HUMAN GENE EVOLUTION

ENCYCLOPAEDIA OF GENE EVOLUTION

Vol. V

HUMAN GENE EVOLUTION

By

Dr. M. Prakash

DPH

DISCOVERY PUBLISHING HOUSE
NEW DELHI-110002

First Published-2006

ISBN 81-8356-169-1 (Set)

Published by

DISCOVERY PUBLISHING HOUSE
4831/24, Ansari Road, Prahlad Street,
Darya Ganj, New Delhi-110002 (India)
Phone: 23279245 • Fax: 91-11-23253475
E-mail:dphtemp@indiatimes.com

Printed at:
Amit Enterprises
Delhi-53

PREFACE

The present title **Encyclopaedia of Gene Evolution** encompasses diverse areas of study from the origin of life and is subsequent record to the molecular and population concepts that explain these events. All biological phenomena derive from evolutionary relationships and past interactions. As the great evolutionary geneticist Theodosius Dobzhansky remarked, "Nothing in biology makes sense except in the light of evolution" yet unifying all biology under an evolutionary theme is still difficult, although the explosive increase in molecular, organismic, and populations information makes the realization of this goal more possible now than ever before.

The aim in writing this text was to bring together the highly dispersed literature on gene structure, function and expression, to integrate this with our emerging knowledge of chromosome and genome structure, and to draw comparison both within and between paralogous and orthologous gene sequences in order to establish the nature of the mutational mechanisms responsible for their evolutionary divergence. Such an approach, encompassing a large number of human genes and their extended families, is essential for any attempt to discern underlying evolutionary principles.

The purpose of this book is to bring together some prevailing knowledge and ideas about evolution, to provide an informed framework of together for under graduates, post-graduate and those of related fields.

Since evolution is the broadest of biological fields, covering the greatest range of disciplines, even the brief survey of evolution offered here has errors and ambiguities. To the extent that this book has been spared many such failings the author owes thanks to many reviewers we commented on one or more sections of this text.

There can be no claim to originality except in the manner of treatment and much of the information has been obtained from the books and scientific journals available in the different libraries.

The author expresses his thanks to his friends and colleagues whose continue inspirations have initiated him to bring out this book.

The author expresses his gratitude to Mr. Wasan and staff of M/s Discovery Publishing House for their whole hearted co-operation in the publication of this book.

Author

CONTENTS

1. Introduction **1–38**

Genomics and Proteomics, Genes, Proteins, Proteomes, Eavesdropping on the Transmission of Genetic Information, Mappings between the Maps, High-Resolution Maps, Picking Out Genes in Genomes, Genomes of Prokaryotes, Genome of the Bacterium Escherichia coli, Genome of the Archaeon Methanococcus jannaschii, Genome of one of the Simplest Organisms: Mycoplasma genitalium, Genomes of Eukaryotes, Genome of Saccharomyces cerevisiae (Baker's yeast), Genome of Caenorhabditis elegans, Genome of Drosophila melanogaster, Genome of Arabidopsis thaliana, Genome of Homo sapiens (Human Genome), Protein Coding Genes, Repeat Sequences, RNA, Single-Nucleotide Polymorphisms (SNPs), Genetic Diversity in Anthropology, Genetic Diversity and Personal Identification, Genetic Analysis of Cattle Domestication, Evolution of Genomes, Please Pass the Genes: Horizontal Gene Transfer, Comparative Genomics of Eukaryotes.

2. Genome Evolution **39–63**

Evolution of Whole Genome, Arrangement of Genes, Gene Clusters, Gene Family, Coding of Genes by DNA, Tandem Repeats, Scattered Repeats, Selfish DNA, Minisatellites (Short Repeats), Scattered Repeats, C-factor Paradox.

3. Pseudogenes (False Genes) **64–86**

Generation of Pseudogenes by Duplication, Generation of Pseudogenes by Retrotransposition, Generation of Pseudogenes by other Means, Origin and Age of Pseudogenes, Patterns of Mutation in Pseudogenes, Pseudogenes and Gene Conversion, Pseudogene Reactivation, Gene Loss/Inactivation in Primates, Urate Oxidase Gene, α-1,3-Galactosyltransferase Gene, Elastase 1 Gene, L-gulono-γ-lactone Oxidase Gene, ADP-Ribosyltransferase I Gene, Haptoglobin Gene, Fertilin-α Gene, T-cell Receptor γ V10 Variable Region Gene, Cytidine Monophospho-N-acetylneuraminic Acid Hydroxylase Gene, Flavin-containing Monooxygenase 2 Gene.

4. Evolution of Split Genes **87—113**

Intron Structure, Function and Evolution, Evolution of Alternative Processing, Alternative Splicing, Aberrant Transcripts, mRNA Surveillance, Ectopic Transcripts, Exon Structure and Evolution, Introns Early or Introns Late? Mechanisms of Intron Insertion and Deletion, Exon Shuffling, Exon Shuffling in the Evolution of Human Genes, Phase Compatibility of Introns, Serine Proteases of Coagulation, Protein Folds, Primordial Exons and the Emergence of Exon Shuffling, Pseudoexons.

5. Gene Families **114—182**

Origin of Human Genes, Genes with a Specifically Human Origin, Human Genes which Originated after the Divergence of Old World Monkeys and New World Monkeys, Human Genes which Originated during Primate Evolution, Human Genes whose Origin Preceded the Divergence of Mammals, Human Genes whose Origin Preceded the Divergence of the Vertebrates, Human Genes whose Origin Preceded the Divergence of the Metazoa, Human Genes whose Origin Preceded the Divergence of Animals and Fungi, Human Genes whose Origin Preceded the Divergence of Plants and Animals, Human Genes whose Origin Preceded the Divergence of Prokaryotes and Eukaryotes, Emergence of Genes and Gene Families has Paralleled Organismal Evolution, Multigene Families, Gene Families, Apolipoprotein genes, Highly Repetitive Multigene Families, Gene Superfamilies, Genes which Undergo Programmed Rearrangement, Convergent Evolution, Coevolution.

6. Regulating Factors **183—220**

Promoters and Enhancers, Evolutionary Conservation of Cis-acting Elements and 'Phylogenetic Footprinting', Nonhomologous Genes Containing Similar Regulatory Elements, Paralogous Genes Containing Dissimilar Regulatory Elements, Orthologous Genes

Containing Dissimilar Regulatory Elements, Bidirectional Promoters, 5' and 3' Untranslated Regions of Genes, Inter-specific Differences in Promoter Selection, Developmental Changes in Gene Expression, Promoter Polymorphisms, Promoter Duplication, Functional Redundancy of Promoter Elements, Recruitment of Repetitive Sequences as Promoter and Silencer Elements, mRNA Editing, Coordinate Regulation, Changes in Expression of Developmentally Significant Gene Products, Transcription Factors, Transcription Factor Families, Functional Conservation of Orthologous Transcription Factors, Functional Redundancy of Paralogous Transcription Factors, Paralogous Transcription Factors, Orthologous Transcription Factors, Alternative Splicing of Transcription Factor Genes, Promoter Shuffling in Transcription Factor Genes, Transcription Factor-binding Site Interactions, Exon Shuffling in the Evolution of Transcription Factors.

7. Molecular Clocks **221–242**

Relative-Rate Test, A Local Clock in Mice, Rats, and Hamsters, Lower Rates in Humans than in Monkeys, Higher Rates in Rodents than in Primates, Male-driven Sequence Evolution, Causes of Rate Variation among Lineages, Overdispersed Clocks, Ratios of Synonymous and Nonsynonymous Rates in Different Lineages.

8. Substituting Nucleotides **243–262**

Noncoding Sequences, Jukes and Cantor's Method, Kimura's Two-Parameter Method, Tajima and Nei's Method, Other Methods, Comparison of Methods, Nonuniform Rates among Nucleotide Sites, Protein-Coding Sequences, Sequence Alignment, Dot-Matrix Method, Similarity and Distance Methods, Multiple Sequences.

9. Evolutionary Reshuffling DNA Sequences **263–283**

Nucleotide Substitution in DNA Sequence, Jukes and Cantor's One-Parameter Model, Kimura's Two-Parameter Model, Other Models of Nucleotide Substitution, Divergence between DNA Sequences, Sequences Similarity, Sequence Dissimilarity, Nonuniform Rates among Nucleotide Sites.

10. Evolution by Gene Duplication **284–322**

Types of DNA Duplication, Domain Duplication and Gene Elongation, α2 Type I Collagen Gene, Apolipoprotein Genes, Prevalence of Domain Duplication, Formation of Gene Families and Acquisition of New Functions, RNA-Specifying Genes, Isozymes, Colour-Sensitive Pigment Proteins, Nonfunctionalization of Redundant Duplicates, Dating Gene Duplications, Globin Superfamily of Genes, Homeobox Genes, Domain and Exon Shuffling, Mosaic Proteins, Phase Limitation on Exon Shuffling, Alternative Pathways for Producing New Functions, Alternative Splicing, Overlapping Genes, Gene Sharing, RNA Editing.

Index **323–326**

1

Introduction

Genomics and Proteomics

The genome of a typical bacterium comes as a single DNA molecule, which, if extended, would be about 2 mm long. (The cell itself has a diameter of about 0.001 mm.) The DNA of higher organisms is organized into chromosomes—normal human cells contain 23 chromosome pairs. The total amount of genetic information per cell—the sequence of nucleotides of DNA—is very nearly constant for all members of a species, but varies widely between species:

Organism	*Genome size*
Epstein-Barr virus	0.172×10^6 base pairs
Bacterium (*E. coli*)	4.6×10^6
Yeast (*S. cerevisiae*)	12.1×10^6
Nematode worm (*C. elegans*)	95.5×10^6
Thale cress (*A. thaliana*)	117.0×10^6
Fruit fly (*D. melanogaster*)	180.0×10^6
Human (*H. sapiens*)	3200×10^6

Not all DNA codes for proteins. Conversely, some genes exist in multiple copies. Therefore, the amount of protein sequence information in a cell cannot easily be estimated from the genome size.

Genes

A single gene coding for a particular protein corresponds to a sequence of nucleotides along one or more regions of a molecule of DNA. The DNA sequence is collinear with the protein sequence. In species for which the genetic material is double-stranded DNA, genes may appear on either strand. Bacterial genes are continuous regions of DNA. Therefore, the functional unit of genetic sequence information from a bacterium is a string of 3*N* nucleotides encoding a string of *N* amino acids, or a string of *N* nucleotides encoding a structural RNA molecule of *N* residues. Such a string, equipped with annotations, would form a typical entry in one of the genetic sequence archives.

In eukaryotes the nucleotide sequences that encode the amino acid sequences of individual proteins are organized in a more complex manner. The relationship between size of gene and size of protein encoded is very different from that in bacteria. Frequently one gene appears split into separated segments in the geniomic DNA. An *exon* is a stretch of DNA retained in the mature messenger RNA that the ribosome translates into protein. An *intron* is an intervening region between two exons. Cellular machinery splices together the proper segments, in RNA transcripts, based on signal sequences flanking the exons in the sequences themselves. Many introns are very long—in some cases substantially longer than the exons.

Regulatory mechanisms organize the expression of genes. Genes may be turned on or off (or more finely regulated) in response to concentrations of nutrients, or to stress, or to unfold complex programs of development during the lifetime of the organism. Many control regions of DNA lie near the segments coding for proteins. They contain sequences that serve as binding sites for the molecules that transcribe the DNA sequence, or sequences that bind regulatory molecules that can *block* transcription. Simple examples occur in bacterial genomes, in which contiguous genes, coding for several proteins that catalyse successive steps in an integrated sequence of reactions, fall under the control of the same regulatory sequence. F. Jacob, J. Monod, and E. Wollman named these *operons*. One can readily understand the utility of a parallel mechanism for control of their expression.

In animals, methylation of DNA provides the signals for tissue-specific expression of developmentally regulated genes. Products of certain genes cause cells to commit suicide—a process called *apoptosis*. Defects in the apoptotic mechanism leading to uncontrolled growth are observed in some cancers, and stimulation of these mechanisms is a general approach to cancer therapy.

Proteins

In principle, a database of amino acid sequences of proteins is inherent in the database of nucleotide sequences of DNA, by virtue of the genetic code. Indeed, new protein sequence data are now being determined by translation of DNA sequences, rather than by direct sequencing of proteins. (Historically, the chemical problem of determining amino acid sequences of proteins directly was solved before the genetic code was established and before methods for determination of nucleotide sequences of DNA were developed. F. Sanger's sequencing of insulin in 1955 first proved that proteins had definite amino acid sequences, a proposition that until then was hypothetical.)

Should any distinction be made between amino acid sequences determined directly from proteins and those determined by translation from DNA? First, we must assume that it is possible correctly to identify within the DNA data stream the regions that encode proteins. The pattern-recognition programs that address this question are subject to three types of errors: a genuine protein sequence may be missed entirely, or an incomplete protein may be reported, or a gene may be incorrectly spliced. Several variations on the theme add to the complexity: Genes for different proteins may overlap, or genes may be assembled from exons in different ways in different tissues. Conversely, some genetic sequences that appear to code for proteins may, in fact, be defective or not expressed. *A protein inferred from a genome sequence is a hypothetical object until an experiment verifies its existence.*

Second, in many cases the expression of a gene produces a molecule that must be modified within a cell, to make a *mature* protein that differs significantly from the one suggested by translation of the gene sequence. In many cases the missing details of *post-translational modifications*—the molecular analogues of body piercing—are quite important, and unlike body piercing, have functional significance. Post-translational modifications include addition of ligands (for instance the covalently-bound haem group of cytochrome *c*), glycosylation, methylation, excision of peptides, and many others. Patterns of disulphide bridges—primary chemical bonds between cysteine residues—cannot be deduced from the amino acid sequence. In some cases, mRNA is edited before translation, creating changes in amino acid sequences that are not inferrable from the genes.

Proteomes

An organism's genome gives a complete but static set of specifications of the potential life of that individual. The state of development of the organism, and its activity at the molecular level at any moment, depend primarily on the amounts and distribution of its proteins. The *proteome project* is a large-scale programme dealing in an integral way with patterns of expression of proteins in biological systems, in ways that complement and extend genome projects.

What kinds of data would we like to measure, and what mature experimental techniques exist to determine them? The basic goal is a spatio-temporal description of the deployment of proteins in the organism. The rates of synthesis of different proteins vary among different tissues and different cell types and states of activity. Methods are available for efficient analysis of transcription patterns of multiple genes. However, because proteins 'turn over' at different rates, it is also necessary to measure proteins directly. The distribution of expressed protein levels is a kinetic balance between rates of protein synthesis and degradation. High-resolution two-dimensional polyacrylamide gel electrophoresis (2D PAGE) shows the pattern of protein content in a sample. Mass-spectroscopic techniques identify the proteins into which the sample has been separated, and their post-translational modifications.

It is also possible to make arrays of proteins, to screen for pairs of proteins that interact. Application of these methods provides a picture of the protein-based activity of an organism, as the genome provides a complete set of potential proteins. R. Simpson has drawn the analogy: if the genome is a list of the instruments in an orchestra, the proteome is the orchestra in the process of playing a symphony.

Eavesdropping on the Transmission of Genetic Information

How hereditary information is stored, passed on, and implemented is perhaps the fundamental problem of biology. Three types of maps have been essential:

1. Linkage maps of genes
2. Banding patterns of chromosomes
3. DNA sequences

These represent three very different types of data. Genes, as discovered by Mendel, were entirely abstract entities. Chromosomes are physical objects, the banding patterns their visible landmarks. Only with DNA sequences are we dealing directly with stored hereditary information in its physical form.

It was the very great achievement of the last century of biology to forge connections between these three types of data. The first steps—and giant strides they were indeed—proved that, for any chromosome, the maps are one-dimensional arrays, and indeed that they are collinear. Any schoolchild now knows that genes are strung out along

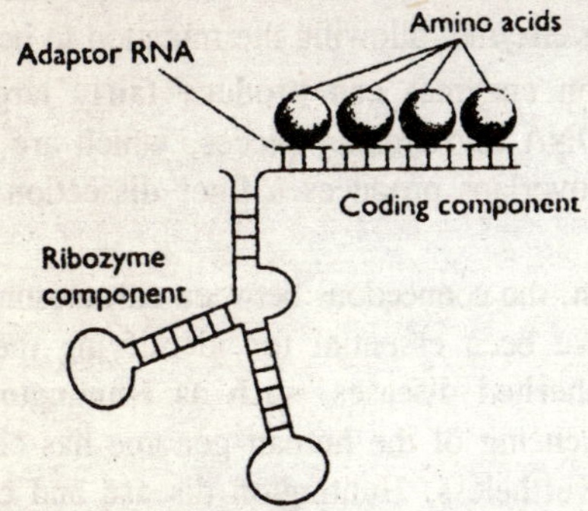

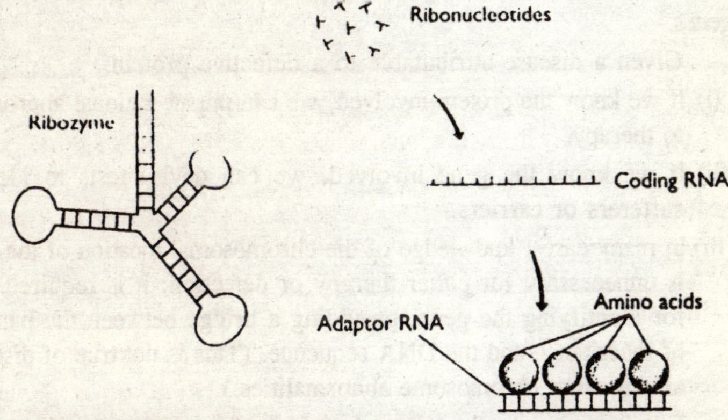

Fig. 1.1. Two scenarios for the evolution of the first coding RNA.

chromosomes, and that each gene corresponds to a DNA sequence. But the proofs of these statements earned a large number of Nobel prizes. Splitting a long molecule of DNA—for example, the DNA in an entire chromosome—into fragments of convenient size for cloning and sequencing requires additional maps to report the order of the fragments, so that the entire sequence can be reconstructed from the sequences of the fragments.

A restriction endonuclease is an enzyme that cuts DNA at a specific sequence, usually about 6 bp long. Cutting DNA with several restriction enzymes with different specificities produces sets of overlapping fragments. From the sizes of the fragments it is possible to construct a *restriction map*, stating the order and distance between the restriction enzyme cleavage sites. A mutation in one of these cleavage sites will change the sizes of the fragments produced by the corresponding enzyme, allowing the mutation to be located in the map.

Restriction enzymes can produce fairly large pieces of DNA. Cutting the DNA into smaller pieces, which are cloned and ordered by sequence overlaps produces a finer dissection of the DNA called *contig map*.

In the past, the connections between chromosomes, genes and DNA sequences have been essential for identifying the molecular deficits underlying inherited diseases, such as Huntington disease or cystic fibrosis. Sequencing of the human genome has changed the situation radically. Nevertheless, Huntington disease and cystic fibrosis result from dysfunction of a single protein encoded by a single gene, and are not correlated with environmental factors or lifestyle. In contrast, many common diseases including diabetes and cancer are affected by many genes.

Given a disease attributable to a defective protein:

(i) If we know the protein involved, we can pursue rational approaches to therapy.

(ii) If we know the gene involved, we can devise tests to identify sufferers or carriers.

(iii) In many cases, knowledge of the chromosomal location of the gene is unnecessary for either therapy or detection; it is required only for identifying the gene, providing a bridge between the patterns of inheritance and the DNA sequence. (This is not true of diseases arising from chromosome abnormalities.)

For instance, in the case of sickle-cell anaemia, we know the protein involved. The disease arises from a single point mutation in haemoglobin. We can proceed directly to drug design. We need the DNA sequence only for genetic testing and counselling. In contrast, if we know neither the protein nor the gene, we must somehow go from the phenotype back to the gene, a process called *positional cloning* or *reverse genetics*. Positional cloning used to involve a kind of Tinker to Evers to Chance cascade from the gene map to the chromosome map to the DNA sequence.

Patterns of inheritance identify the type of genetic defect responsible for a condition. They show, for example, that Huntington disease and cystic fibrosis are caused by single genes. To find the

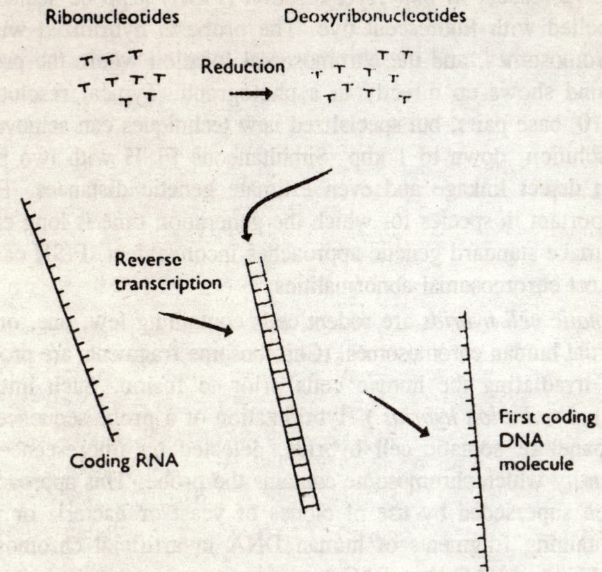

Fig. 1.2. Conversion of a coding RNA molecule into the progenitor of the first DNA genome.

gene associated with cystic fibrosis it was necessary to begin with the gene map, using linkage patterns of heredity in affected families to localize the affected gene to a particular region of a particular chromosome. Knowing the general region of the chromosome, it was then possible to search the DNA of that region to identify candidate genes, and finally to pinpoint the particular gene responsible and sequence it. In contrast, many diseases do not show simple inheritance, or, even if only a single gene is involved, heredity creates only a predisposition the clinical consequences of which depend on environmental factors. The full human genome sequence, and measurements of expression patterns, will be essential to identify the genetic components of these more complex cases.

Mappings between the Maps

A gene linkage map can be calibrated to chromosome banding patterns through observation of individuals with deletions or translocations of parts of chromosomes. The genes responsible for phenotypic changes associated with a deletion must lie within the deletion. Translocations are correlated with altered patterns of linkage and recombination.

There have been several approaches to coordinating chromosome banding patterns with individual DNA sequences of genes:

(i) In Fluorescent *In Situ* Hybridization (FISH) a probe sequence is labelled with fluorescent dye. The probe is hybridized with the chromosomes, and the chromosomal location where the probe is bound shows up directly in a photograph. Typical resolution is $\sim 10^5$ base pairs, but specialized new techniques can achieve high resolution, down to 1 kbp. Simultaneous FISH with two probes can detect linkage and even estimate genetic distances. This is important in species for which the generation time is long enough to make standard genetic approaches inconvenient. FISH can also detect chromosomal abnormalities.

(ii) *Somatic cell hybrids* are rodent cells containing few, one, or even partial human chromosomes. (Chromosome fragments are produced by irradiating the human cells prior to fusion. Such lines are called *radiation hybrids*.) Hybridization of a probe sequence with a panel of somatic cell hybrids, detected by fluorescence, can identify which chromosome contains the probe. This approach has been superseded by use of clones of yeast or bacteria or phage containing fragments of human DNA in artificial chromosomes (YACs and BACs and PACs).

Of course, for human DNA these methods are obsolete, now that the sequence is known (but mind the gaps). Given a DNA sequence one would just look it up.

High-Resolution Maps

Formerly, genes were the only visible portions of genomes. Now, markers are no longer limited to genes with phenotypically observable effects, which are anyway too sparse for an adequately high-resolution map of the human genome. Now that we can interrogate DNA sequences directly, any features of DNA that vary among individuals can serve as markers, including:

Variable number tandem repeats

Variable number tandem repeats (VNTRs), also called minisatellites. VNTRs contain regions 10-100 base pairs long, repeated a variable number of times-same sequence, different number of repeats. In any individual, VNTRs based on the same repeat motif may appear only once in the genome; or several times, with different lengths on different chromosomes. The distribution of the sizes of the repeats is the marker. Inheritance of VNTRs can be followed in a family and mapped to a disease phenotype like any other trait. VNTRs were the first genetic sequence data used for personal identification—genetic fingerprints — in paternity and in criminal cases.

Formerly, VNTRs were observed by producing *restriction fragment length polymorphisms* (RFLPs) from them. VNTRs are generally flanked by recognition sites for the same restriction enzyme, which will neatly excise them. The results can be spread out on a gel, and detected by Southern blotting. Note the distinction: VNTRs are characteristics of genome sequences; RFLPs are artificial mixtures of short stretches of DNA created in the laboratory in order to identify VNTRs.

It is much easier and more efficient to measure the sizes of VNTRs by amplifying them with PCR, and this method has replaced the use of restriction enzymes.

Short tandem repeat polymorphisms

Short tandem repeat polymorphisms (STRPs), also called microsatellites. STRPs are regions of only 2-5 bp but repeated many times; typically 10-30 consecutive copies. They have several advantages as markers over VNTRs, one of which is a more even distribution over the human genome.

There is no reason why these markers need lie within expressed genes, and usually they do not. (The CAG repeats in the gene for huntingtin and certain other disease genes are exceptions.)

Panels of micro satellite markers greatly simplify the identification of genes. It is interesting to compare a recent project to identify a disease gene now that the human genome sequence is available, with such classic studies as the identification of the gene for cystic fibrosis.

Additional mapping techniques deal more directly with the DNA sequences, and can short-circuit the process of gene identification:

Contig or Contiguous clone map

A *contig*, or *contiguous clone map*, is a series of overlapping DNA clones of known order along a chromosome from an organism of interest—for instance, human—stored in yeast or bacterial cells as YACs (Yeast Artificial Chromosomes) or BACs (Bacterial Artificial Chromosomes). A contig map can produce a very fine mapping of a genome. In a YAC, human DNA is stably integrated into a small extra chromosome in a yeast cell. A YAC can contain up to 10^6 base pairs. In principle, the entire human genome could be represented in 10 000 YAC clones. In a BAC, human DNA is inserted into a plasmid in an *E. coli* cell. (A plasmid is a small piece of double-stranded DNA found in addition to the main genome, usually but not always circular.) A BAC can carry about 250 000 bp. Despite their smaller capacities, BACs are preferred to YACs because of their greater stability and ease of handling.

Sequence tagged site

A *sequence tagged site* (STS) is a short, sequenced region of DNA, typically 200—600 bp long, that appears in a unique location in the genome. It need not be polymorphic. An STS can be mapped into the genome by using PCR to test for the presence of the sequence in the cells containing a contig map.

One type of STS arises from an *expressed sequence tag* (EST), a piece of cDNA (complementary DNA; i.e., a DNA sequence derived from the messenger RNA of an expressed gene.) The sequence contains only the exons of the gene, spliced together to form the sequence that encodes the protein. cDNA sequences can be mapped to chromosomes using FISH, or located within contig maps.

How do contig maps and sequence tagged sites facilitate identifying genes? If you are working with an organism for which the full genome sequence is not known, but for which full contig maps are available for all chromosomes, you would identify STS markers tightly linked to your gene, then locate these markers in the contig maps.

Picking Out Genes in Genomes

Computer programs for genome analysis identify *open reading frames* or *ORFs*. An ORF is a region of DNA sequence that begins with an initiation codon (ATG) and ends with a stop codon. An ORF is a potential protein-coding region.

Approaches to identifying protein-coding regions choose from or combine two possible approaches:

1. *Detection of regions similar to known coding regions from other organisms*. These regions may encode amino acid sequences similar to known proteins, or may be similar to Espressed Sequence Tags (ESTs). Because ESTs are derived from messenger RNA, they correspond to genes known to be expressed. It is necessary to sequence only a few hundred initial bases of cDNA to give enough information to identify a gene: characterization of genes by ESTs is like indexing poems or songs by their first lines.
2. *Ab initio methods, that seek to identify genes from the properties of the DNA sequences themselves*. Computer-assisted annotation of genomes is more complete and accurate for bacteria than for eukaryotes. Bacterial genes are relatively easy to identify because they are contiguous—they lack the introns characteristic of eukaryotic genomes—and the intergene spaces are small. In higher organisms, identifying genes is harder. Identification of exons is

one problem, assembling them is another. Alternative splicing patterns present a particular difficulty.

A framework for ab initio gene identification in eukaryotic genomes includes the following features:

- The initial (5') exon starts with a transcription start point, preceded by a core promotor site such as the TATA box typically ~30bp upstream. It is free of in-frame stop codons, and ends immediately before a GT splice signal. (Occasionally a non-coding exon precedes the exon that contains the initiator codon.)
- Internal exons, like initial exons, are free of in-frame stop codons. They begin immediately after an AG splice signal and and end immediately before a GT splice signal. The flanking introns contain consensus branch point sequences and polypyrimidine tracts that interact with the splicing apparatus.
- The final (3') exon starts immediately after an AG splice signal and ends with a stop codon, followed by a polyadenylation signal sequence. (Occasionally a noncoding exon follows the exon that contains the stop codon.)

All coding regions have non-random sequence characteristics, based partly on codon usage preferences. Empirically, it is found that statistics of hexanucleotides perform best in distinguishing coding from non-coding regions. Starting from a set of known genes from an organism as a training set, pattern recognition programs can be tuned to particular genomes.

Accurate gene detection is a crucial component of genome sequence analysis. This problem is an important focus of current research.

Genomes of Prokaryotes

Most prokaryotic cells contain their genetic material in the form of a large single circular piece of double-stranded DNA, usually less than 5 Mb long. In addition they may contain plasmids.

The protein-coding regions of bacterial genomes do not contain introns. In many prokaryotic genomes the protein-coding regions are partially organized into *operons*—tandem genes transcribed into a single messenger RNA molecule, under common transcriptional control. In bacteria, the genes of many operons code for proteins with related functions. For instance, successive genes in the trp operon of *E. coli* code for proteins that catalyse successive steps in the biosynthesis of tryptophan. In archaea, a metabolic relationship between genes in operons is less frequently observed.

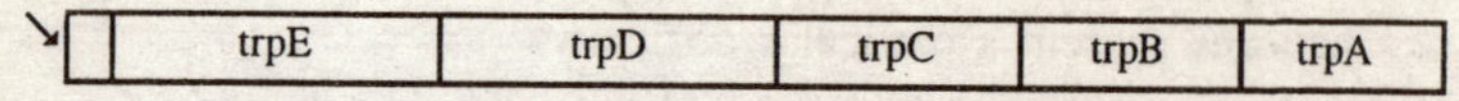

Fig. 1.3. The trp operon in E. coli.

The typical prokaryotic genome contains only a relatively small amount of non-coding DNA (in comparison with eukaryotes), distributed throughout the sequence. In *E. coli* only ~11 % of the DNA is non-coding.

Genome of the Bacterium *Escherichia coli*

E. coli, strain K-12, has long been the workhorse of molecular biology. The genome of strain MG1655, published in 1997 by the group of F. Blattner at the University of Wisconsin, contains 4 639 221 bp in a single circular DNA molecule, with no plastids. Approximately 89% of the sequence codes for proteins or structural RNAs. An inventory reveals:

- 4285 protein-coding genes
- 122 structural RNA genes
- non-coding repeat sequences
- regulatory elements
- transcription/translation guides
- transposases
- prophage remnants
- insertion sequence elements
- patches of unusual composition, likely to be foreign elements introduced by horizontal transfer.

Analysis of the genome sequence required identification and annotation of protein-coding genes and other functional regions. Many *E. coli* proteins were known before the sequencing was complete, from the many years of intensive investigation: 1853 proteins had been described before publication of the genome sequence. Other genes could be assigned functions from identification of homologues by searching in sequence databanks. The narrower the range of specificity of the function of the homologues, the more precise could be the assignment. Currently, over 60% of proteins can be assigned at least a general function. Other regions of the genome are recognized as regulatory sites, or as mobile genetic elements, also on the basis of similarity to homologous sequences known in other organisms.

The distribution of protein-coding genes over the genome of *E. coli* does not seem to follow any simple rules, either along the DNA or on different strands. Indeed, comparison of strains suggests that genes are mobile.

The *E. coli* genome is relatively gene dense. Genes coding for proteins or structural RNAs occupy ~89% of the sequence. The average size of an ORF is 317 amino acids. If the genes were evenly distributed, the average intergenic region would be 130 bp; the observed average distance between genes is 118 bp. However, the sizes of intergenic regions vary considerably. Some intergenic regions are large. These contain sites of regulatory function, and repeated sequences. The longest intergenic region, 1730 bp, contains non-coding repeat sequences.

Approximately three-fourth of the transcribed units contain only one gene; the rest contain several consecutive genes, or operons. It is estimated that the *E. coli* genome contains 630—700 operons. Operons vary in size, although few contain more than five genes. Genes within operons tend to have related functions.

In some cases, the same DNA sequence encodes parts of more than one polypeptide chain. One gene codes for both the τ and γ subunits of DNA polymerase III. Translation of the entire gene forms the τ subunit. The γ subunit corresponds approximately to the N-terminal two-thirds of the τ subunit. A frame shift on the ribosome at this point leads to chain termination 50% of the time, causing a 1: 1 ratio of expressed τ and γ subunits. There do not appear to be any overlapping genes in which different reading frames both code for expressed proteins.

In other cases, the same polypeptide chain appears in more than one enzyme. A protein that functions on its own as lipoate dehydrogenase is also an essential subunit of pyruvate dehydrogenase, 2-oxoglutarate dehydrogenase and the glycine cleavage complex.

Having the complete genome, we can examine the protein repertoire of *E. coli*. The largest class of proteins are the enzymes — approximately 30% of the total genes. Many enzymatic functions are shared by more than one protein. Some of these sets of functionally similar enzymes are very closely related, and appear to have arisen by duplication, either in *E. coli* itself or in an ancestor. Other sets of functionally similar enzymes have very dissimilar sequences, and differ in specificity, regulation, or intracellular location.

Several features of *E. coli's* generous endowment of enzymes give it an extensive and versatile metabolic competence, which allow it to grow and compete under varying conditions:

1. It can synthesize all components of proteins and nucleic acids (amino acids and nucleotides), and cofactors.
2. It has metabolic flexibility: Both aerobic and anaerobic growth are possible, utilizing different pathways of energy capture. It can grow on many different carbon and nitrogen sources. Not all metabolic pathways are active at any time; the alternatives allow response to changes in conditions.
3. A wide range of transporters allows uptake of required substrates.
4. Even for specific metabolic reactions there are many cases of multiple enzymes. These provide redundancy, and contribute to an ability to tune metabolism to varying conditions. Complex regulatory mechanisms integrate the protein expression pattern.
5. However, *E. coli* does not possess a complete range of enzymatic capacity. It cannot fix CO_2 or N_2.

We have described here some of the static features of the *E. coli* genome and its protein repertoire. Current research is moving into dynamic aspects, to investigations of protein expression patterns.

Genome of the Archaeon *Methanococcus jannaschii*

S. Luria once suggested that to determine common features of all life one should not try to survey everything, but rather identify the organism most different from us and see what we have in common with it. The assumption was that the way to do this would be to find an organism adapted to the most different environment.

Deep-sea exploration has revealed environments as far from the familiar as those portrayed in science fiction. Hydrothermal vents are underwater volcanoes emitting hot lava and gases through cracks in the ocean floor. They create niches for communities of living things disconnected from the surface, depending on the minerals exuded from the vent for their sources of inorganic nutrients. They support living communities of microorganisms that are the only known forms of life not dependent on sunlight, directly or indirectly, for their energy source.

The microorganism *Methanococcus jannaschii* was collected from a hydrothermal vent 2600 m deep off the coast of Baja California, Mexico, in 1983. It is a thermophilic organism, surviving at temperatures from 48—94°C, with an optimum at 85°C. It is a strict anaerobe, capable of self-reproduction from inorganic components. Its overall metabolic equation is to synthesize methane from H_2 and CO_2.

M. jannaschii belongs to the archaea, one of the three major divisions of life along with the bacteria and eukarya. The archae a comprise groups of prokaryotes, including organisms adapted to extreme environmental conditions such as high temperature and pressure, or high salt concentration.

The genome of *M. jannaschii* was sequenced in 1996 by The Institute for Genomic Research (TIGR). It was the first archaeal genome sequenced. It contains a large chromosome containing a circular double-stranded DNA molecule 1,664,976 bp long, and two extrachromosomal elements of 58,407 and 16,550 bp, respectively. There are 1743 predicted coding regions, of which 1682 are on the chromosome, and 44 and 12 on the large and small extrachromosomal elements respectively. Some RNA genes contain introns. As in other prokaryotic genomes there is little non-coding DNA.

M. jannaschii would appear to satisfy Luria's goal of finding our most distant extant relative. Comparison of its genome sequence with others shows that it is distantly related to other forms of life. Only 38% of the ORFs could be assigned a function, on the basis of homology to proteins known from other organisms. (Now over 50% can be assigned a function.) However, to everyone's great surprise, archaea are in some ways more closely related to eukaryotes than to bacteria! They are a complex mixture. In archaea, proteins involved in transcription, translation, and regulation are more similar to those of eukaryotes. Archaeal proteins involved in metabolism are more similar to those of bacteria.

Genome of one of the Simplest Organisms: Mycoplasma genitalium

Mycoplasma genitalium is an infectious bacterium, the cause of non-gonococcal uretheitis. Its genome was sequenced in 1995 by a collaboration of groups at TIGR, The Johns Hopkins University and The University of North Carolina. The genome is a single DNA molecule containing 580,070 bp. This is the smallest cellular genome yet sequenced. So far, *M. genitalium* is the closest we have to a *minimal organism*, the smallest capable of independent life. (Viruses, in contrast, require the cellular machinery of their hosts.)

The genome is dense in coding regions. Four hundred and sixty-eight genes have been identified as expressed proteins. Some regions of the sequence are gene rich, others gene poor, but overall 85% of the sequence is coding. The average length of a coding region is 1040 bp. As in other bacterial genomes, the coding regions do not contain introns. Further compression of the genome is achieved by overlapping genes. It appears that many of these have arisen through loss of stop codons.

The gene repertoire of *M. genitalium* includes some that encode proteins essential for independent reproduction, such as those involved in DNA replication, transcription, and translation, plus ribosomal and transfer RNAs. Other genes are specific for the infectious activity. These include adhesins that mediate binding to infected cells, other molecules for defence against the host's immune system, and many transport proteins. As adaptation to the parasitic lifestyle of the organism, there has been widespread loss of metabolic enzymes, including those responsible for amino acid biosynthesis—indeed, one amino acid is absent from all *M. genitalium* proteins.

Genomes of Eukaryotes

It is rare in science to encounter a completely new world containing phenomena entirely unsuspected. The complexity of the eukaryotic genome is such a world.

In eukaryotic cells, the majority of DNA is in the nucleus, separated into bundles of nucleoprotein, the chromosomes. Each chromosome contains a single double-stranded DNA molecule. Smaller amounts of DNA appear in organelles—mitochondria and chloroplasts. The organelles originated as intracellular parasites. Organelle genomes usually have the form of circular double-stranded DNA, but are sometimes linear and sometimes appear as multiple circles. The genetic code by which organelle genes are translated differs from that of nuclear genes.

Nuclear genomes of different species vary widely in size. The correlation between genome size and complexity of the organism is very rough. It certainly does not support any preconception that humans stand on a pinnacle. In many cases differences in genome size reflect different amounts of simple repetitive sequences, often referred to as 'junk DNA.'

In addition to variation in DNA content, eukaryotic species vary in the number of chromosomes and distribution of genes among them. Some differences in the distribution of genes among chromosomes involve translocations, or chromosome fragmentations or joinings. For instance, humans have 23 pairs of chromosomes; chimpanzees have 24. Human chromosome 2 is equivalent to a fusion of chimpanzee chromosomes 12 and 13. The difficulty of chromosome pairing during mitosis in a zygote after such an event can contribute to the reproductive isolation required for species separation.

Other differences in chromosome complement reflect duplication or hybridization events. The wheat first used in agriculture, in the

Middle East at least 10,000—15,000 years ago, is a diploid called *einkorn* (*Triticum monococcum*), containing 14 pairs of chromosomes. *Emmer* wheat (*T. dicoccum*), also cultivated since palaeolithic times, and *durum* wheat (*T. turgidum*), are merged hybrids of relatives of einkorn with other wild grasses, to form tetraploid species. Additional hybridizations, to different wild wheats, gave hexaploid forms, including *spelt* (*T. spelta*), and modern common wheat *T. aestivum*. *Triticale*, a robust crop developed in modern agriculture and currently used primarily for animal feed, is an artificial genus arising from crossing durum wheat (*Triticum turgidim*) and rye (*Secale cereale*). Most triticale varieties are hexaploids.

All these species are still cultivated—some to only minor extents—and have their individual uses in cooking. Spelt, or *farro* in Italian, is the basis of a well-known soup; pasta is made from durum wheat; and bread from *T. aestivum*.

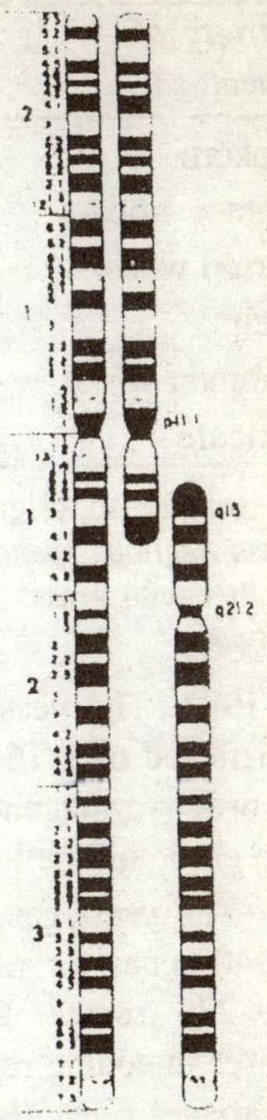

Fig. 1.4. Left: human chromosome 2. Right: matching chromosomes from a chimpanzee.

Even within single chromosomes, gene families are common in eukaryotes. Some family members are *paralogues*—related genes that have duplicated within the same genome, and have in many cases diverged to provide separate functions in descendant species. Changes in expression pattern may precede development of novel function. (*Orthologues*, in contrast, are homologues in different species that often perform the same function. For instance, human α and β globin are paralogues, and human and horse myoglobin are orthologues.) Other related sequences may be pseudogenes, which may have arisen by duplication, or by retrotransposition from messenger RNA, followed by the accumulation of mutations to the point of loss of function. The human globin gene cluster is a good example.

Genome of *Saccharomyces cerevisiae* (Baker's yeast)

Yeast is one of the simplest known eukaryotic organisms. Its cells, like our own, contain a nucleus and other specialized intracellular compartments. The sequencing of its genome, by an unusually effective international consortium involving ~100 laboratories, was completed

Variety of wheat	Classification	Chromosome complement
Einkorn	*Triticum monococcum*	AA
emmer wheat	*Triticum turgidum*	AABB
durum wheat	*Triticum turgidum*	AABB
spelt	*Triticum spelta*	AABBDD
common wheat	*Triticum aestivum*	AABBDD
triticale	*Triticosecale*	AABBRR

A—genome of original diploid wheat or a relative, B—genome of a wild grass *Aegilops speltoides* or *Triticum speltoides* or a relative, D—genome of another wild grass, *Triticum tauschii* or a relative, R—genome of rye *Secale cereale*.

in 1992. The yeast genome contains 12057500 bp of nuclear DNA, distributed over 16 chromosomes. The chromosomes range in size over an order of magnitude, from the 1352 kbp chromosome IV to the 230 kbp chromosome I.

The yeast genome contains 5885 predicted protein-coding genes, ~140 genes for ribosomal RNAs, 40 genes for small nuclear RNAs, and 275 transfer RNA genes. In two respects, the yeast genome is denser in coding regions than the known genomes of the more complex eukaryotes *Caenorhabditis elegans*, *Drosophila melanogaster*, and human: (1) Introns are relatively rare, and relatively small. Only 231 genes in yeast contain introns. (2) There are fewer repeat sequences compared with more complex eukaryotes.

A duplication of the entire yeast genome appears to have occurred ~150 mil. lion years ago. This was followed by translocations of pieces of the duplicated DNA and loss of one of the copies of most (~92%) of the genes.

Of the 5885 potential protein-coding genes, 3408 correspond to known proteins. About 1000 more contain some similarity to known proteins in other species. Another ~800 are similar to ORFs in other genomes that correspond to unknown proteins. Many of these homologues appear in prokaryotes. Only approximately one-third of yeast proteins have identifiable homologues in the human genome.

In taking censuses of genes, it has been useful to classify their functions into broad categories. The following classification of yeast protein functions is taken from http://www.mips.biochem.mpg.de/proj/yeast/catalogues/funcat/:

- Metabolism
- Energy
- Cell growth, cell division and DNA synthesis
- Transcription
- Protein synthesis
- Protein destination
- Transport facilitation
- Cellular transport and transport mechanisms
- Cellular biogenesis
- Cellular communication/signal transduction
- Cell rescue, defence, cell death and ageing
- Ionic homeostasis
- Cellular organization
- Transposable elements, viral and plasmid proteins
- Unclassified

Yeast is a testbed for development of methods to assign functions to gene products. The search for homologues has been exhaustive and it continues. Collections of mutants exist containing a knockout of every gene. (A unique sequence 'bar code' introduced into each mutant facilitates identification of the ones that grow under selected conditions.) Cellular localization and expression patterns are being investigated. Several types of measurements, including those based on activation of transcription by pairs of proteins that can form dimers, are producing catalogues of interprotein interactions.

Genome of *Caenorhabditis elegans*

The nematode worm *Caenorhabditis elegans* entered biological research in the 1960s, at the express invitation of Sydney Brenner. He recognized its potential as a sufficiently complex organism to be interesting but simple enough to permit complete analysis, at the cellular level, of its development and neural circuitry.

The *C. elegans* genome, completed in 1998, provided the first full DNA sequence of a multicellular organism. The *C. elegans* genome contains ~97 Mbp of DNA distributed on paired chromosomes I, II, III, IV, V and X. There is no Y chromosome. Different genders in *C. elegans* appear in the XX genotype, a self-fertilizing hermaphrodite; and the XO genotype, a male.

The *C. elegans* genome is about eight times larger than that of yeast, and its 19099 predicted genes are approximately three times the number in yeast. The gene density is relatively low for a eukaryote, with ~1 gene/5 kb of DNA. Exons cover ~27% of the genome; the genes contain an average of 5 introns each. Approximately 25% of the genes are in clusters of related genes.

Many *C. elegans* proteins are common to other life forms. Others are apparently specific to nematodes. Forty-two percent of proteins have homologues outside the phylum, 34% are homologous to proteins of other nematodes and 24% have no known homologues outside *C. elegans* itself. Many of the proteins have been classified according to structure and function.

Several kinds of RNA genes have been identified. The *C. elegans* genome contains 659 genes for tRNA, almost half of them (44%) on the X chromosome. Spliceosomal RNAs appear in dispersed copies, often identical. (Spliceosomes are the organelles that convert pre-mRNA transcripts to mature mRNA by excising introns.) Ribosomal RNAs appear in a long tandem array at the end of chromosome I. 5S RNAs appear in a tandem array on chromosome V. Some RNA genes appear in introns of protein-coding genes.

The *C. elegans* genome contains many repeat sequences. Approximately 2.6% of the genome consists of tandem repeats. Approximately 3.6% of the genome contains inverted repeats; these appear preferentially within introns, rather than between genes. Repeats of the hexamer sequence TTAGGC appear in many places. There are also simple duplications, involving hundreds to tens of thousands of kilobases.

Genome of *Drosophila melanogaster*

Drosophila melanogaster, the fruit fly, has been the subject of detailed studies of genetics and development for almost a century. Its genome sequence, the product of a collaboration between Celera Genomics and the Berkeley Drosophila Genome Project, was announced in 1999.

The chromosomes of *D. melanogaster* are nucleoprotein complexes. Approximately one-third of the genome is contained in heterochromatin, highly coiled and compact (and therefore densely staining) regions flanking the centromeres. The other two-thirds is euchromatin, a relatively uncoiled, less-compact form. Most of the active genes are in the euchromatin. The heterochromatin in *D. melanogaster* contains many tandem repeats of the sequence AATAACATAG, and relatively few genes.

The total chromosomal DNA of *D. melanogaster* contains about ~180 Mbp. The released sequence consists of the euchromatic portion, about 120 Mbp.

The genome is distributed over five chromosomes: three large autosomes, a Y chromosome, and a fifth tiny chromosome containing only ~1 Mb of euchromatin. The fly's 13601 genes are approximately double the number in yeast, but are fewer than in *C. elegans*, perhaps a surprise. The average density of genes in the euchromatin sequence is 1 gene/9 kb; much lower than the typical 1 gene/kb densities of prokaryotic genomes.

Despite the fact that insects are not very closely related to mammals, the fly genome is useful in the study of human disease. It contains homologues of 289 human genes implicated in various diseases, including cancer, and cardiovascular, neurological, endocrinological, renal, metabolic, and haematological diseases. Some of these homologues have different functions in humans and flies. Other human disease-associated genes can be introduced into, and studied in, the fly. For instance, the gene for human spinocerebellar ataxia type 3, when expressed in the fly, produces similar neuronal cell degeneration. There are now fly models for Parkinson disease and malaria.

The non-coding regions of the *D. melanogaster* genome must contain regions controlling spatiotemporal patterns of development. The developmental biology of the fly has been studied very intensively. It is, therefore, an organism in which the study of the genomics of development will prove extremely informative.

Genome of *Arabidopsis thaliana*

As a flowering plant *Arabidopsis thaliana* is a very distant relative of the other higher eukaryotic organisms for which genome sequences are available. It invites comparative analysis to identify common and specialized features.

The *Arabidopsis thaliana* genome contains ~125 Mbp of DNA, of which 115.4 Mbp was reported in 2000 by The *Arabidopsis* Genome Initiative. There are five pairs of chromosomes, containing 25 498 predicted genes. The genome is relatively compact, with 1 gene/4.6 kb on the average. This figure is intermediate between prokaryotes and *Drosophila*, and roughly similar to *C. elegans*. The genes of *Arabidopsis* are relatively small. Exons are typically 250 bp long, and introns relatively small, with mean length 170 bp. Typical of plant genes is an enrichment of coding regions in GC content.

Most *Arabidopsis* proteins have homologues in animals, but some systems are unique, among higher organisms, to plants, including cell wall production and photosynthesis. Many proteins shared with animals have diverged widely since the last common ancestor. Typical of another difference between plants and animals, 25% of the nuclear genes in plants have signal sequences governing their transport into organelles—mitochondria and chloroplasts—compared to 5% of mitochondrial-targeted nuclear genes in animals.

Traces of large- and small-scale duplication events appear in the *Arabidopsis* genome. Fifty-eight per cent of the genome contains 24 duplicated segments, ≥100 kb in length. The genes in the duplicated segments are usually not identifiable homologues. However, 4140 genes (17% of all genes) appear in 1528 tandem arrays of gene families, containing up to 23 members. The pattern of duplications suggests that an ancestor of *Arabidopsis* underwent a whole-genome duplication, to form a tetraploid species. That this happened a long time—estimated at 112 million years—ago is shown by the very high amount of loss of one copy of duplicated genes, and wide divergence of those that remain.

Higher plants must integrate the effects of three genomes— nuclear, chloroplast and mitochondrial. The organelle genomes are much smaller:

Gene distribution in *A. thaliana* between nucleus and organelles

	Nuclear	*Chloroplast*	*Mitochondrion*
Size (kb)	125,100	154	367
Protein genes	25,498	79	58
Density (kb/protein gene)	4.5	1.2	6.25

Many genes for proteins synthesized by nuclear genes and transported to organelles appear to have originated in the organelles and been transferred to the nucleus.

GENOME OF *HOMO SAPIENS* (HUMAN GENOME)

In February 2001, the International Human Genome Sequencing Consortium and Celera Genomics published, separately, drafts of the human genome. The sequence amounted to $\sim 3.2 \times 10^9$ bp, thirty times larger than the genomes of *C. elegans* or *D. melanogaster*. One reason for this disparity in size is that coding sequences form less than 5% of the human genome; repeat sequences over 50%. Perhaps the most surprising feature was the small number of genes identified. The finding of only about 30,000-40,000 genes suggests that alternative splicing patterns make a very significant contribution to our protein repertoire. It is estimated that ~35% of genes have alternative splicing patterns.

The human genome is distributed over 22 chromosome pairs plus the X and Y chromosomes. The DNA contents of the autosomes range from 279 Mbp down to 48 Mbp. The X chromosome contains 163 Mbp and the Y chromosome only 51 Mbp.

The exons of human protein-coding genes are relatively small compared to those in other known eukaryotic genomes. The introns are relatively long. As a result many protein-coding genes span long stretches of DNA. For instance, the dystrophin gene, coding for a 3 685 amino acid protein, is >2.4 Mbp long.

Protein Coding Genes

Analysis of the human protein repertoire implied by the genome sequence has proved difficult because of the problems in reliably detecting genes, and alternative splicing patterns. The International Human Genome Sequencing Consortium has estimated that the total count will be ~32000 genes in all.

The top categories in a functional classification are:

Function	*Number*	*%*
Nucleic acid binding	2207	14.0
DNA binding	1656	10.5
DNA repair protein	45	0.2
DNA replication factor	7	0.0
Transcription factor	986	6.2
RNA binding	380	2.4
Structural protein of ribosome	137	0.8
Translation factor	44	0.2
Transcription factor binding	6	0.0
Cell cycle regulator	75	0.4
Chaperone	154	0.9
Motor	85	0.5
Actin binding	129	0.8
Defense/immunity protein	603	3.8
Enzyme	3242	20.6
Peptidase	457	2.9
Endopeptidase	403	2.5
Protein kinase	839	5.3

Protein phosphatase	295	1.8
Enzyme activator	3	0.0
Enzyme inhibitor	132	0.8
Apoptosis inhibitor	28	0.1
Signal transduction	1790	11.4
Receptor	1318	8.4
Transmembrane receptor	1202	7.6
G-protein linked receptor	489	3.1
Olfactory receptor	71	0.4
Storage protein	7	0.0
Cell adhesion	189	1.2
Structural protein	714	4.5
Cytoskeletal structural protein	145	0.9
Transporter	682	4.3
Ion channel	269	1.7
Neurotransmitter transporter	19	0.1
Ligand binding or carrier	1536	9.7
Electron transfer	33	0.2
Cytochrome P450	50	0.3
Tumour suppressor	5	0.0
Unclassified	4813	30.6
Total	15683	100.0

A classification based on structure revealed the most common types:

Most common types of proteins

Protein	*Number*
Immunoglobulin and major histocompatibility complex domain	591
Zinc finger, C2H2 type	499
Eukaryotic protein kinase	459
Rhodopsin-like GPCR superfamily	346
Serine/Threonine protein kinase family active site	285
EGF-like domain	259
RNA-binding region RNP-1 (RNA recognition motif)	214
G-protein beta WD-40 repeats	196

Src homology 3 (SH3) domain	194
Pleckstrin homology (PH) domain	188
EF-hand family	185
Homeobox domain	179
Tyrosine kinase catalytic domain	173
Immunoglobulin V-type	163
RING finger	159
Proline rich extensin	156
Fibronectin type III domain	151
Ankyrin-repeat	135
KRAB box	133
Immunoglobulin subtype	128
Cadherin domain	118
PDZ domain (also known as DHR or GLGF)	117
Leucine-rich repeat	113
Serine proteases, trypsin family	108
Ras GTPase superfamily	103
Src homology 2 (SH2) domain	100
BTB/POZ domain	99
TPR repeat	92
AAA ATPase superfamily	92
Aspartic acid and asparagine hydroxylation site	91

Repeat Sequences

Repeat sequences comprise over 50% of the genome:

- Transposable elements, or interspersed repeats — almost half the entire genome! These include the LINEs and SINEs.
- Retroposed pseudogenes.
- Simple 'stutters' —repeats of short oligomers. These include the minisatellites and microsatellites. Trinucleotide repeats such as CAG, corresponding to glutamine repeats in the corresponding protein, are implicated in numerous diseases.
- Segmental duplications, of blocks of ~10-300 kb. Interchromosomal duplications appear on non-homologous chromosomes, sometimes at multiple sites. Some intrachromosomal duplications include closely spaced duplicated regions many kilobases long of very similar sequence implicated in genetic diseases; for example Charcot-Marie-Tooth syndrome type 1A, a progressive peripheral

neuropathy resulting from duplication of a region containing the gene for peripheral myelin protein 22.

- Blocks of tandem repeats, including gene families

RNA

RNA genes in the human genome include:

1. 497 transfer RNA genes. One large cluster contains 140 tRNA genes within a 4 Mb region on chromosome 6.
2. Genes for 28S and 5.8S ribosomal RNAs appear in a 44-kb tandem repeat unit of 150—200 copies. 5S RNA genes also appear in tandem arrays containing 200—300 genes, the largest of which is on chromosome 1.
3. Small nucleolar RNAs include two families of molecules that cleave and process ribosomal RNAs.
4. Spliceosomal snRNAs, including the U1, U2, U4, U5, and U6 snRNAs, many of which appear in clusters of tandem repeats of nearly-identical sequences, or inverted repeats.

Single-Nucleotide Polymorphisms (SNPs)

A single-nucleotide polymorphism or SNP (pronounced 'snip') is a genetic variation between individuals, limited to a single base pair which can be substituted, inserted or deleted. Sickle-cell anaemia is an example of a disease caused by a specific SNP: an A→T mutation in the β-globin gene changes a Glu→Val, creating a sticky surface on the haemoglobin molecule that leads to polymerization of the deoxy form.

Web resource: Human genome information

Interactive access to DNA and protein sequences:
http://www.ensembl.org/
Images of chromosomes, maps, loci:
http://www.ncbi.nlm.nih.gov/genome/guide/
Gene map 99:
http://www.ncbi.nlm.nih.gov/genemap99/
Overview of human genome structure:
http://hgrep.ims.u-tokyo.ac.jp
Single-nucleotide polymorphisms:
http://snp.cshl.org/
Human genetic disease:
http://www.ncbi.nlm.nih.gov/Omim/
http://www.geneclinics.org/profiles/all.html
Ethical, legal, social issues:
http://www.nhgri.nih.gov/ELSI/

SNPs are distributed throughout the genome, occurring on the average every 2000 base pairs. Although they arose by mutation, many positions containing SNPs have low mutation rates, and provide stable markers for mapping genes. A consortium of four major academic genome centres and 11 private companies is creating a high-quality, high-density human SNP map, with a commitment to open public access. A recently published version contains 1.42 million SNPs.

Not all SNPs are linked to diseases. Many are not within functional regions (although the density of SNPs is higher than the average in regions containing genes). Some SNPs that occur within exons are mutations to synonymous codons, or cause substitutions that do not significantly affect protein function. Other types of SNPs can cause more than local perturbation to a protein: (1) A mutation from a sense codon to a stop codon, or *vice versa*, will cause either premature truncation of protein synthesis or 'readthrough.' (2) A deletion or insertion will cause a phase shift in translation.

The A, B and O alleles of the genes for blood groups illustrate these possibilities. A and B alleles differ by four SNP substitutions. They code for related proteins that add different saccharide units to an antigen on the surface of red blood cells.

Allele	Sequence	Saccharide
A	...gctggtgacccctt...	N-acetylgalactosamine
B	...gctcgtcaccgcta...	galactose
O	...cgtggt-acccctt...	—

The O allele has undergone a mutation causing a phase shift, and produces no active enzyme. The red blood cells of type O individuals contain neither the A nor the B antigen. This is why people with type O blood are universal donors in blood transfusions. The loss of activity of the protein does not seem to carry any adverse consequences. Indeed, individuals of blood types Band O have greater resistance to smallpox.

Strong correlation of a disease with a specific SNP is advantageous in clinical work, because it is relatively easy to test for affected people or carriers. But if a disease arises from dysfunction of a specific protein, there ought to be many sites of mutations that could cause inactivation. A particular site may predominate if (1) all bearers of the gene are descendants of a single individual in whom the mutation occurred, and/or (2) the disease results from a *gain* rather than loss of a specific property, such as in the ability of sickle-cell haemoglobin to polymerize, and/or (3) the mutation rate at a particular site is unusually high, as in the Gly380→Arg mutation in the fibroblast growth receptor gene FGFR3, associated with achondroplasia (a syndrome including short stature).

In contrast, many independent mutations have been detected in the BRCA1 and BRCA2 genes, loci associated with increased disposition to early-onset breast and ovarian cancer. The normal gene products function as tumour suppressors. Insertion or deletion mutants causing phase shifts generally produce a missing or inactive protein. But it cannot be deduced a priori whether a novel substitution mutant in BRCA1 or BRCA2 confers increased risk or not.

Treatments of diseases caused by defective or absent proteins include:

1. *Providing normal protein.* We have mentioned insulin for diabetes, and Factor VIII in the most common type of haemophilia. Another example is the administration of human growth hormone in patients with an absence or severe reduction in normal levels. Use of recombinant proteins eliminates the risk of transmission of AIDS through blood transfusions or of Creutzfeldt-Jakob disease from growth hormone isolated from crude pituitary extracts.
2. *Life-style adjustments that make the function unnecessary.* Phenylketonuria (PKU) is a genetic disease caused by deficiency in phenylalanine hydroxylase, the enzyme that converts phenylalanine to tyrosine. Accumulation of high levels of phenylalanine causes developmental defects, including mental retardation. The symptoms can be avoided by a phenylalanine-free diet. Screening of newborns for high blood phenylalanine levels is legally required in the United States and many other countries.
3. Gene therapy to replace absent proteins is an active field of research.

Other clinical applications of SNPs reflect correlations between genotype and reaction to therapy (pharmacogenomics). For example, an SNP in the gene for N-acetyl transferase NAT—2 is correlated with peripheral neuropathy—weakness, numbness and pain in the arms, legs, hands or feet—as a side effect of treatment with isoniazid (isonicotinic acid hydrazide), a common treatment for tuberculosis. Patients with this SNP are given alternative treatment.

Genetic Diversity in Anthropology

SNP data are of great utility in anthropology, giving clues to historical variations in population size, and migration patterns.

Degrees of genetic diversity are interpretable in terms of the size of the founding population. Founders are the original set of individuals from whom an entire population is descended. Founders can be either original colonists, such as the Polynesians who first settled in New Zealand, or merely the survivors of a near-extinction. Cheetahs show the effects of a population bottleneck, estimated to have occurred 10 000 years ago. All living cheetahs are as closely related to one another as siblings. Extrapolations of mitochondrial DNA variation in contemporary humans suggest a single maternal ancestor who lived 140,000—200,000 years ago. Calling her Eve suggests that she was the first woman. But fossil evidence for humans goes back much longer. Mitochondrial Eve was in fact the founder of a surviving population following a near-extinction.

Population-specific SNPs are informative about migrations. Mitochondrial sequences provide information about female ancestors and Y chromosome sequences provide information about male ancestors. For example, it has been suggested that the population of Iceland—first inhabited about 1100 years ago—is descended from Scandinavian males, and from females from both Scandinavia and the British Isles. Medieval Icelandic writings refer to raids on settlements in the British Isles. A fascinating relationship between human DNA sequences and language families has been investigated by L.L. Cavalli-Sforza and colleagues. These studies have proved useful in working out interrelationships among American Indian languages. They confirm that the Basques, known to be a linguistically isolated population, have been genetically isolated also.

In the discovery of isolated populations, anthropological genetics provides data useful in medicine, for mapping disease genes is easier if the background variation is low. Genetically isolated populations in Europe include, in addition to the Basques, the Finns, Icelanders, Welsh, and Lapps. In Iceland, good genealogical and medical records are available. In a controversial move, the government of Iceland passed a law in 1998 authorizing the creation of a database of the medical records, family history and genetic sequences of its 275,000 citizens. A company, Decode Genetics, will collaborate with the worldwide pharmaceutical industry in the analysis and application of these resources.

Genetic Diversity and Personal Identification

Variations in our DNA sequences give us individual fingerprints, useful for identification and for establishment of relationships, including but not limited to questions of paternity. The use of DNA analysis as evidence in criminal trials is now well-established.

Genetic fingerprinting techniques were originally based on patterns of VNTRs, but have been extended to include analysis of other features including mitochondrial DNA sequences.

For most of us, all our mitochondria are genetically identical, a condition called homoplasmy. However, some individuals contain mitochondria with different DNA sequences; this is heteroplasmy. Such sequence variation in a disease gene can complicate the observed inheritance pattern of the disease.

The most famous case of heteroplasmy involved Tsar Nicholas II of Russia. After the revolution in 1917 the Tsar and his family were taken to exile in Yekaterinburg in Central Russia. During the night of 16-17 July 1918, the Tsar, Tsarina Alexandra, at least three of their five children, plus their physician and three servants who had accompanied the family, were killed, and their bodies buried in a secret grave. When the remains were rediscovered, assembly of the bones and examination of the dental work suggested—and sequence analysis confirmed—that the remains included an expected family group. The identity of the remains of the Tsarina were proved by matching the mitochondrial DNA sequence with that of a maternal relative, Prince Philip, Chancellor of the University of Cambridge, Duke of Edinburgh, and grandnephew of the Tsarina.

However, comparisons of mitochondrial DNA sequences of the putative remains of Nicholas II with those of two maternal relatives revealed a difference at base 16169: the Tsar had a C and the relatives a T. The extreme political and even religious sensitivities mandated that no doubts were tolerable. Further tests showed that the Tsar was heteroplasmic; T was a minor component of his mitochondrial DNA at position 16169. To confirm the identity beyond any reasonable question, the body of Grand Duke Georgij, brother of the Tsar, was exhumed, and was shown to have the same rare heteroplasmy.

Genetic Analysis of Cattle Domestication

Animal resources are an integral and essential aspect of human culture. Analysis of DNA sequences sheds light on their historical development and on the genetic variety characterizing modern breeding populations.

Contemporary domestic cattle include those familiar in Western Europe and North America, *Bos taurus*; and the zebu of Africa and India, *Bos indicus*. The most obvious difference in external appearance is the humpback of zebu. It has been widely believed that the domestication of cattle occurred once, about 8000-10,000 years ago, and that the two species subsequently diverged.

Analysis of mitochondrial DNA sequences from European, African and Asian cattle suggest however that (1) all European and African breeds are more closely related to each other than either is to Indian breeds, and (2) the two groups diverged about 200,000 years ago, implying recent independent domestications of different species. The similarity in physical appearance of African and Indian zebu (and other similarities at the molecular level; for instance, VNTR markers in nuclear DNA) must then be attributable to importation of cattle from India to East Africa.

Evolution of Genomes

The availability of complete information about genomic sequences has redirected research. A general challenge in analysis of genomes is to identify 'interesting events'. A background mutation rate in coding sequences is reflected in *synonymous* nucleotide substitutions: changes in codons that do not alter the amino acid. With this as a baseline one can search for instances in which there are significantly higher rates of *non-synonymous* nucleotide substitutions: changes in codons that cause mutations in the corresponding protein. (Note, however, that synonymous changes are not necessarily selectively neutral.)

Given two aligned gene sequences, we can calculate K_a = the number of non-synonymous substitutions, and K_s = the number of synonymous substitutions. (The calculation involves more than simple counting because of the need to estimate and correct for possible multiple changes.) A high ratio of K_a/K_s identifies pairs of sequences apparently showing positive selection, possibly even functional changes.

The new field of comparative genomics treats questions that can only now be addressed, such as:

- What *genes* do different phyla share? What genes are unique to different phyla? Do the arrangements of these genes in the genome vary from phylum to phylum?
- What homologous *proteins* do different phyla share? What proteins are unique to different phyla? Does the integration of the activities of these proteins vary from phylum to phylum? Do the mechanisms of control of expression patterns of these proteins vary from phylum to phylum?
- What *biochemical functions* do different phyla share? What biochemical functions are unique to different phyla? Does the

integration of these biochemical functions vary from phylum to phylum? If two phyla share a function, and the protein that carries out this function in one phylum has a homologue in the other, does the homologous protein carry out the same function?

The same questions could be asked about different species within each phylum.

The human genome may have first claim on our attention. But the ability of different genomes to illuminate one another is of the utmost importance.

M.A. Andrade, C. Ouzounis, C. Sander, J. Tamames, and A. Valencia compared the protein repertoire of species from the three major domains of life: *Haemophilus influenzae* represented the bacteria, *Methanococcus jannaschii* the eukarya, and *Saccharomyces cerevisiae* (yeast) the eukarya. Their classification of protein functions contained as major categories processes involving energy, information, and communication and regulation:

General functional classes

- Energy
  - — Biosynthesis of cofactors, amino acids
  - — Central and intermediary metabolism
  - — Energy metabolism
  - — Fatty acids and phospholipids
  - — Nucleotide biosynthesis
  - — Transport
- Information
  - — Replication
  - — Transcription
  - — Translation
- Communication and regulation
  - — Regulatory functions
  - — Cell envelope/cell wall
  - — Cellular processes

The number of genes in the three species are:

Species	*Number of genes*
Haemophilus influenzae	1680
Methanococcus jannaschii	1735
Saccharomyces cerevisiae	6278

Are there, among these, shared proteins for shared functions? In the category of Energy, proteins are shared across the three domains. In the category of Communication, proteins are unique to each domain. In the categories of Regulation and Information, archaea share some proteins with bacteria and others with eukarya.

Analysis of shared functions among all domains of life has led people to ask whether it might be possible to define a *minimal organism*—that is, an organism with the smallest gene complement consistent with independent life based on the central DNA→RNA→protein dogma (i.e. excluding protein-free life forms based solely on RNA). The minimal organism must have the ability to reproduce, but not be required to compete in growth and reproductive rate with other organisms. One may reasonably assume a generous nutrient medium, relieving the organism of biosynthetic responsibility, and dispensing with stress-response functions including DNA repair.

The smallest known independent organism is *Mycoplasma genitalium*, with 468 predicted protein sequences. In 1996, A.R. Mushegian and E.V. Koonin compared the genomes of *M. genitalium* and *H. influenzae*. (At the time, these were the only completely sequenced bacterial genomes.) The last common ancestor of these widely diverged bacteria lived about two billion years ago. Of 1703 protein-coding genes of *H. influenzae*, 240 are homologues of proteins in *M. genitalium*. Mushegian and Koonin reasoned that all of these must be essential, but might not be sufficient for autonomous life—for some essential functions might be carried out by unrelated proteins in the two organisms. For instance, the common set of 240 proteins left gaps in essential pathways, which could be filled by adding 22 enzymes from *M. genitalium*. Finally, removing functional redundancy and parasite-specific genes gave a list of 256 genes as the proposed necessary and sufficient minimal set.

What is in the proposed minimal genome? Functional classes include:

- Translation, including protein synthesis
- DNA replication
- Recombination and repair—a second function of essential proteins involved in DNA replication.
- Transcription apparatus
- Chaperone-like proteins
- Intermediary metabolism—the glycolytic pathway

- No nucleotide, amino acid, or fatty acid biosynthesis
- Protein-export machinery
- Limited repertoire of metabolite transport proteins

It should be emphasized that the viability of an organism with these proteins has not been proven. Moreover, even if experiments proved that some minimal gene content—the proposed set or some other set—is necessary and sufficient, this does not answer the related question of identifying the gene complement of the common ancestor of *M. genitalium* and *H. influenzae*, or of the earliest cellular forms of life. For only 71% of the proposed set of 256 proteins have recognizable homologues among eukaryotic or archaeal proteins.

Nevertheless, identification of functions necessary common to all forms of life allows us to investigate the extent to which different forms of life accomplish these functions in the same ways. Are similar reactions catalyzed in different species by homologous progeins? Genome analysis has revealed families of proteins with homologoues in archaea, bacteria, and eukarya. The assumption is that these have evolved from an individual ancestral gene through a series of speciation and duplication events, although some may be the effects of horizontal transfer. The challenge is to map common functions and common proteins.

Several thousand protein families have been identified with homologues in archaea, bacteria and eukarya. Different species contain different amounts of these common families. In bacteria, the range is from *Aquifex aeolicus*, 83% of the proteins of which have archaeal and eukaryotic homologues, to *Borrelia burgdorferi*, in which only 52% of the proteins have archaeal and eukaryotic homologues. Archaeal genomes have somewhat higher percentage (62-71%) of proteins with bacterial and eukaryotic homologues. But only 35% of the proteins of yeast have bacterial and archaeal homologues.

Does the common set of proteins carry out the common set of functions? Among the proteins of the minimal set identified from *M. genitalium*, only ~30% have homologues in all known genomes. Other essential functions must be carried out by unrelated proteins, or in some cases possibly by unrecognized homologues. The protein families for which homologues carry out common functions in archaea, bacteria and eukarya are enriched in those involved in translation.

The picture is emerging that evolution has explored the vast potential of proteins to different extents for different types of functions. It has been most conservative in the area of protein synthesis.

Protein functional class	*Number of families appearing in all known genomes*
Translation, including ribosome structure	53
Transcription	4
Replication, recombination, repair	5
Metabolism	9
Cellular processes: (chaperones, secretion, cell division, cell wall biosynthesis)	9

Please Pass the Genes: Horizontal Gene Transfer

Learning that *S. griseus* trypsin is more closely related to bovine trypsin than to other microbial proteinases, Brian Hartley commented in 1970 that, '... the bacterium must have been infected by a cow'. It was a clear example of lateral or horizontal gene transfer—a bacterium picking up a gene from the soil in which it was growing, which an organism of another species had deposited there. The classic experiments on pneumococcal transformation by O. Avery, C. MacLeod, and M. McCarthy that identified DNA as the genetic material were an even earlier example. In general, horizontal gene transfer is the acquisition of genetic material by one organism from another, by natural rather than laboratory procedures, through some means other than descent from a parent during replication or mating. Several mechanisms of horizontal gene transfer are known, including direct uptake, as in the pneumococcal transformation experiments, or via a viral carrier.

Analysis of genome sequences has shown that horizontal gene transfer is not a rare event, but has affected most genes in microorganisms. It requires a change in our thinking from ordinary 'clonal' or parental models of heredity. Evidence for horizontal transfer includes (1) discrepancies among evolutionary trees constructed from different genes, and (2) direct sequence comparisons between genes from different species:

- In *E. coli*, 755 ORFs (a total of 547.8 kb, ~18% of the genome) appear to have entered the genome by horizontal transfer after divergence from the *Salmonella* lineage 100 million years ago.
- In microbial evolution, horizontal gene transfer is more prevalent among operational genes—those responsible for 'housekeeping' activities such as biosynthesis—than among informational genes—

those responsible for organizational activities such as transcription and translation. For example, *Bradyrhizobium japonicum*, a nitrogen-fixing bacterium symbiotic with higher plants, has two glutamine synthetase genes. One is similar to those of its bacterial relatives; the other 50% identical to those of higher plants. Rubisco (ribulose1,5-bisphosphate carboxylase/oxygenase), the enzyme that first fixes carbon dioxide at the entry to the Calvin cycle of photosynthesis, has, been passed around between bacteria, mitochondria and algal plastids, as well as undergoing gene duplication. Many phage genes appearing in the *E. coli* genome provide further examples and point to a mechanism of transfer.

Nor is the phenomenon of horizontal gene transfer limited to prokaryotes. Both eukaryotes and prokaryotes are chimeras. Eukaryotes derive their informational genes primarily from an organism related to *Methanococcus*, and their operational genes primarily from proteobacteria with some contributions from cyanobacteria and methanogens. Almost all informational genes from *Methanococcus* itself are similar to those in yeast. Nor is gene transfer necessarily limited to ancient ancestors. It has been suggested (but is now disputed) that numerous bacterial genes have entered the human genome. Conversely, at least eight human genes appeared in the *M. tuberculosis* genome.

The observations hint at the model of a 'global organism', a genetic common market, or even a World Wide DNA Web from which organisms download genes at will! How can this be reconciled with the fact that the discreteness of species has been maintained? The conventional explanation is that the living world contains ecological 'niches' to which individual species are adapted. It is the discreteness of niches that explains the discreteness of species. But this explanation depends on the stability of normal heredity to maintain the fitness of the species. Why would not the global organism break down the lines of demarcation between species, just as global access to pop culture threatens to break down lines of demarcation between national and ethnic cultural heritages? Perhaps the answer is that it is the informational genes, which appear to be less subject to horizontal transfer, that determines the identity of the species.

It is interesting that although evidence for the importance of horizontal gene transfer is overwhelming, it was dismissed for a long time as rare and unimportant. The source of the intellectual discomfort is clear: Parent-to-child transmission of genes is at the heart of the Darwinian model of biological evolution whereby selection (differential reproduction) of parental phenotypes alters gene frequencies in the next generation. For offspring to gain genes from elsewhere than their parents smacks of Lamarck and other discredited alternatives to the paradigm. The evolutionary tree as an organizing principle of biological relationship is a deeply ingrained concept: scientists display an environmentalist-like fervour in their commitment to trees, even when trees are not an appropriate model of a network of relationships. Perhaps it is well to recall that Darwin knew nothing of genes, and the mechanism that generated the variation on which selection could operate was a mystery to him. Maybe he would have accepted horizontal gene transfer more easily than his followers!

Comparative Genomics of Eukaryotes

A comparison of the genomes of yeast, fly, worm and human revealed 1308 groups of proteins that appear in all four. These form a conserved core of proteins for basic functions, including metabolism, DNA replication and repair, and translation.

These proteins are made up of individual protein domains, including single-domain proteins, oligomeric proteins, and modular proteins containing many domains (the biggest, the muscle protein titin, contains 250-300 domains.) The proteins of the worm and fly are built from a structural repertoire containing about three times as many domains as the proteins of yeast. Human proteins are built from about twice as many as those of the worm and fly. Most of these domains appear also in bacteria and archaea, but some are specific to (probably, invented by) vertebrates (see following Table). These include proteins

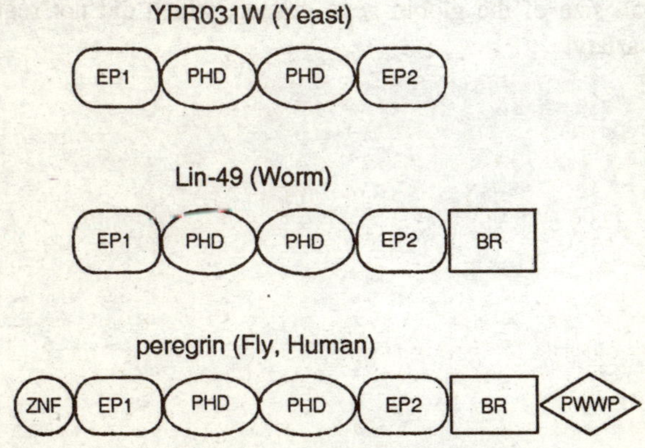

Fig. 1.5. Evolution by accretion of domains, of molecules related to perigrin, a human protein that probably functions in transcription regulation.

that mediate activities unique to vertebrates, such as defence and immunity proteins, and proteins in the nervous system; only one of them is an enzyme, a ribonuclease.

Distribution of probable homologues of predicted human proteins

Vertebrates only	22%
Vertebrates and other animals	24%
Animals and other eukaryotes	32%
Eukaryotes and prokaryotes	21%
No homologues in animals	1%
Prokaryotes only	1%

To create new proteins, inventing new domains is an unusual event. It is far more common to create different combinations of existing domains in increasingly complex ways. A common mechanism is by accretion of domains at the ends of modular proteins. This process can occur independently, and take different courses, in different phyla.

Gene duplication followed by divergence is a mechanism for creating protein families. For instance, there are 906 genes + pseudogenes for olfactory receptors in the human genome. These are estimated to bind ~10000 odour molecules. Homologues have been demonstrated in yeast and other fungi (some comparisons are odorous), but it is the need of vertebrates for a highly-developed sense of smell that multiplied and specialized the family to such a great extent. Eighty percent of the human olfactory receptor genes are in clusters. Compare the small size of the globin gene cluster, which did not require such great variety.

2

Genome Evolution

The organization of genes into genomes is a large question, and this chapter looks at only two aspects of this issue. Both involve non-Mendelian inheritance, in which genes are copied laterally through the DNA rather than vertically from parent to offspring. We first consider concerted evolution by looking at the definition of a gene family, and how such families originate by duplication. We then see that gene families often show an evolutionary pattern that is difficult or impossible to explain if mutations are independent in each gene of the family; it can be explained if mutations move laterally, however. Next, we move on to the evolution of selfish DNA—that is, functionless sequences of DNA that are passively copied from generation to generation. We see that large quantities of non-coding DNA are present in eukaryotic species, and that (at least in some species) much of it is repetitive, being made up of duplications of certain unit sequences. We look at the kinds of repetitive DNA and the genetic processes that operate in them, such as slipping and unequal crossing over among tandem repeats, and transposition between scattered repeats. If non-coding repetitive DNA is selfish DNA, it would explain the apparently unnecessary excess of DNA in eukaryotes.

Evolution of Whole Genome

The theory of population genetics has been concerned with discrete genetic loci that are inherited according to Mendel's laws. The theory shows how mutation, selection, drift, migration, and linkage determine changes in gene frequencies. It was first developed in the 1920s and 1930s, before the dawn of molecular genetics. Since the 1960s, however, it has become increasingly clear that only a part—and probably a small part, at that—of the DNA in an organism consists of discrete genes, each coding for a protein or having some regulatory function. Large amounts of non-coding DNA are present as well, which are often arranged in the form of repeating unit sequences. The genes themselves, moreover, can have distinctive arrangements on the chromosomes; related genes, for instance, may be distributed in clusters in the DNA. These discoveries do not contradict the theory of population genetics. Indeed, the theory of population genetics is the only theory we have to make sense of them. However, additional concepts are needed to explain this system. In this chapter we will concentrate on two such concepts: concerted evolution and selfish DNA. They are related in that both depend on non-Mendelian heredity.

Any theory of how a genetic (or any other) phenomenon evolves must specify how the phenomenon originated, and then how it spread through the population. In the standard theory of evolution, new forms originate as mutations at a locus and then spread by either natural selection or random drift. The ideas in this chapter describe mechanisms for the origin of variants, rather than for their spread through a population. Once the new genetic variant of the types discussed in this chapter has originated, it spreads or does not spread by the same processes of drift and selection as any other genes.

Arrangement of Genes

How are the many genes of an organism arranged in its DNA? The theory of population genetics would make good sense if single genes were scattered arbitrarily. Some genes do exist by themselves as single copies, but many genes are arranged along the chromosomes in groups of related genes. The groups are called *gene clusters*. Related genes may be arranged in more than one physical cluster, and a whole family of related genes is called a *gene family*. Gene clusters and gene families may vary in importance in different taxonomic groups. They seem to be much rarer, for example, in insects than in mammals, so most of this chapter will be concerned with mammalian and vertebrate genes. Gene clusters, where they exist, have peculiar evolutionary properties, which we can introduce by means of two well-studied cases: the ribosomal RNA genes and the globin gene family.

Eukaryotic ribosomes are made up of three classes of ribosomal RNA: 5S, 18S and 28S. The 18S and 28S parts are formed from a large 45S precursor molecule that is transcribed from the DNA as a whole. Along the DNA, the 45S ribosomal genes are arranged in rows of multiple copies of a recognizable unit; hundreds of these units are

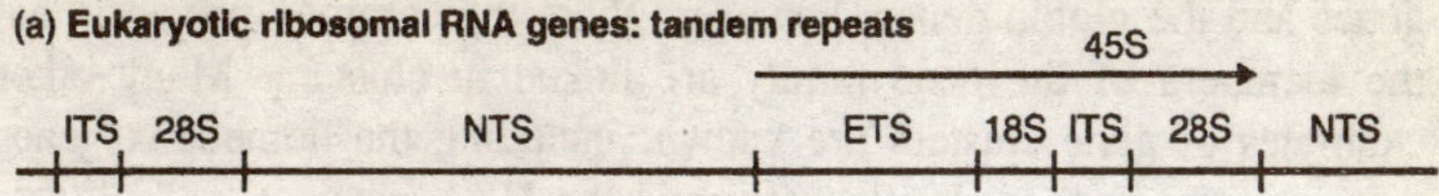

(b) Globin gene family: gene clusters

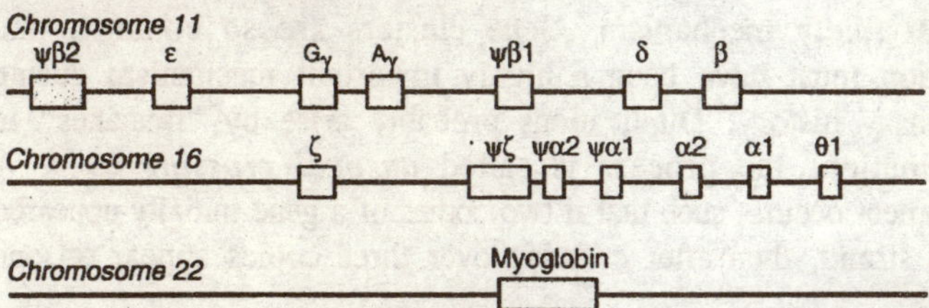

Fig. 2.1. Eukaryotic ribosomal RNA genes consist of long rows of tandem repeats of a unit genic organization. (b) The globin gene family consists of a number of clusters of related genes.

strung out in a row of *tandem repeats*. In the African clawed toad *Xenopus laevis*, the 400-600 tandem repeats all appear on one chromosome; in humans, approximately 300 repeats are scattered through five chromosomes in tandem repetitive blocks of various sizes. The organism presumably needs multiple copies of the gene to synthesize large quantities of the gene product. Other functional genes also exist in tandem repeats. The five histone genes, at least in *Drosophila* and sea urchins, have this structure. Although the histone genes of vertebrates are also a multiple gene family, they are scattered about, rather than appearing side by side.

The globin gene family has a slightly different arrangement. The hemoglobin molecule that transports oxygen in the blood comprises a number of components. Human hemoglobin is a tetramer. In an adult, it is made up of two α-globins and two β-globins; in the fetus, of two α- and two γ-globins; and in the embryo, two ∈- and two ζ-globins. The sequences of these five globin types are similar and the genes that code for them occur in clusters on the chromosomes. In humans, two main clusters of globin genes exist: the α-globin cluster on chromosome 16 and the β-globin cluster on chromosome 11; the myoglobin gene on chromosome 22 is also part of the same family. Unlike the ribosomal RNA genes, the globins are not simple tandem repeats of an identical set of genes. Instead, the different globin genes represent distinct genes with distinct sequences. The ribosomal RNA genes and the globin genes, however, share the common property that the members of the gene family are linked in clusters. Many other examples of gene clusters are known, including the homeobox genes that are important in development and the HLA genes.

GENE CLUSTERS

How could a cluster of genes, like the globin gene family, originate? For related genes strung out along a chromosome, duplication seems the most likely mechanism. Gene clusters are so common that duplication must have been a highly important mechanism during evolutionary history. Duplications probably arise by "mistakes" in recombination; the process is called *unequal crossing over*. A misalignment occurs, such that if two copies of a gene initially appeared on each strand, then after crossing over three copies appear on one

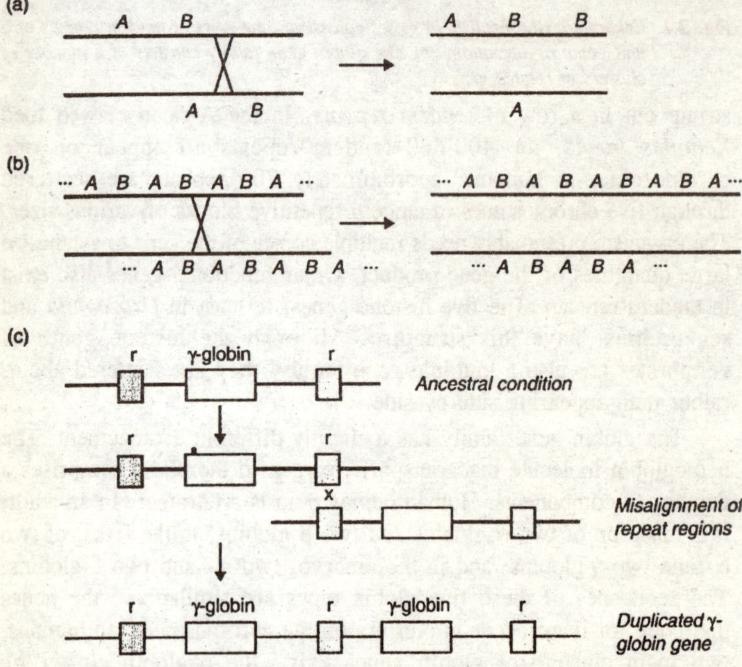

Fig. 2.2. Unequal crossing over occurs when the sequences of the two chromosomes are misaligned at recombination. (a) In a simple case, chromosomes with three copies and one copy of a gene could be generated from two chromosomes with two genes each. (b) In practice, misalignment is more likely if a long series of copies of similar sequences is present. (c) The human γ-globin genes include a repeat sequence (solid box) on either side of the gene itself (open box); this repeat could have been mismatched, causing duplication.

strand and one copy appears on the other strand. It is particularly easy to envisage a misalignment if a number of copies of a gene already occur in a row. The sequences of the two strands are matched up when they align for recombination and a gene could easily align with a non-complementary copy in the series of repeats.

In the human globin genes, duplication may have been facilitated by a non-coding repeat sequence between the globin genes. For example, the two γ-globin genes on chromosome 11 each have a short repeat sequence on either side. If this repeated sequence existed before the duplication, it could have been the target for the mismatching that produced the duplication. Certain variants of the human α-globin genes provide even more evidence for the importance of unequal crossing over among the globins. A normal human has two α-globin genes, but individuals can have three, or a single, α-globin gene. The obvious explanation is that these copies were generated by unequal crossing over.

Once a duplication has originated in a single copy, it could spread through the population by drift or selection. After the duplication became fixed, the two genes could retain the same sequence (like in ribosomal RNA genes) if selection favoured multiple copies of the gene. Alternatively, they could diverge to form a gene family (like the globins). A third possibility is that if only one of the copies of the gene is needed, the other might be switched off or decay by mutation and drift.

Heterozygous advantage at a locus may set up selection for gene duplication. Suppose that selection at a locus favours heterozygotes because an individual is better off with a two versions of a protein—perhaps because each protein is adapted to slightly different conditions. While these two versions are supplied by a single locus, disadvantageous homozygotes are generated when two heterozygotes breed together. If the gene were duplicated, the two versions could be supplied by homozygotes for each at the two loci, and the disadvantageous segregational load disappears.

The human globin gene family, as we saw, is arranged in two (or three, including myoglobin) clusters on two (or three) chromosomes. The kind of duplication may well explain the origin of many genes within each cluster, but is less likely to explain how more than one such cluster originates. Multiple clusters are more likely to have originated by translocation, by the duplication of part or all of a chromosome, or by polyploidy (i.e., the doubling, or multiplying, of the whole genome). For instance, suppose that one pair of chromosomes did not segregate equally at meiosis. One of the daughter cells would acquire a double set of the chromosome and the other would acquire none. The latter offspring would probably die, because it is missing the genes of the chromosome. Thus, the result is ultimately a mutation to a double set of the chromosome. As with gene duplication, after a chromosome has duplicated, its new set of genes could be maintained, diverge, or be suppressed, according to the outcome of drift or selective circumstances.

The globin gene family in two species of *Xenopus* supports this interpretation. In *Xenopus tropicalis*, unlike mammals, the α- and β-globin clusters are on the same chromosome. *X. laevis* (a tetraploid species), however, has two sets of the linked α- and β-globin gene cluster. If the α set were suppressed on one chromosome and the β set on the other, the mammalian arrangement would have evolved.

The sequences of the human globin genes can be arranged in a phylogeny. The phylogeny suggests that at least seven duplications have occurred in the history of the molecule. Once we have the tree, it is easy to use other molecular clock evidence to calibrate this scheme and estimate the times of the duplications. The difference in the sequence of the α- and β-globins, for instance, suggests they are derived from a duplication that happened approximately 500 million years ago.

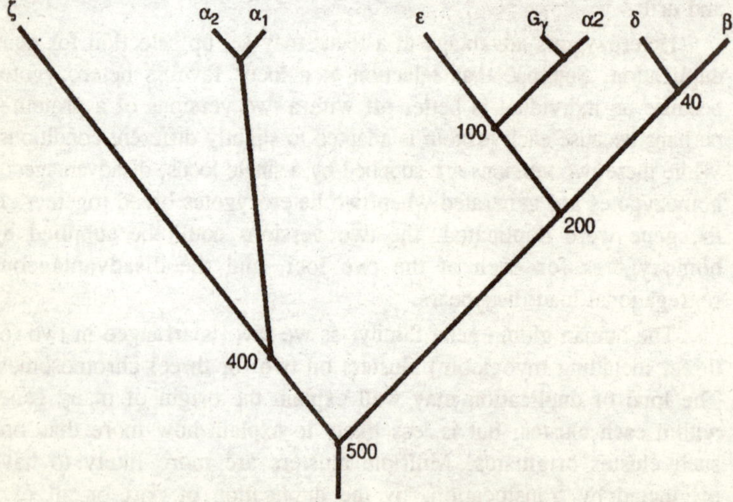

Fig. 2.3. A phylogeny of the human globin genes.

These inferred dates pose an interesting paradox. If we compare the sequences of the human α_1- and α_2-globins, the split apparently happened very recently. In fact, it may post-date the split between humans and the great apes and it certainly post-dates the split between the great apes and the rest of the primates. In this case, the primates outside the great apes should have only one α-globin. The prediction, however, is false: all the primates have two α-globin molecules. The full taxonomic distribution of the α-globins suggests that the gene duplicated at least before the origin of the mammals (about 85 million years ago), and probably before the split between mammals and birds (about 300 million years ago). Yet the similarity of the α_1- and α_2-globins in humans suggests the genes duplicated about 1 million years ago. Which of the figures is correct? This paradox, which is a common characteristic of evolution in gene families, is called *concerted evolution*.

Gene Family

If each tandem repeat of the ribosomal RNA gene evolved independently by mutation, drift, and selection, the different genes might gradually diverge from one another over time. Different mutations arise in different genes occasionally in the lineages of two species and become fixed. Because the evolutionary changes accumulate independently both in the different genes and in the different species, we might expect that the similarity between the genes in a gene family within a species would be approximately the same as the similarity between copies of the same gene in different species. In the modern species A and B at the top of Fig 2.4, if you examine one gene in the gene family, such as the gene at the extreme left, reveals one difference between it in the two species; there is also one difference between this gene and the fifth and seventh gene within species A.

This structure provides only a rough prediction of how evolution could proceed. The different genes in a gene family within a species have been separate since the duplication, whereas copies of the same gene in different species have been separate since the species split apart. If the duplication occurs in both species, then the speciation event will be more recent than the duplication. Other things being equal, the copies of the same gene in different species should then be more similar than the duplicates within a species. On the other hand, correlated selection pressures may operate within a species, tending to make the duplicated copies within a species more similar.

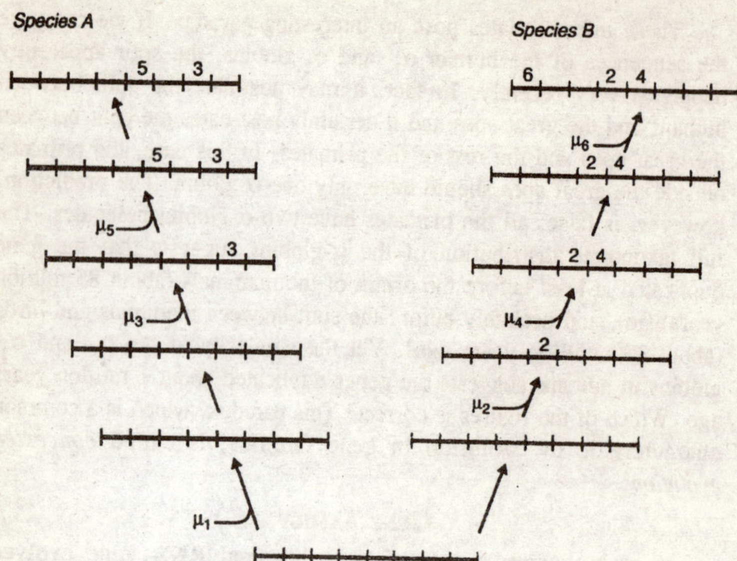

Fig. 2.4. Evolution of a gene family in two evolutionary lineages (in separate species).

And what about the facts? The first study, by Brown et al (1972), was made for the genes that code for ribosomal RNA in *Xenopus*. The 18S and 28S regions both fit the expected pattern. The sequences were almost identical in all tandem repeats of both *X. borealis* and *X. laevis*, suggesting strong stabilizing selection. For the non-transcribed spacer region (NTS), however, the pattern was different. Again, within a species the sequence was similar between tandem repeats, but this sequence differed in the two species. The NTS regions of *X. borealis* and *X. laevis* showed no more similarity than would two quite different genes—whereas within a species the NTS regions all had the same sequence. Thus, the NTS region showed concerted evolution.

The α-globin genes of primates described earlier can be used to illustrate the same principle. All primates, we saw, have two α-globins; we can therefore assume that the common ancestor of primates had two α-globin genes. The sequence of each α-globin gene differs between primate species. In the great apes, for example, any two species differ by about 2.5 amino acid substitutions in each gene. However, within a species, the α_1- and α_2-globins differ by only about one-tenth of that amount (this difference has been shown to be true not only for humans, but also for chimpanzees, gorillas, and rhesus monkeys). If one gene accumulates about 2.5 amino acid changes in the time between two species, then two different genes (α_1 and α_2) that have been separated for 300 million years should have accumulated many more changes—if they have been evolving independently. The conclusion is that they have not evolved independently; they have evolved in concert.

Why do gene families show concerted evolution? The concept is very difficult to explain if the genes evolve by mutations that occur independently at each gene locus. Indeed, independent mutations combined with neutral drift can be ruled out as an explanation. If the α_1- and α_2- genes drifted at random independently, they should be much more different than observation shown them to be within each species. Independent mutations and selection can, in principle, explain concerted

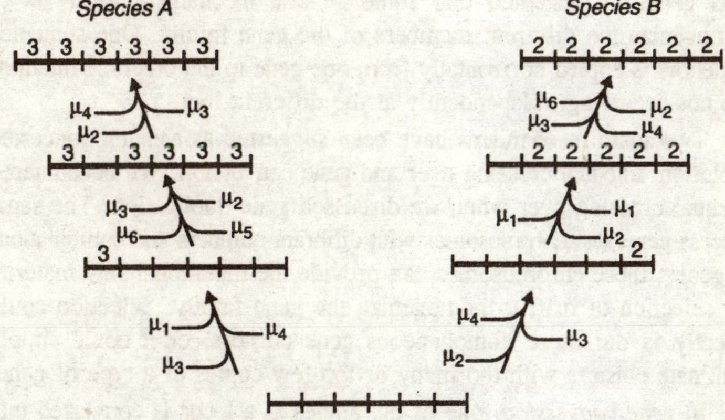

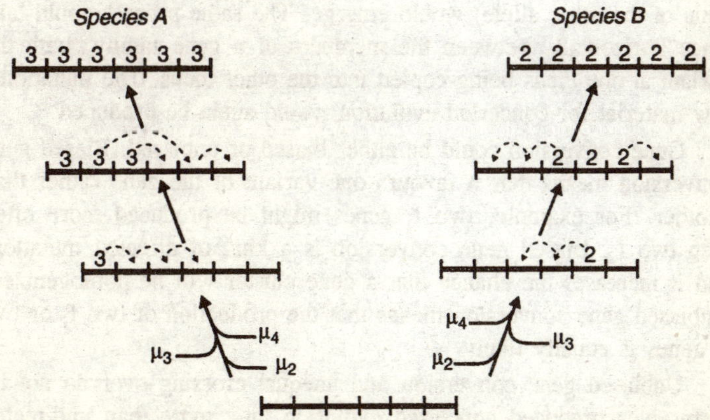

Fig. 2.5. Concerted evolution.

evolution. The same mutations would have to arise independently in both α_1- and α_2-globins in each species and then be fixed by selection, perhaps because one variant of the two genes was favoured in the conditions under which humans evolved, whereas another variant of both was favoured in the conditions under which chimpanzees evolved.

For the two globins in a number of primate species, this theory is a possibility. However, as the gene family showing concerted evolution grows larger, the explanation becomes increasingly incredible. When we reach the ribosomal RNA genes, with their innumerable tandem repeats, it is beyond belief that the same mutation could have occurred independently at each locus and been fixed by selection. In this case, it is generally accepted that some genetic mechanism must act to homogenize the different members of the gene family. One sequence somehow is copied horizontally from one gene to the others. Mutations are not occurring independently at the different loci.

Two main mechanisms have been suggested as causing concerted evolution: unequal crossing over and gene conversion. We encountered unequal crossing over when we discussed gene duplication. The same process generates chromosomes with different numbers and combinations of genes; these chromosomes can provide the mutational raw material for selection or drift to homogenize the gene family. Selection could directly favour more homogeneous gene clusters, or it could simply eliminate clusters with too many or too few copies of a type of gene.

In *gene conversion*, one of the alleles at a locus is converted into the other allele. Thus, when the heterozygote f_1f_2 segregated, instead of the Mendelian proportions of one f_1 for one f_2, two f_1 or two f_2 (and none of the other allele) would emerge. The same process could take place horizontally between the members of a gene family, with the variant at one locus being copied into the other locus. The mutational raw material for concerted evolution would again be produced.

Gene conversion could be either biased or unbiased. Biased gene conversion means that it favours one variant of the gene rather than another. For example, two f_1 genes might be produced more often than two f_2. Biased gene conversion is a kind of directed mutation, and it increases the chance that a gene cluster will be homogenized. Unbiased gene conversion means that the production of two f_1 or two f_2 genes is equally likely.

Unbiased gene conversion and unequal crossing over do not by themselves produced concerted evolution, any more than undirected mutation alone produces other types of evolution. They account only for the origin of variants—that is, they cause some individuals of a population to have more homogeneous gene families than other individuals. Selection and drift are still needed to explain how one of the gene family variants becomes fixed in the population. In the case of drift, a lateral "march to homozygosity" would occur between loci much like the process for a single locus.

The relative importance of unequal crossing over and gene conversion is unknown, by Hillis et al. (1991) found that the genes for ribosomal RNA showed concerted evolution even in an asexually reproducing lizard, which does not have crossing over. Gene conversion was therefore presumably responsible for the concerted evolution. Moreover, one sequence variant was favoured, suggesting that gene conversion may be biased in this case.

So far we have considered concerted evolution in a single gene cluster. The process is not confined to gene clusters on one chromosome, however. The ribosomal RNA genes in primates appear in clusters on five chromosomes, but the genes on different chromosomes show concerted evolution like genes on the same chromosome. This evolution is probably made possible when the regions of the chromosomes encoding the ribosomal RNA genes physically come into association. Unequal crossing over, or gene conversion, could then take place, just like that hypothesized to allow concerted evolution among genes on the same chromosome.

Concerted evolution is usually confined to only some of the genes in a gene family. For example, neither the α- and β-, nor the β- and γ-globins evolve in concert, even though they belong to the same gene family. The reason is that gene conversion takes place only between genes with some degree of sequence similarity. The chance of gene conversion is probably roughly proportional to the sequence similarity between two genes. A newly duplicated gene, therefore, might undergo one of two evolutionary courses. If gene conversion happens before the two genes have diverged dramatically, they may evolve in concert. If, however, they diverge too far for gene conversion to be possible, they will "escape" from one another, and their future evolution will happen independently. At a further stage, when they have diverged far enough, we should cease to classify these genes as members of the same gene family. Which of these two fates a pair of gene undergoes will be determined by the relative rates of gene conversion and of the mutations that break down sequence similarity. If the former is high relative to the latter, concerted evolution is likely; if the opposite is true, the genes will escape and diverge from one another.

The chance of gene conversion is probably not simply proportional to sequence similarity. The insertion of a mobile genetic element into a gene can prevent future conversion of that gene, or of part of it to one side of the insertion. Schimenti and Duncan have suggested that the insertion of an *Alu* element has prevented gene conversion, and concerted evolution, between certain genes in the β-globin families of cows and goats. The chance of concerted evolution is, therefore, controlled by both the sequence similarity of two genes and the insertion of a mobile genetic element in them.

In summary, the genes in gene clusters evolve in parallel. This concerted evolution is probably not caused by selection and drift operating on independent mutations at each locus; rather, it requires some mechanism for the same mutation to arise at the different loci. Gene conversion or unequal crossing over allow mutations to move between loci, and selection or drift could then fix the same mutation at all loci. In the case of biased gene conversion, the mutational mechanism itself will help to fix the favoured variant. The result would be the observed pattern of concerted evolution.

Coding of Genes by DNA

So far we have been concerned with genes that code for proteins, or at least for RNA. However, much of the DNA in the organisms of many species probably does not code for anything. Consider humans as an example. The human genome contains about 3×10^9 nucleotide pairs, and geneticists disagree about what proportion of it is *coding DNA* (i.e., DNA that is transcribed to produce proteins or regulate the production of proteins). Estimates are fallible, but the figure may be about 10-25%; the highest estimates go up to 50%. Even if 50% of the genome does code for genes, a large proportion of the genome still does something other than coding for, or controlling the production of, proteins. DNA that does not code for genes is called *non-coding* DNA.

DNA re-association experiments allow an estimate of the proportion of non-coding DNA. In these experiments, the DNA is first heated and melted into single strands, and the single-stranded DNA is then allowed, in a cooler solution, to join together (re-associate). The rate at which the strands re-associate is controlled by their sequence similarity. Early work identified three classes of DNA. A part of the DNA re-associated quickly; it is mainly highly repetitive DNA, made up of large numbers of repeats of simple sequences. At the other extreme, "single-copy" DNA joined together relatively slowly; this part of the DNA probably codes for most of the genes. In between is a class of middle-repetitive DNA, which is repeated but not as much as the highly repetitive DNA. On the assumption that the single-copy DNA serves as the coding part of the genome, we can use the proportion of single-copy DNA (i.e., slowly re-associating DNA) as a first estimate of the coding DNA in the genome.

Table 2.1. list the genome sizes and proportion of single-copy DNA for a number of species. These numbers suggest, in two ways, that a large quantity of non-coding DNA exist.

First, the figures for single-copy DNA are often much less than 100%. Only one-fifth of the DNA of the toad *Bufo bufo* is single-copy DNA, for example. If only single-copy DNA acts as the coding part, then four-fifths of *B. bufo*'s DNA is non-coding. This observation is generally true in eukaryotes, but not in prokaryotes. Bacterial DNA appears to be much more economically organized.

Second, large differences appear between species, and these differences can hardly all be due to variations in the numbers of coding genes. Fifteen times as many genes may be needed to build a human as to build a protozoan. In the words of Orgel and Crick, however, "it seems implausible that the number of radically different genes needed in a salamander is 20 times that in a man." The obvious deduction is that the differences derive mainly from non-coding DNA. Thus, a genome consists of more than simply genes.

The experiments provide a second clue about the nature of the non-coding DNA—namely, that it is often repetitive. Geneticists distinguish a number of kinds of repetitive DNA. A first distinction is between *tandem* and *scattered repeats*—that is, whether the repeat sequences appear next to one another (in tandem) or are scattered through the genome. Repetitive DNA can also be classified by the length and number of the repeat units.

No generally agreed classification exists for repetitive DNA, although the suggested classifications are all similar. We will use the following classification developed by Charlesworth et al. (1994). There may be other kinds of repetitive DNA that are not readily classifiable in one of these categories—or there may not be: the question is a topic of current research. (The word "locus" features in the descriptions below. It is used here in a slightly different way from traditional genetics: a locus for a tandem repeat is the site in the chromosome where a whole series of repeats is present. Thus, it refers to the whole array, not just one unit within it.)

Table 2.1. Amount of DNA percentage of single copy DNA in various animal species.

Invertebrates			*Vertebrates*		
Species	*1C (pg)*	*% sc DNA*	*Species*	*1C (pg)*	*% sc DNA*
Protozoa			Protochordata		
Tetrahymena pyriformis	0.2	90	*Ciona intestinalis*	0.2	70
Coelenterata			Pisces		
Aurelia aurita	0.7	70	*Scyliorhinus stellatus*	6.1	39
Nemertini (Rhyncocoela)			*Leuascus cephalus*	5.5	44
Cerebratulus	1.4	60	*Raja montagui*	3.4	47
Mollusca			*Rutilus rutilus*	4.8	54
Aplysia california	1.8	55	Amphibia		
Crassostrea virginica	0.7	60	*Necturus masculosus*	83.0	12
Spisula solidissima	1.2	75	*Bufo bufo*	7.0	20
Loligo loligo	2.8	75	*Triturus cristatus*	21.0	47
Arthropoda			*Xenopus laevis*	3.1	75
Prosimulium multidentatum	0.18	56	Reptilia		
Drosophila melanogaster	0.18	60	*Natrix natrix*	2.5	47
Limulus polyphemus	2.8	70	*Terrapene carolina*	4.1	54
Musca domestica	0.9	90	*Caiman crocodylus*	2.6	66
Chironomous tentans	0.21	95	*Python reticulatus*	1.7	71
Echinodermata			Aves		
Strongylocentrotus purpuratus	0.9	75	*Gallus domesticus*	1.2	80
			Mammalia		
			Homo sapiens	3.5	64
			Mus musculus	3.5	70

Tandem Repeats

1. *Microsatellites*. Repeats of short (2-5 base pairs) nucleotide sequences. The number of repeats varies between loci, but the average is on the order of approximately 100 repeats. Many microsatellite loci are scattered through the genome; an average human, for example, contains about 30,000 microsatellite loci. Microsatellites have also been detected in other vertebrates, insects, and plants.
2. *Minisatellites*. Repeats of longer (approximately 15 base pairs) nucleotide sequences. The number of repeats varies minisatellite loci, and many loci are scattered through the genome. The average length of any one locus is usually about 500-2000 nucleotides. Minisatellites have been studied in humans and other vertebrates, fungi, and plants.
3. *Satellite DNA*. The size of the repeated unit varies in different cases, with some being as small (5-15 base pairs) as micro- and minisatellites, while others are larger (about 100 base pairs). They are often found in large blocks of 1000 or more repeats of the unit sequence in regions of the chromosome near the centromere or the telomere. They do not seem to be as variable in the number of repeats per site as are micro- and minisatellites.

Scattered Repeats

Longer sequences (on the order of approximately 100 base pairs) that are distributed throughout the genome, usually in single copies bounded by other sequences rather than in tandem repeats. Three examples from humans are the sequences called *Alu*, *Kpn*, and poly (C-A); the three together make up about 20% of the human genome.

A unit sequence of any of the three kinds of tandem repetitive DNA can be found at more than one place in the genome. At all those sites they are found in the form of tandem repeats. Scattered repeats are also found at many sites, but usually with one a single copy of the sequence appearing at each site.

Two main evolutionary hypotheses have been suggested to explain the presence of this repetitive, non-coding DNA. One hypothesis says that such DNA is functional, even though it does not actually encode genes. It may be needed for some regulatory or structural reason, for example, or perhaps it keeps the genes apart or correctly configured in the DNA molecule's three-dimensioned shape. Alternatively, the repetitive DNA may be selfish DNA—either neutral "junk" DNA, or parasitic DNA.

We will concentrate on the "selfish DNA" hypothesis here, for two reasons. First, it is widely accepted as the explanation for much of the repetitive non-coding DNA. Second, it is a new evolutionary concept. If the repetitive DNA is functional, it will evolve just like genic DNA and requires no special discussion. If the DNA is non-functional for the organism that contains it, something different is occurring. It is also worth noting that comparisons of non-coding repetitive DNA made between different species have often shown the presence of concerted evolution, as in the gene families of coding DNA discussed earlier. The mechanisms presumably operate in much the same way in the two cases, and we will not discuss concerted evolution here.

Selfish DNA

In 1980, Doolittle and Sapienza, and Orgel and Crick, made more explicit an idea that many other geneticists had also pondered—the idea that some, or all, repetitive DNA may have no use for the organism. It may instead be *selfish DNA*. Selfish DNA is non-transcribed and non-coding, and contributes nothing to the well-being of the organism. In most cases, such DNA is selectively neutral except for the energetic burden of replicating it. If it was excised, the organism would suffer no disadvantage. After a stretch of selfish DNA originates, it is passively replicated and passed on from parent to offspring. Changes in its frequency in the population would occur by drift. If a non-coding sequence was not neutral—perhaps because it interfered with the construction or metabolism of the organism—it would be selected against. Likewise, if it accumulated to such an extent that the cell cycle was slowed by the need to replicate the entire DNA sequence, selection would probably act to reduce it. Provided the quantity of a particular sequence is not excessive, is not transcribed, and accumulates in parts of the genome where is does not interfere with genic regulation and transcription, there is no reason why selfish DNA should not evolve.

Selfish DNA might be either "active" or "passive". The sequence itself might not influence the chance that it spreads in the DNA and is retained. It is then a passive kind of selfish DNA and could accumulate as "junk DNA" in the genome. Alternatively, a particular sequence might, for some reason related to its sequence, have a better than average chance of spreading through the DNA. These sequences would be a more active, parasitic kind of selfish DNA, and they would proliferate until checked by natural selection because they interfered with vital DNA functions or became so abundant that their replication imposed a significant cost on the organism. For the possible examples of selfish DNA we shall discuss here, we do not know for certain to which class they belong. This distinction should be kept in mind, however.

How might selfish DNA originate? In theory, any kind of mutation could give rise to selfish DNA. A point mutation could inactivate a gene, converting it into a pseudogene. If the inactivated gene was redundant and the new pseudogene was not in a position to interfere with any important function, the mutation could be fixed by drift and the pseudogene retained as passive selfish DNA. For repetitive DNA, other kinds of mutations would be needed.

We will discuss tandem and scattered repetitive DNA separately, because they are believed to originate by different mechanisms. In this section, we will examine three ways of creating tandem repetitive DNA briefly.

The three processes that have been suggested to create tandem repeats are called slippage, unequal crossing over, and rolling circle replication.

Slippage occurs when one DNA strand is being created by replication from another strand. The two strands may slip relative to one another by a few nucleotides, such that either a few nucleotides

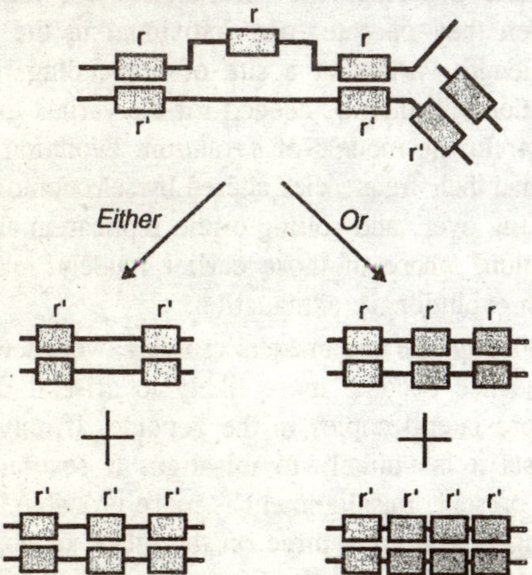

Fig. 2.6. Slippage occurs when a sequence of DNA is copied twice, generating a new strand with an extra copy of that sequence, or missed out generating a new strand with one copy less.

are missed or copied twice. Such an event is most probable in a region of short repeats, and the result is a new strand that has either more or less repeats than the original strand. Slippage occurs only over short distances, and is the most likely mechanism by which new length variants arise for microsatellite sequences. It may produce new length variants in minisatellites as well, although unequal crossing over is also important in this case.

Unequal crossing over can produce a strand with a greater or lesser number of repeats, and successive rounds of unequal crossing over can build up or break down long sequences of repeats. Unequal crossing over has not yet been confirmed to operate in minisatellites, but it is strongly suspected to play a role based on indirect evidence. While slippage could operate in the short unit satellites, unequal crossing over could operate in all of them.

The third mechanism, *rolling circle replication*, has also been suggested to affect all types of satellites. This method of replication has been well studied in bacterial plasmids and some viruses; the idea that it operates in eukaryotes is more speculative. It would require the excision of a sequence from the chromosome. The sequence would then have to form a circle, after which it could multiply by rolling circle replication. The replicated sequences would then have to be reinserted into the chromosome.

All of these processes are mechanisms for the origin of new variants. When they operate, one individual in the population will have a new length variant at a site of non-coding repetitive DNA. Drift or selection will then be needed for the variant to spread through the population. In the models of evolution, evolution occurred when rare variants had their frequencies altered by selection or drift. Slippage, unequal crossing over, and rolling circle replication are all analogous to the "mutation" phase of those earlier models, and are thus only one part of an evolutionary explanation.

Under both slippage and unequal crossing over, new length variants of a unit sequence become more likely to arise if that sequence is present in more lateral copies in the genome. If only one copy of a sequence exists it is unlikely to misalign. If several repeats of the sequence are present, misalignment is more probable, as copy two on one strand aligns with copy three on the other, or copy eight on one strand aligns with copy six on the other. The more copies of a sequence that exist, the more likely this misalignment becomes. If, therefore, a genome contained two sorts of neutral, functionless sequence—one in single copy and the other in multiple repeats—then (other things being equal) the repeated sequence would be more likely to grow laterally through the genome during evolution. If the highly repetitive rows of tandem repeats are actually selfish DNA, their lengths will fluctuate through evolutionary time. Unequal crossing over, slippage, and drift will constantly be causing them to shrink and grow. At any one time, some very long regions of repeats will be found, and these regions will be recognized as highly repetitive DNA. After a few million years, a currently long repeat could potentially have shrunk and some other, previously shorter series lengthened. The frequency distribution of lengths will likely remain probabilistically constant.

Minisatellites (Short Repeats)

While studying the human myoglobin gene in the early 1980s, Jeffreys discovered a short sequence of repeated DNA within an intron. He used the short sequence as a "probe" to see whether the same sequence was present anywhere else in the genome. The probe hybridized at several regions. Jeffreys then extracted the DNA from all these regions, and found that each DNA sample consisted of a unit sequence in a row of tandem repeats. The number of repeats at each site was highly variable between individuals; one site, for example, might contain 25 repeats in one individual, 5 in another individual, and 47 in another. Although the unit sequence itself is variable, a common core sequence has been characterized and is 16 (or perhaps 8) bases long. The particular sequence is not the only sequence that can form this sort of scattered short set of repeats, but it is the most studied. Jeffreys called these sequences minisatellites; the name (like

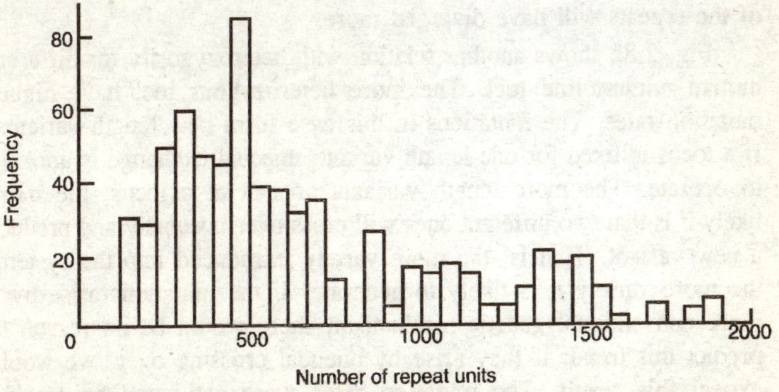

Fig. 2.7. Frequency distribution of numbers of repeats at one minisatellite locus in a total of 344 individuals (688 gametes).

that of microsatellites and satellites) derives from the way they can be detected when DNA is centrifuged.

Why do minisatellites vary so much in the number of repeats? We saw earlier that slippage and unequal crossing over are thought to generate new length variants, and we can now look at two predictions of this idea. Both concern the heterozygosities of different minisatellite loci. The alleles at a minisatellite locus are the different length variants, which differ in their number of repeats of the unit sequence. We can measure the frequencies of the different length variants at each locus, and calculate the heterozygosity for each by the standard formula.

The first relation to consider exists between the heterozygosity of a locus and the similarity of the unit sequence. So far we have described repetitive DNA as if the unit sequence were identical in all repeats. In practice the sequence can differ slightly in terms repeats, with greater differences appearing at some loci. When Stephan and Cho plotted the heterozygosity against the sequence similarity for 10 minisatellite loci in humans, they found a strong negative relation. The loci that are most variable in numbers of length variants possess the most similar unit sequences. The relation is expected if new variants arise by unequal crossing over, for two reasons. First, unequal crossing over is most likely if the unit sequences are more similar. Second, unequal crossing over causes the same variant to be spread through the DNA at a site. If unequal crossing over has happened more frequently at one locus than at another, the unit sequences should look more similar at those loci where the process is more common. If unequal crossing over has not occurred much at a locus, the sequences of the repeats will have diverged more.

Fig. 2.8b shows another relation with heterozygosity for different human minisatellite loci. The more heterozygous loci have higher mutation rates. The mutations in this case form new length variants. If a locus is fixed for one length variant, unequal exchange is unlikely to operate. The more length variants present at a locus, the more likely it is that two different ones will cross over unequally and produce a new variant. That is, the more variety introduced into the system, the more variety it is likely to generate. If the mutations arise by a more conventional genetic mechanism, there would be no reason to predict this trend; if they arise by unequal crossing over, we would expect this result. The mutation rates were estimated by tracing minisatellites through human pedigrees, to see when new variants arose. Notice that the rate is high, up to 5% per gamete.

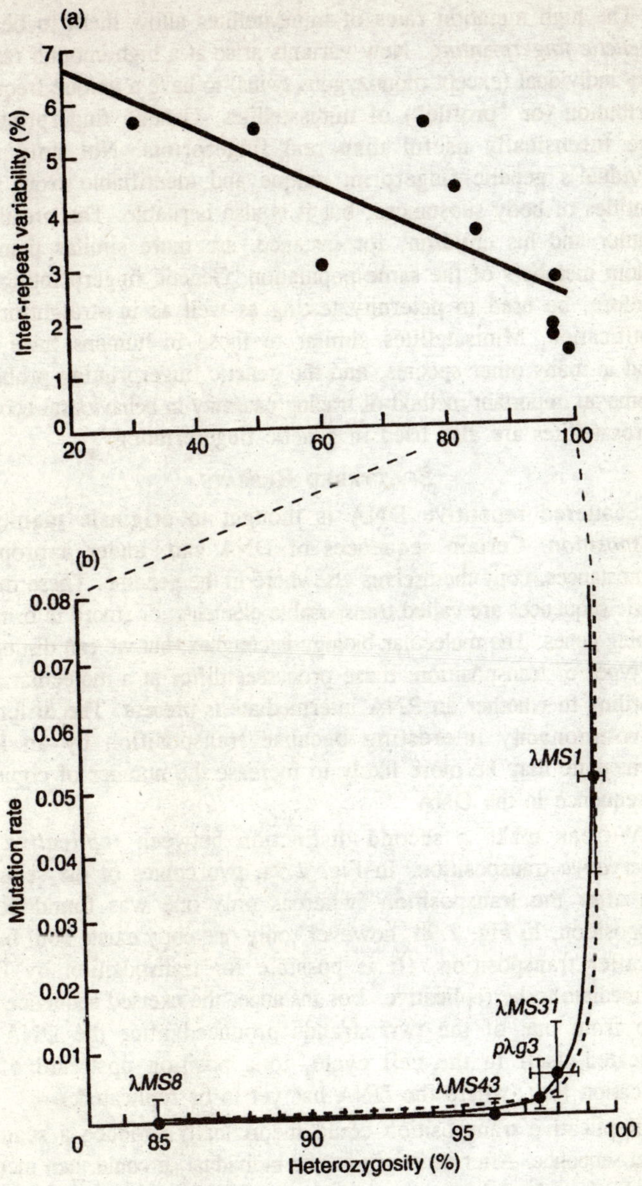

Fig. 2.8. Two relations of the variation of different minisatellite loci in humans. (a) The sequences of the units are more similar at loci that are more heterozygous. (b) Mutation rates are higher at more variable loci. Both relationships make sense if length variants at minisatellites mainly originate by unequal crossing over.

The high mutation rates of minisatellites allow them to be used in *genetic fingerprinting*. New variants arise at a high enough rate for every individual (except monozygous twins) to have a unique frequency distribution (or "profile") of minisatellites. Genetic fingerprints are more forensically useful than real fingerprints. Not only is an individual's genetic fingerprint unique and identifiable from small quantities of body substances, but it is also heritable. The profiles of a father and his children, for instance, are more similar than two random members of the same population. Genetic fingerprinting can, therefore, be used in paternity testing as well as in straightforward identification. Minisatellites similar to those in humans have been found in many other species, and the genetic fingerprinting probe has become an important method of tracing paternity in behavioural ecology. Microsatellites are also used in genetic fingerprinting.

Scattered Repeats

Scattered repetitive DNA is thought to originate mainly by *transposition*. Certain sequences of DNA can, under appropriate circumstances, copy themselves elsewhere in the genome. These mobile genetic sequences are called transposable elements, or (more informally) jumping genes. The molecular biology is complex, but we can distinguish two types of transposition; these processes differ at a molecular level according to whether an RNA intermediate is present. The difference is evolutionarily interesting because transposition by an RNA intermediate may be more likely to increase the number of copies of the sequence in the DNA.

We can make a second distinction between *replicative* and *conservative* transposition. In Fig. 2.9a, two copies of the sequence exist after the transposition, whereas only one was found before transposition. In Fig. 2.9b, however, only one copy exists both before and after transposition. (It is possible for transposition by DNA intermediate to be replicative. For instance, the excised sequence may jump from one of the two strands produced after the DNA has replicated itself in the cell cycle, to a position upstream of the replication fork, where the DNA has yet to be replicated.)

Replicative transposition could theoretically produce a scattered repeat sequence. After its origin in one individual, it could then increase in frequency (by selection or drift) to result in the pattern of scattered repeats we now observe. Did real scattered sequences evolve in this way? The best evidence that a scattered repeat sequence had originated by transposition would be found by observing the sequence transposing.

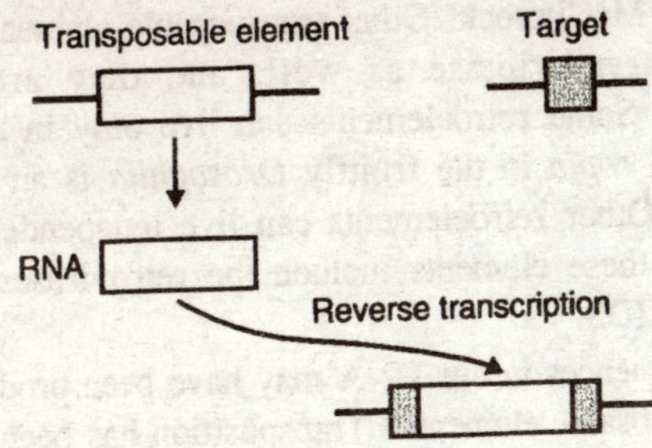

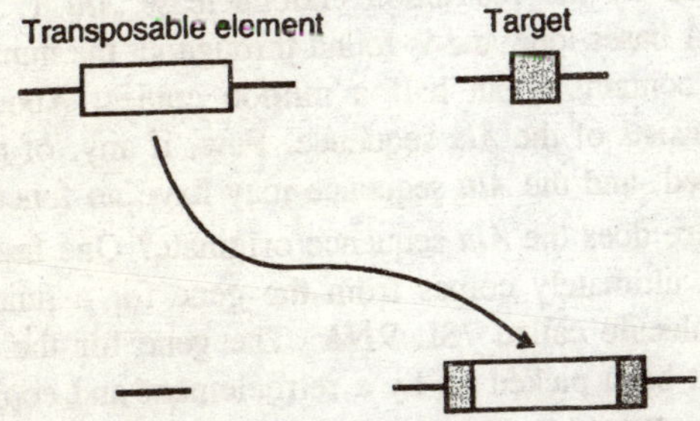

Fig. 2.9. Two main kinds of transposable elements. (a) Retroelements, which transpose by the reverse transcription of an RNA intermediate. (b) Transposable elements, which transpose by excising from one site and inserting elsewhere, without an DNA intermediate.

A more general method is to study the characteristics of the DNA sequences that are known to transpose, and of the scattered repeats, to see whether the latter resemble the former. Transposable elements have some characteristics features, and the similarity between these and at least some well-studied scattered repeat sequences is enough to make a compelling case for their origin by transposition. Let us look first at some known (or strongly suspected) transposable elements, and then at an example of a scattered repeat.

The simplest transposable elements lack the enzyme reverse transcriptase. They possess only a "transposase" enzyme together with one or more other genes, and are transposed by a DNA intermediate. Such elements usually have inverted repeat sequences at either end; the same sequence is found at both ends, with one sequence forming a mirror image of the other. The most famous examples are the first transposable elements to have been described—the ones in maize discovered by McClintock. Other transposable elements have the gene for reverse transcriptase as well, and they are classified as retroelements. Some retroelements can live only in the genome; the element called *copia* in the fruitfly *Drosophila* is an example of this phenomenon. Other retroelements can live independently or within a host genome; these elements include the retroviruses, such as HIV, the agent of AIDS.

Which sequences in our DNA may have been produced from these kinds of transposable elements? Transposition has been most studied in bacteria, and our knowledge for multicellular organisms is less certain, but some likely examples have been identified. The *Alu* sequence is a hot candidate in mammals. *Alu* is the name for a characteristic sequence recognized by the restriction endonuclease *Alu* 1. The sequence is about 281 bases long and is found throughout the human genome; each genome contains about half a million copies. About 5% of human DNA consists of the *Alu* sequence. Few, if any, of the sequences are transcribed, and the *Alu* sequence may have no function.

Where does the *Alu* sequence originate? One favoured hypothesis is that it ultimately comes from the gene for a functional ribosomal RNA molecule called 7SL RNA. The gene for the 7SL RNA could have first been picked up by a retroelement and copied back into the DNA by reverse transcription; the many *Alu* sequences are believed to have been formed as second-order derivatives of the 7SL RNA gene, by recurrent rounds of retroelemental reverse transcription from a number of reverse-transcribed copies of the gene itself. The *Alu* sequence itself is not strictly a retroelement because it lacks the gene for reverse transcription, but it looks like a sequence that is derived from reverse transcription. Perhaps it is picked up by a retroelement, and the gene for reverse transcription is lost when it is copied back into the DNA. Although the sequence has not been proved to be functionless, a strong suspicion exists that it is an example of selfish DNA. The large numbers of copies of *Alu* suggest that something about its sequence makes *Alu* particularly likely to be copied; it would then be an example of active selfish DNA.

Transposable elements can have interesting evolutionary consequences, even if they evolved originally by neutral drift. The insertion of an *Alu* sequence can prevent gene conversion between a pair of genes. Concerted evolution between the two genes then becomes impossible. As a consequence, duplicated genes in which an *Alu* sequence is inserted would be more likely to show independent evolution than genes without such an insertion. A second suggested consequence involves chromosomal evolution. Chromosomes may be particularly likely to rearrange around *Alu* sequences, and the numbers and positions of the sequence may, therefore, influence the rate and form of chromosomal evolution. Finally, at the level of the DNA sequence, when a transposon inserts into a new gene, it causes a mutation. The recipient gene has a new sequence after receiving the transposon, which may (or may not) cause a mutation at the phenotypic level. McClintock first showed that mutations can be caused by transposition in maize. Transposition probably accounts for a substantial proportion of observed mutations. It may differ from other kinds of mutation in that it may occur more at some sites in the genome than at others. Mutations caused by nucleotide changes may be more randomly distributed.

C-Factor Paradox

Let us return to the observation that more DNA is present in the cells of some species, particularly eukaryotes, than is needed to code for their genes. The degree of excess is controversial, but no one dc that at least some excess DNA exists in at least some species. (Likewise, no one doubts that at least some of the species differences in DNA content, such as between humans and bacteria, are due to differences in phenotypic complexity.) The apparent excess of DNA is sometimes called the C-factor paradox. What purpose does the excess DNA serve?

Selfish DNA is an attractive, if hypothetical, solution to this question. The excess DNA would then have no use or function, but would be selectively neutral, and passively replicated from generation to generation. The assumptions of this hypothesis are plausible—unequal crossing over, gene conversion, transposition, and neutral drift all do happen. Moreover, if the hypothesis is correct, it would explain why this apparently functionless DNA actually is functionless. It could also explain why the degree of excess DNA differs so much among species, because the amount of repetitive DNA will fluctuate through time as it evolutionarily grows by mutation and drift (or selection) and shrinks when selection acts against excessive accumulations of repeats. Under this scenario, different species would represent the still-frames in a moving picture. While this idea is simply a hypothesis, it is, however, a theoretically attractive hypothesis. At the present time, the best available explanation for the apparent excess of DNA in our cells is that it results from the passive accumulation of selfish DNA.

3

PSEUDOGENES (FALSE GENES)

Pseudogenes may be regarded as 'floating hulks', gene-derived DNA sequences that are not longer capable of being expressed as protein products. Some are the remnants of once active genes that have acquired inactivating mutations which either preclude their transcription or at the very least prevent mRNA translation. Others are retrotransposed copies of expressed mRNAs which, since they almost always lack promoter sequences, incapable of being expressed and therefore also tend to accumulate deleterious mutations. Pseudogenes ought not, however, to be regarded entirely as evolutionary *cul-de-sacs*. Although it is unlikely that reactivating (reverse) mutations will occur so as to restore their function, pseudogenes may nevertheless influence the evolution of other functionally significant sequences by for example mediating recombination events or acting as sequence donors in gene conversion.

Generation of Pseudogenes by Duplication

A considerable number of pseudogenes have been descried in the human genome that have retained the exon-intron structure of their functional source genes. One assumes that such sequences arose by simple duplication of functional gene sequences but became inactivated since their intrinsic redundancy prevented selection from maintaining their potential to be expressed. Many of the examples have therefore accumulated nonsense mutations, frameshift deletions and insertions, or single base-pair substitutions within splice sites, any one of which would have been sufficient to render the expression of th sequences impossible. Perhaps the archetypal example is that of the human pseudogene (CYP21P) for cytochrome P450c21 which is closely linked

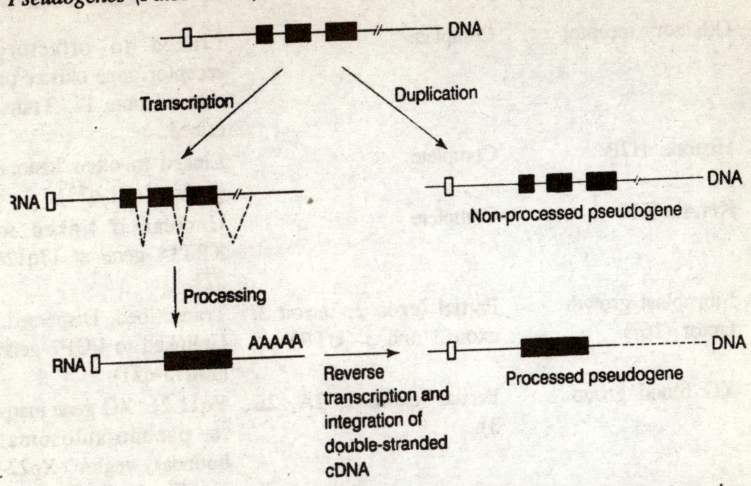

Fig. 3.1. Schematic diagram depicting the different mechanisms of pseudogene formation.

to its cognate gene (CYP21; 6921.3) as a result of a duplication event which occurred before the separations of apes from Old World monkeys >23 Myrs ago. This pseudogene contains three inactivating mutations: an 8 bp deletion in exon 3, the insertion of a T in exon 7 and a nonsense mutation in exon 8. The 8 bp deletion is present in both humans and chimpanzees whereas the other two mutations are human-specific consistent with the 8 bp deletion being the original inactivating mutation in the human-chimpanzee lineage. In gorilla and orangutan, the extra Cyp21 gene copies are inactivated by various other mutations.

Table 3.1. Examplex of human pseudogenes (ψ) generated by duplication

Pseudogene	Complete/partial	Comments
Cytochrome P450c21	Complete	Linked to CYP21 (6p21.3) gene
Amylase	Partial (lacks exons 1-3)	Linked to AMY gene cluster (1p21)
Apolipoprotein C1	Complete	Linked to APOC1 (19q13.2) gene
Tyrosinase	Partial (exons 4 and 5 and non-coding)	11p11.2-cen. Unlinked to TYR gene (11q14-q21). Not transcribed
Dopamine D5 receptor (2)	Complete (intronless ψ derived from intronless gene)	ψ on chromosomes 1 and 2. Unlinked to DRD5 gene (4p15-p16)

Olfactory receptor	Complete	Linked to olfactory receptor gene cluster on chromosome 17. Transcribed.
Histone H2B	Complete	Linked to other histone genes (1q21-q23)
Keratin K14	Complete	Unclear if linked to KRT14 gene at 17q12-q21
Fibroblast growth factor (16?)	Partial (exon 2, intron 2, exon 3 and 3' UTR)	Transcribed. Dispersed. Unlinked to FGF7 gene (15q15-q21)
XG blood group	Partial (exons 1, 2A, 2B, 3)	Yq11.21. XG gene maps to pseudo-autosomal boundary region, Xp22-pter Transcribed.
Interleukin-8 receptor	Complete	2q34-q35. Linked to IL8RA and ILR8B genes
Kallmann syndrome gene	Complete	Yq11. Unlinked to KAL1 gene (Xp22.3)
α2-Macroglobulin	Complete	12p12-p13. Linked to A2M gene
von Willebrand factor	Partial (exons 23-28)	Chromosome 22. Unlinked to VWF gene (12p13)
Mannose-binding protein	Complete	10q22. Probably linked 10 MBP gene (10q11-q21)
Serum amyloid A (SAA3)	Complete	Not transcribed. Linked to SAA gene cluster (11p15).
Carbonic anhydrase V	Partial (exons 3-7, introns 3-6)	16p11.2-p12. Unlinked to CA5 gene (16q24.3)
Calcitonin	Partial (exons 2 and 3)	Linked to CALCA gene (11p15)
Protein S	Partial (exons 2-15 present)	Linked to PROS1 gene (3p11-q11). Not transcribed.
β-Glucocerebrosidase	Complete	Linked to GBA gene (1q21). Transcribed.
Metaxin	Complete	Linked to MTX gene (1q21). Not transcribed.

β-Glucuronidase (2)	Complete and partial	5p13 and 5q13. Unlinked to GUSB gene (7q11).
α-Globin	Complete	Linked to HBA1 (16p13.3-pter) gene. Transcribed.
ζ-Globin	Complete	Linked to HBA1 (16p13.3-pter) gene. Not transcribed.
Immunoglobulin Cε2	Partial	Linked to immunoglobulin Cε1 IGHE gene (14q32)
Immunoglobulin γ	Complete	Linked to IGHG gene cluster (14q32).
Immunoglobulin Vk(3)	Complete	Chromosomes 1, 15, 22. Unlinked to IGKV gene cluster (2p12)
Immunoglobulin V_H	Complete	Linked to immunoglobulin V_H (IGHV) gene cluster (14q32.3)
HLA class I (9)	Complete	Linked to HLA-A, HLA-E, HLA-G and HLA-F genes (6p21.3)
HLA class II	Complete	Linked to HLA-DOB1 genes (6p21.3)
Prostaglandin EP4 receptor (2)	Complete	Unclear if linked to PTGER4 gene (5p13)
Adrenoleukodystrophy	Partial (exons 7-10)	Unclear if linked to ALD gene at Xq28.
β-Globin (1984)	Complete	Linked to HBB gene (11p15)
Aldehyde dehydrogenase (ALDH8)	Complete	Transcribed
Iduronate-2-sulfatase	Partial (exons 2 and 3, introns 2 and 7)	Linked to IDS gene (Xq28). Not transcribed.

From inspection of Table 3.1, it is apparent that such pseudogenes are essentially of two types: either complete copies of the source gene or partial copies of that gene. Complete copies tend to be linked to the source gene whereas partial copies are often (although not always) unlinked to the gene from which they originated. This may reflect mechanistic deferences in pseudogene generation. Since many of the pseudogenes are duplicated copies of entire genes that include their

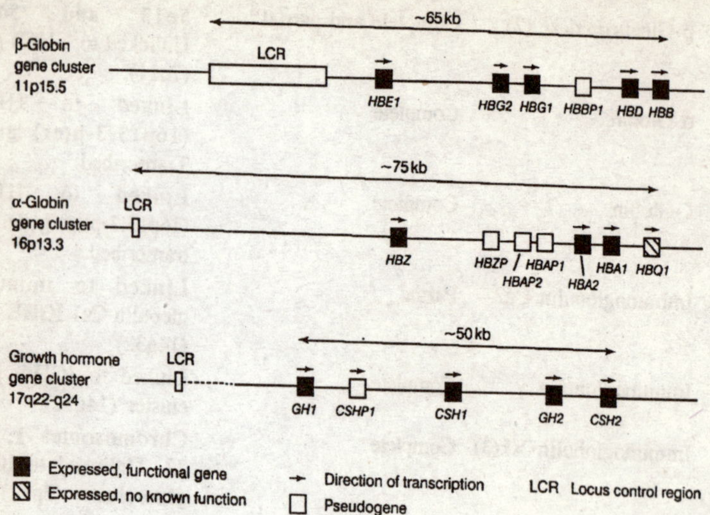

Fig. 3.2. Human β-globin, α-globin, and growth hormone gene clusters illustrating relative locations of nonprocessed pseudogenes.

original promoters, it is not surprising that some are capable of being transcribed. However, some of the partial, truncated pseudogenes also appear to be transcribed. We may nevertheless assume that the nonsense mutations and frameshift insertions and deletions acquired by these sequences serve to preclude the translation of any mRNA synthesized.

A particularly good example of the generation of multiple partial pseudogenes is that of the numerous sequences related to the neurofibromatosis type 1 (NF1; 17q11.2) gene identified on human chromosomes 2, 12, 14, 15, 18, 20, 21 and 22. These NF1-homologous sequences contain numerous nucleotide substitutions, small deletions and insertions (some inactivating) but still exhibit >90% homology to the corresponding sequences of the NF1 gene. They appear to have arisen by partial duplication of the NF1 gene between 22 Myrs and 33 Myrs ago with subsequent rounds of duplication generating new copies which were then transposed to the pericentromeric regions of the other chromosomes.

In cases where the parental source gene is intronless [e.g. some argininosuccinate synthetase pseudogenes and the dopamine D5 receptor pseudogene], the pseudogenes may have been secondarily created by the genomic duplication of pre-existing processed pseudogenes.

The occasional human pseudogene manifests a presence/absence polymorphism, for example the T-cell receptor β v6.10 pseudogene which has been found in the genome of most but not all individuals tested.

Generation of Pseudogenes by Retrotransposition

The second main category of pseudogenes probably accounts for the majority of inactivated gene sequences found in the human genome. They are termed processed pseudogenes since they are thought to have originated by retrotransposition of a correctly processed mRNA intermediate lacking intervening sequences. Th mRNA origin is evidenced by the lack of introns and the frequent presence of poly(A) tracts at their 3' ends. Flanking direct repeats, commonly of 9-17 bp are usually present and have probably been acquired during the process of integration. Many processed pseudogenes correspond to the entire length of the coding region of their source genes but others are truncated (e.g. CFTR exon 9 pseudogenes fragments). The vast majority of processed pseudogenes are located on different chromosomes from their functional source genes. This is not very surprising in view of the fact that retrotransposition necessarily requires a mobile mRNA intermediate.

Many processed pseudogenes have, like their duplication-derived counterparts, acquired multiple genetic lesions that preclude their expression. This is not however an invariant finding as evidenced by the human metallothionein processed pseudogene which has no such lesions and retains an open reading frame. This sequence is nevertheless probably not transcribed since, owing to its mRNA origin, it is necessarily lacks the promoter elements necessary for transcription to occur. In general, retrotransposed pseudogenes are rarely associated with active promoter elements and are not therefore usually expressed. Again there are always exceptions, for example the human glutamine synthetase pseudogene which contains a functional promoter/enhancer in its 5' flanking sequence that allows it to be transcribed although not expressed (owing to the presence of numerous frameshift mutations). Those processed 'pseudogenes' which possess both open reading frames and some inherent or acquired promoter activity such that they can be expressed are really not pseudogenes at all but rather examples of gene creation by retrotransposition.

To have been inherited, processed pseudogenes must have originated in the germline. It therefore follows that their functional source genes must have been expressed in the germline. Consistent with this assumption, many processed pseudogenes correspond either to ubiquitously expressed 'housekeeping' genes (e.g. snRNA genes) or to genes that are known to be expressed in germ cells (e.g. β-tubulin).

Table 3.2. Examples of human pseudogenes generated by retrotransposition

Pseudogene	*Chromosomal location/comments*
Dihydrofolate reductase (2)	Unlinked to functional source gene DHFR (5q11-q13)
ADP-ribosyltransferase, NAD$^+$ (2)	Chromosomes 13 and 14. Unlinked to functional source gene ADPRT (1q42). Present in gorilla (98% homologous). Originated ~27 Myrs ago.
Ribosomal protein L7	Located within intron 1 of c-fms (CSF1R) gene (5q33)
High mobility group protein	Flanked by 15 bp direct repeat. Originated only ~1 Myrs ago.
Topoisomerase 1	Chromosome 1. Truncated. Unlinked to functional source gene TOP1 (20q12-q13)
Glutamine synthetase	Flanked by 9 bp direct repeat. Transcribed.
FAU proto-oncogene	Chromosome 18. Integrated within sequence homologous to promoter of islet amyloid polypeptide (IAPP) gene. 317 amino acid open reading frame. Not transcribed. Unlinked to FAU gene (11q13)
28S ribosomal RNA	Flanked by 16 bp direct repeats
Ribosomal protein L23A	Chromosome 13. Unlinked to functional source gene RPL23A (17q11)
Ribosomal protein L38	Located in promoter region of type-1 angiotensin II receptor gene (AGTR1; 3q21-q25)
Ferritin L (5)	Unlinked to functional source gene FTL (19q13).
Ferritin H (18)	Highly dispersed (1p, 1q, 2q, 3q, 4, 6p, 31q, 14, 17p, X). Unlinked to functional source gene FTH1 (11q12-q13).
Glyceraldehyde-3-phosphate dehydrogenase (5)	Dispersed including Xp21-p11. Not linked to functional source gene GAPD (12p13).
Cytochrome c oxidase subunit Vib (3)	Unlinked to functional source gene COX6B (19q13)

Cytochrome b_5 (2)	Partial. Transcribed. 14q31-q32 and 20p11. Unlinked to functional source gene CYB5 (18q23).
Prothymosin α (5)	Unlinked to functional source gene PTMA (chromosome 2). Not transcribed.
Argininosuccinate synthetase (14)	Highly dispersed. Originated during last 40 Myrs. Unlinked to functional source gene ASS (9q34)
Immunoglobulin J	Chromosome 8q13-q21. Unlinked to functional source gene IGJ (4q21). Originated ~40-50 Myrs ago.
Immunoglobulin ε	Present in Old World monkeys and gorilla but appears to have been lost in chimpanzee.
Immunoglobulin λ	Unlinked to functional source immunoglobulin λ genes IGLC1 (22q11).
Calmodulin (5)	Unlinked to functional source gene CALM1 (14q24-q31). Originated ~21-49 Myrs ago.
Phosphoglycerate kinase	
Metallothionein	
β-Actin (3)	Chromosomes 6, 15 and 18. Unlinked to functional source gene ACTB (7p22)
γ-Actin (3)	Chromosomes 3q23, 20p13, 6p21. Unlinked to functional source gene ACTG1 (17q25)
β-Tubulin	8q21-8pter. Unlinked to functional source gene TUBB (6p21-pter).
Laminin receptor	Flanked by 19 bp direct repeats. Not transcribed. Originated 3.5-5 Myrs ago.
Enolase 1	Chromosome 1q41-q42. Unlinked to functional source gene ENO1 (1p36-pter)
Uracil-DNA glycosylase (2)	Located on chromosomes 14 and 16. Unlinked to functional source gene UNG (12q23-q24)
Tropomyosin, non-muscle	
Zinc finger protein ZNF75	12q13. Unlinked to functional source gene ZNF75 (Xq26)

Adenylate kinase 3	Located within intron 10 of NF1 gene at 17q11.2
Nucleotide excision repair protein RAD52	Chromosome 2. Unlinked to functional source gene RAD52 (12p12-q13)
Serotonin 5-hydroxytryptamine 1D receptor	Transcribed. Capable of being translated into 28 amino acid polypeptide. Unlinked to functional source gene HTR1D (1p34-p36)
Fatty acid-binding protein 3	13q13-q14. Unlinked to functional source gene FABP3 (1p32-p33).
Cysteine and glycine-rich protein 2	3q21. Unlinked to functional source gene CSRP2 (12q21).
c-Ki-ras 1 proto-oncogene	6p12-p11. Unlinked to functional source gene KRAS2 (12p12).
c-elk-1 proto-oncogene (2)	14q32 within IGHV locus. Unlinked to functional source gene ELK1 (Xp11.2). Insertion occurred 30-60 Myrs ago prior to IGHV locus duplication.
Interleukin-6 signal transducer	17p11. Unlinked to functional source gene IL6ST (5q11).
SHC transforming protein	Xq12-q13. Unlinked to functional source gene SHC1 (1q21).
Mitochondrial-3-glycerol phosphate dehydrogenase 2	Chromosome 17. Unlinked to functional source gene GPD2 (2q24)
Mitochondrial elongation factor Tu	17q11. Unlinked to functional source gene TUFM (16p11)
Mitochondrial NADH: ubiquinone oxidoreductase 24-kDa subunit	19q13.3-qter. Unlinked to functional source gene NDUFV2 (18p11.2-p11.31). Partially processed.
NADH: ubiquinone oxido reductase B13 subunit	11p15. Unlinked to functional source gene NDUFA5 (7q32)
Solute carrier family 9 (NHE3)	Chromosome 10. Unlinked to functional source gene SLC9A3 (5p15)
B-raf	Xq13. Unlinked to functional source gene BRAF (7q34)
GTP-binding protein Gαq	2q14-q21. Unlinked to functional source gene GNAQ (9q21)
αE-catenin	5q22. Unlinked to functional source gene CTNNA1 (5q11)

Acyl-CoA binding protein	Chromosome 6. Unlinked to functional source gene DBI (2q12-q21).
Methylthioadenosine phosphorylase	3q28. Unlinked to functional source gene MTAP (9p21).
Serine hydroxymethyltransferase	1p32-23. Unlinked to functional source gene SHMT1 (17p11-p12).
Hexokinase II	X chromosome. Unlinked to functional source gene HK2 (2p12). Originated 14-16 Myrs ago. Integrated into LINE element.
Dihydrolipoamide succinyl-transferase	1p31. Unlinked to functional source gene DLST (14q24)
S-adenosylmethionine decarboxylase	Xq28. Unlinked to functional source gene AMD1 (6q21-q22).
Dual specificity phosphatase 8	10q11.2. Unlinked to functional source gene DUSP8 (11p15).
G protein-coupled receptor kinase GRK6	Chromosome 13. Unlinked to functional source gene GPRK6 (5q35)
Estrogen-related receptor ERRα	13q12. Unlinked to functional source gene ESRRA (11q12-q13)
Phosphatidylinositol glycan class F	Chromosome 5q35. Unlinked to functional source gene PIGF (2p16-p21)
Nucleophosmin (7)	4 full-length, 3 truncated. Unlinked to functional source gene NPM1 (5q35)
RNA helicase, DDX10	9q21-q22. Unlinked to functional source gene DDX10 (11q22-q23)
Transcriptional elongation factor TFIIS	Unlinked to functional source gene TCEA1 (3p21-p22)
U1 snRNA	
U3 snRNA	
U4 snRNA	
U6 snRNA	
U10 snRNA	

However, some processed pseudogenes are derived from transcripts of tissue-specific genes such as those encoding the immunoglobulins. These pseudogenes may have originated from ectopic transcripts, the consequence of the 'leaky' transcription which appears to be a property of every gene in every cell. In a very few cases, processed pseudogenes have been shown to be derived from antisense transcripts.

Some genes possess only a single processed pseudogene copy whereas others [e.g. the U1 and U6 small nuclear RNA (snRNA genes] can possess hundreds. In some cases, the processed pseudogenes greatly outnumber their functional counterparts. One example is that of the human ribosomal protein multigene family whose members are composed predominantly of multiple processed pseudogenes. The proportion of processed pseudogenes to functionally active genes in any one gene family may reflect the level of transcription of the source gene in the germline. It is also likely to be influenced by the private sequence characteristics of the source gene which will determine both the efficiency of priming of reverse transcriptase and of their eventual integration of reverse transcripts. Thus, for example the plethora of U3 snRNA pseudogenes may be explicable in terms of the inherent ability of U3 snRNA to act as a self-priming template for reverse transcriptase. Both human and murine cells possess an endogenous reverse transcriptase activity although it is unclear whether this is derived from retroviral infection, endogenous sources of reverse transcriptase such as retroviral elements, or perhaps more likely, LINE elements.

Some human processed pseudogenes have been inserted into the introns of other genes, for example a ribosomal protein L7 pseudogene in intron 1 of the *c-fms* (CSF1R; 5q33)) proto-oncogene, a phosphoglycerate mutase pseudogene in intron 1 of the Menkes disease (ATP7A; Xq13.3) gene, an L5 ribosomal protein pseudogene in an intron of the small nuclear ribonucleoprotein N (SNRPN; 15q12) gene, an L21 ribosomal protein pseudogene in intron 13 of the breast cancer (BRCA1; 17q21) gene and an adenylate kinase 3 pseudogene in intron 10 of the neurofibromatosis type 1 (NF1; 17q11.2) gene. If integrated upstream of a gene, the pseudogene sequence may affect expression of that gene. For instance, a processed ribosomal protein S25 pseudogene, which has become integrated into the promoter region of a rat class I alcohol dehydrogenase gene, inactivated a glucocorticoid response element and contributed a novel suppressor element. Similarly, the human L38 ribosomal protein processed pseudogene has become integrated into the promoter of the type-1 angiotensin II receptor (AGTR1; 3q21-q25) gene although the effect, if any, on the expression of the gene is unclear. In general, the sites into which processed pseudogenes integrate appear to be AT-rich and various models have been proposed which attempt to describe the process of integration. Co-retrotransposition sometimes occurs with other sequences e.g. endogenous retroviral elements, LINE elements and possibly *Alu* sequences.

Some pseudogenes may be duplicated copies of pre-existing processed pseudogenes, for example the tandemly arranged proliferating cell nuclear antigen pseudogenes on chromosome 4q34, the L7a ribosomal protein pseudogenes associated with the closely linked lymphotactin α (LTNA) and β (LTNB) genes on chromosome lq23 or the *elk-1* pseudogene associated with the immunoglobulin heavy chain locus (IGHV) on 14q32.

A highly unusual presence/absence polymorphism has been noted for a human dihydrofolate reductase processed pseudogene. This pseudogene is present in only a proportion of individuals and this proportion varies between racial groups. However, it is unclear whether the polymorphism is a consequence of the recent acquisition of the pseudogene or whether it instead results from its occasional loss.

Generation of Pseudogenes by other Means

Although the vast majority of known pseudogenes are either inactivated genomic copies (partial or complete) of functional genes or processed pseudogenes, there are a few unusual examples which do not fit easily into either of these two grouping. One such case is that of the hybrid β-δ-globin pseudogene found in lemurs. The first and second exons plus intron 1 appear to be derived from the ψβ-globin pseudogene whilst intron 2, exon 3, and the 3' UTR originated from the δ-globin gene. The hybrid pseudogene probably originated from an unequal crossover which fused sequences derived from the ψβ-globin pseudogene and the δ-globin gene.

The cadherin pseudogene on human chromosome 5q13 described by Selig et al. (19950 is also unusual. It contains one exon derived from the source gene encoding cadherin 12 (CDH12; 5p13-p14) plus flanking introns, an open reading frame of 794 amino acids, and is transcribed at a 10-fold higher rate than the source gene. Selig et al. (1995) proposed that this pseudogene had been only partially processed in so far as the intron sequences had not been completely removed from the mRNA before reverse transcription and integration had taken place. Other examples of *semi-processed* human pseudogenes include one at lq24-q25 derived from the MADS box transcription factor 2 (MEF2A; 5q26) gene and another at 19q13.3-qter derived from the mitochondrial NADH: ubiquinone oxidoreductase (NDUFV2; 18p11.2-p11.31)) gene and at least 13 pseudogenes which have independently originated from an identical mis-spliced transcript of the protein geranylgeranyltransferase type Iβ (PGGT1B) gene.

Origin and Age of Pseudogenes

Some indication of the relative age of a pseudogene can be obtained by the extent of homology between the pseudogene and its parent or source gene. Thus, in the human α-globin gene cluster on chromosome 16, the ψζ-globin pseudogene is >99.5% homologous to its parental ζ-globin (HBZ; 16p13.3-pter) gene whereas the rather older ψα-globin pseudogene is only 75-80% homologous to its parental α-globin gene.

The approximate age of a pseudogene can be estimated by determining whether or not an orthologous sequence is present at an identical location in other related species of known phylogeny. Thus, the XG blood group pseudogene is present in the great apes but not in Old World monkeys, New World monkeys or prosimians whilst multiple copies of the keratinocyte growth factor pseudogene are present in chimpanzee, gorilla but not in gibbons or Old World monkeys. Similarly, the ψβ1-globin pseudogene has been found in prosimians and New World monkeys as well as the anthropoid apes and human and is therefore thought to have originated very early in primate evolution. Rouquier et al. (1998) noted distinct human-and gorilla specific inactivating mutations in the olfactory receptor pseudogene (termed 912-93) that is located on human chromosome 11q11-q12 and which exhibits synteny in the hominoid primates. Finally, the serine hydroxymethyltransferase processed pseudogene, derived from the SHMT1 gene on chromosome 17p11.2, arose during the evolution of the primates but different lesions have occurred in different phylogenetic branches of the family making this pseudogene a potentially useful marker in molecular studies.

Attempts have also been made to estimate the age of pseudogenes (time since duplication and/or inactivation) by comparing the proportion of silent and replacement changes in the pseudogene sequence to that exhibited by the active parent gene. Such estimates assume that following inactivation, pseudogenes will accumulate mutations at the same rate as silent changes in active genes. However, what is often ignored is that there is still some selection against silent changes in functional genes. Moreover, such estimates are extremely sensitive to the confounding effects of gene conversion whose homogenizing influence serves continually to restart the molecular clock leading to the underestimation of the age of the pseudogene.

Patterns of Mutation in Pseudogenes

Pseudogenes represent extremely useful tools for the study of mutation because, since they lack a biological function and are not subject to selective constraints, all mutations that occur in pseudogenes are selectively neutral and will become fixed in the population with equal probability.

Pseudogenes exhibit a very high rate of nucleotide substitution with the CpG dinucleotide being a hotspot for mutation. By contrast, deletions occur about once every 40 nucleotide substitutions whilst insertions occur about once every 100 nucleotide substitutions. The age of the pseudogene, however, is not always linearly related to the numbers of deletions and insertions present, and this may be due at least in part to gene conversion. The rate of DNA loss through deletion from processed pseudogenes appear to be considerably higher in rodents than in humans. Interpretation of such interspecies comparisons can however be confounded by possible differences in the frequency of gene conversion between species.

The mutation rate within processed pseudogenes appears to be higher than in regions flanking the site of insertion. This is potentially explicable if one considers that those sites which are inherently the most mutable have been maintained by selection within the retrotransposed sequence up until pseudogene formation whereas such sites have been removed in the flanking regions which have been unconstrained by the effects of natural selection for much longer periods of evolutionary time.

Pseudogenes and Gene Conversion

Gene conversion is the modification of one of two alleles by the other. It involves the nonreciprocal correction of an 'acceptor' gene or DNA sequence by a 'donor' sequence which remains physically unchanged. In most known cases, gene conversion has occurred between highly homologous and closely linked gene sequences. Examples of gene conversion that have occurred during primate evolution. Gene conversion has also served as a mutational mechanism causing human inherited disease; probable examples involve the genes and pseudogenes for steroid 21-hydroxylase (CYP21; 6p21.3), polycystic kidney disease, (PKD1; 16p13), neutrophil cytosolic factor p47-*phox*, (NCF1; 7q11.2) immunoglobulin l-like polypeptide 1, (IGLL1; 22q11), glucocerebrosidase (GBA; 1q21), von Willebrand factor (VWF; 12p13), and phosphomannomutase (PMM2; 16p13). These gene/pseudogenes pairs are all closely linked with the exception of the VWF gene (12p13) and its pseudogene (22q11-q13) and the PMM2 gene (16p13) and its pseudogene (18p). Together these two exceptions would seem to establish a precedent for the occurrence of gene conversion between unlinked loci in the human genome.

Could pseudogenes (both the highly dispersed processed variety and/or the closely linked duplication-derived pseudogenes) also have templated advantageous changes in their single copy functional source genes over evolutionary time? Certainly the converse is true since sequence changes in pseudogenes can be templated by their functional homologues. If pseudogene-templated changes in functional genes were not deleterious, they could eventually have become fixed in populations or even species. Thus, in principle, pseudogenes, whether processed or non-processed, could act as a reservoir of sequence variation which could at some stage be transferred back to the functional gene. In this way, different mutational combinations might be put together within the pseudogene, all the time being shielded from selective pressure, and then functionally tested after simultaneous transfer to the expressed gene copy. For any one gene, the contribution of pseudogene-mediated gene conversion to the process of evolutionary change might be expected to depend on:

(i) The number of homologous pseudogenes (processed or unprocessed) in the genome that are homologous to the gene in question.

(ii) *Whether the pseudogenes are linked or unlinked to the functional gene*. Gene conversion does occur between unlinked homologous sequences. Gene conversion between multi-copy, dispersed retrotransposons such as *Alu* sequences and LINE elements has also been reported. However, interchromosomal gene conversion events are predicted to be much less frequent than intra-chromosomal events.

(iii) *The length of the sequence involved in the gene conversion event*. Were this to be too large, inactivating mutations present in the pseudogene would be more likely to be transferred to the functional gene thereby inactivating it rather than altering it. In practice, however, gene conversion events are often quite localized, one example being the HLA-DQB1 gene (6p21.3) in which gene conversion events have been confined to exon 2 without extending into the adjacent introns. Some gene conversion events can nevertheless involve rather longer stretches of DNA, for example over the entire rhesus monkey $\gamma 1$ and $\gamma 2$ globin genes and over the entire hominoid immunoglobulin Cα genes. The interpretational problem in such studies lies in the fact that it is difficult to distinguish a single large gene conversion event from multiple overlapping short gene conversions.

(iv) *The age of the pseudogenes*. The degree of sequence homology to the functional gene will decay with time and this could serve to reduce the frequency of gene conversion as well as the length of sequence involved. Further, the number of inactivating mutations acquired by the pseudogene will increase with time and this would have the effect of reducing the likelihood of a functionally productive gene conversion event.

At the present time, definitive evidence for pseudogene-mediated gene conversion is fairly sparse. Since by its very nature, gene conversion tends to cover its own traces, such evidence may prove hard to obtain. Further, although the donor sequence might be a pseudogene in an extant genome, this does not mean that it was necessarily a pseudogene at the time of gene conversion event. The demonstration that the sequence in question has been accumulating mutations over a considerable period of time would provide evidence in favour of an ancient inactivating event. Again, however, gene conversion could confound estimates of the length of evolutionary time elapsed. The above notwithstanding, one possible example of pseudogene-mediated gene conversion involves a human immunoglobulin V_H pseudogene (V4-55P) which may have served as a donor sequence in the conversion of two functional V_H (IGHV; 14q32) genes, V4-4b and V4-28. Similarly, a human pseudogene in the human growth hormone gene cluster may have templated sequence changes in a functional source gene: the growth hormone 1 (GH1; 17q22-q24) gene promoter region contains five single base polymorphic alleles (at −57, −1, +3, +16, and +26) which are identical to the bases present at the homologous locations in the promoter of the closely linked and evolutionarily related pseudogene (CSHL1) but are different to those present in the more distal CSH1, GH2 and CSH2 genes. A gorilla HlA-A gene has been shown to be similar to an HLA-AR pseudogene only in exon 2 with the remainder of the gene being closely related to other primate A locus genes. Finally, immunoglobulin V gene diversity in chickens is known to be increased by gene conversion during B-cell development; the germline pool of donor sequence information for somatic gene conversion is found in the families of V pseudogenes located 5' to the single functional V gene at each locus.

Pseudogene Reactivation

Pseudogene may be by definition inactivation but this does not mean that they can never regain their activity. Indeed, Marshall et al. (1994) calculated that the resurrection of pseudogenes is probabilistically feasible within about 6 Myrs of formation but that is unlikely after more than 10 Myrs have elapsed owing to the accumulation of multiple inactivating mutations.

One putative example of the natural resurrection of a pseudogene is that of a bovine seminal ribonuclease gene which appears to have become reactivated after the divergence of the kudu but before the divergence of the ox, between ~5 Myrs and 10 Myrs ago. Since the seminal ribonuclease genes are highly homologous to the pancreatic ribonuclease gene family (the seminal proteins emerged about 35 Myrs ago by a process of duplication and divergence from a pancreatic protein gene), the repair of the seminal ribonuclease gene may have been effected by gene conversion, involving the transfer of genetic information from a pancreatic ribonuclease gene.

An example of the natural reactivation of a pseudogene in humans is that of a truncated γE crystallin gene (CRYGEP1) found in association with Coppock-like cataract. *Crystallin genes*, the γ-crystallin gene cluster at 2q33-q35 comprises four gene and two pseudogenes, γE and γF, both of which contain in-frame stop codons. In the above mentioned patient, a number of sequence changes occurred within an around the TATA box in the promoter region of the γE pseudogenes which appear to have resulted in a 10-fold increase in promoter activity. The predicted product of the γE pseudogene is a 6kDa N-terminal γ-crystallin fragment which could be responsible for the increased opacity of the lens in the patient.

In an evolutionary context, the reactivation of the human ψζ1-globin pseudogene (HBZP; 16p13) by removal of its sole inactivating mutation appears to have occurred by gene conversion. This change, which has probably been templated by the neighbouring ζ-globin (HBZ; 16p13) gene, is a common polymorphism in a number of different populations.

The above examples of the reactivation of pseudogenes were naturally occurring. Reactivation of an inactivated pseudogene can however also be effected artificially by correction of the inactivating mutation(s) and restoration of the sequence originally present prior to inactivation. One example of this is the artificial reactivation by *in vitro* mutagenesis of the human T-cell receptor γ variable V10 pseudogene which was originally inactivated by a single base-pair substitution in a donor splice site. Such experimental studies serve to demonstrate that the corrected lesion was the sole impediment to expression and therefore likely to be the original inactivating mutation.

Gene Loss/Inactivation in Primates

Gene can be created but they may also be destroyed. As we have seen, one form of gene inactivation involves pseudogene formation. This process involves the creation of a copy of a specific gene (whether cDNA or genomic) followed by its inactivation by the sudden acquisition, and subsequent steady accumulation of, deleterious mutations. Owing to the redundancy of genetic information available upon gene duplication, there is often no selective pressure to retain the integrity of the new sequence because the sequence from which it was derived still exists and functions normally.

By contrast, the loss of inactivation of a single copy essential gene is likely to have a dramatic phenotype effect. Many of the lesions responsible will be evident by their clinical manifestations (assuming that they are not embryonic lethal) and we should expect them to be rapidly removed from the population by natural selection. On the other hand, the inactivation of a nonessential gene could in principle have relatively few deleterious consequences and might even have a beneficial effect under the appropriate circumstances. One possible example of this is the C4b-binding protein β-chain gene (C4BPB; 1q32) gene which occurs as a single copy functional gene in the human genome but has been inactivated in the mouse. That the loss of this gene has been fixed evolutionarily in the murine lineage suggests that it is at least neutral with respect to fitness. It's loss may even have conferred a selective advantage (antithrombotic effect?) in which case, it could have become fixed within a relatively short period of evolutionary time. Several similar examples of the inactivation of single copy genes from primate genomes have also been documented and these *single copy pseudogenes* will now be described in some detail.

Urate Oxidase Gene

Urate oxidase is a copper binding enzyme found in most vertebrates which catalyses the conversion of uric acid to allantoin. Although found in Old World monkeys and in the majority of New World monkeys, urate oxidase activity has been lost in the hominoid line. The human urate oxidase (UOX; 1p22) 'gene' was isolated by screening a genomic library with a porcine cDNA probe. It was found to contain two nonsense (CGA→TGA) mutations at codons 33 and 187. These lesions were also found to be present in chimpanzee and gorilla but only the codon 33 mutation was detected in orangutan. This implies that the codon 33 mutation was the original inactivating mutation and that it must have occurred before the divergence of orangutan from the chimpanzee/human line, between 7 Myrs and 13 Myrs ago. Urate oxidase activity is also absent in the gibbon but this has been shown to be due to a13 bp deletion in exon 2 of the *Uox* gene. The loss of urate oxidase activity in primates has therefore been due to the occurrence of at least two distinct gene lesions in different lineages. That this gene could have been lost in primates as a result of at least two independent mutational events is consistent with its loss being advantageous. Although somewhat far-fetched perhaps, it has been suggested that since uric acid is a potent antioxidant, an increased uric acid concentration might have contributed to a lowering of the somatic mutation rate with a consequent increase in hominoid lifespan. The loss of urate oxidase may however be responsible for at least some cases of renal stones and gouty arthritis in humans.

α-1,3-Galactosyltransferase Gene

α–1,3–Galactosyltransferase (α-1,3 GT) is responsible for the synthesis of the α-galactosyl epitope present in the cell surface receptors of most mammals including prosimians and New World monkeys. However, the catarrhines (Old World monkeys, apes and humans) appear to lack this epitope and produce large amounts of antibodies against it. Two inactivating single base-pair deletions (Del C822 and del G904) were found in the human α-1,3, GT (GGTA1) 'gene' located at chromosome 9q34. Since chimpanzees possess both deletions whereas orangutan and gorilla only have del904, it was reasoned that del904 was the original inactivating mutation. Old world monkeys were found not to possess either lesion and no other inactivating mutation was obvious. Clearly, as with urate oxidase, at least two distinct inactivating mutations have occurred in apes and Old World monkeys. Selective pressure to suppress α-galactosyl transferase expression may have been mediated by a pathogen which infected primates via cell surface receptors containing the α-galactosyl epitope. Alternatively, the pathogen might have expressed α-galactosyl epitopes and an effective host immune response would have required the suppression of autologous α-galactosyl epitope synthesis.

Elastase I Gene

The human pancreatic elastase I (ELA1; 12q13) 'gene' is transcriptionally silent even though the coding sequence and splice junctions appear to be intact. Studies of the corresponding (transcriptionally active) rat gene have demonstrated that some 205 bp immediately upstream of the transcriptional start site are required for pancreatic expression. This regulatory region comprises an enhancer between –205 bp immediately upstream of the transcriptional start site are required for pancreatic expression. This regulatory region comprises an enhancer between –205 and –72 that contains all the information necessary for tissue-specific expression, and a minimal promoter element between –71 and +8. The enhancer contains three functional elements (A, B, and C), two of which (A and B) bind pancreas-specific transcription factors (PTF1 and PDX1) that mediate the cell type specificity of the enhancer. The 5' flanking region of the human ELA1 'gene' was studied by Rose and MacDonald (1997). Its nucleotide sequence differs by 28% from its rat counterpart whilst its enhancer/promoter strength (as measured by reporter gene assays) was 2.6% that of the homologous rat sequence. Comparison of the rat and human gene 5' flanking regions revealed a total of 13 nucleotide differences within the A, B, and C elements. A combination of the three human enhancer elements, 'repaired' by reference to the rat sequence, together with the rat promoter, partially restored the activity of the human enhancer. These authors went on to demonstrate that two mutations in the A element and four mutations in the B element served to abolish the binding of the cognate transcription factors. The degree to which these lesions individually exert a deleterious effect on enhancer function will not be apparent without further functional studies. In addition, without phylogenetic studies of the ELA1 genes from other mammals/primates, it is unclear in which order these mutations occurred during evolution. This notwithstanding, we may still conclude that the silencing of the human ELA1 'gene' has come about through the mutational inactivation of the upstream enhancer rather than through the alteration of the coding region.

L-gulono-γ-lactone Oxidase Gene

Primates (including humans) and guinea pigs have at least one thing in common: they are, unlike other mammals, unable to synthesize L-ascorbic acid from D-glucose owing to a deficiency of the enzyme L-gulono-γ-lactone oxidase. Using a rat cDNA as a probe, Nishikimi et al. (1994) isolated the once active human L-gulono-γ-lactone oxidase (GULOP) 'gene' which is located at chromosome 8p21.1. A considerable number of inactivating mutations were found including the deletion of exon VIII, two 1 bp deletions, one 1 bp insertion and two nonsense mutations. The presence of two non-consensus dinucleotides at donor splice sites (GC and GG instead of GT) also suggested the introduction of potentially inactivating mutations at splice sites. Since both Old World and New World monkeys are deficient in L-gulono-γ-lactone oxidase whilst prosimians possess this enzyme, its loss must have preceded the divergence of New World from Old World monkeys (~45 Myrs ago) but occurred after the divergence of the prosimians from the simian line (50-60 Myrs ago).

ADP-Ribosyltransferase 1 Gene

ADP-ribosyltransferase 1 (RT6) is a glycosyl phosphatidylinositol-anchored cell membrane protein expressed in peripheral T cells and intra-epithelial lymphocytes. In the rat, expression of RT6 has been used to distinguish subsets of T cells and a defect in RT6 expression has been associated with susceptibility to autoimmune type I diabetes. Haag et al. (1994) have shown that the human (ART2P; 11p15) and chimpanzee 'genes' contain three nonsense mutations, at codon 47, 141, and 193 respectively, which serve to inactivate them. No further copies of the ART2P 'gene' were detectable in the human genome consistent with its presumed status of single copy pseudogene. It is possible that inactivation of the ART2P gene conferred a selective advantage if, for example, it resulted in the loss of a membrane receptor for a pathogenic virus. It is as yet unclear if loss of the RT6 protein confers increased susceptibility to disease.

Haptoglobin Gene

Haptoglobin, a hemoglobin-binding protein, is encoded by three closely linked gene in chimpanzee (HP, HPR, and HPP) but only two (HP, HPR; 16q22.1) in human. The loss of the HPP 'gene' in human occurred after the separation of the human and chimpanzee lineages by an unequal homologous crossover that deleted most of the HPP gene.

Fertilin-α Gene

In most mammals including macaques and baboons, α-fertilin contributes a sub unit to a heterodimeric membrane glycoprotein on the sperm surface. The macaque and baboon possess two α-fertilin genes which appear to lack introns (indicative perhaps of a retrotranspositional origin). Only a single copy homologue (FTNA) is present in the human genome and, although this sequence is transcribed in the testis, it represents a nonfunctional pseudogene since it contains numerous mutations which render translation of the transcript impossible. What is puzzling is that although the 5' half of the human FTNA 'gene' is highly homologous to the macaque *Ftna1* gene, the 3' half differs markedly from both *Ftna1* and *Ftna2* genes. Jury et al. (1997) attempted to explain the creation of a solitary nonfunctional human gene in terms of a recombination event between the two ancestral primate *Ftna* genes.

T-cell Receptor γ V10 Variable Region Gene

The 'gene' encoding the V10 variable region of the T-cell receptor γ (TCRG; 7p15) has a single base-pair substitution in a donor splice site which serves to inactivate the gene in humans. The counterparts of this gene in chimpanzee and gorilla contain functional splice sites and therefore this lesion must have occurred in the human lineage. Restoration of the defective splice site by *in vitro* mutagenesis has been shown to generate a correctly spliced product

Cytidine Monophospho-N-acetylneuraminic Acid Hydroxylase Gene

Humans differ from the great apes in that they lack the enzyme CMP-*N*-acetyl-neuraminic acid hydroxylase which, as the name suggests, hydroxylates *N*-acetyl-neuraminic acid. This cell surface sialic acid molecule is thought to be involved in cell-cell recognition and in cell-pathogen interactions. The loss of enzyme activity in human is caused by a 92 bp deletion in the coding region of the CMP-*N*-acetylneuraminic acid hydroxylase (CMAH; 6p22-p23) gene. This deletion results in a frameshift leading to premature termination of translation. It is specific to the human lineage since it is not found in any of the great apes but it is not of recent origin since it has been found to be present in Caucasians, African Americans, Japanese, Kung bushmen and Khwe pigmies. Chou et al. (1998) speculated that the loss of N-acetylneuraminic acid hydroxylation may have had implications for susceptibility to infectious disease since some pathogens utilize sialic acids as specific binding sites on mammalian cells.

Flavin-containing Monooxygenase 2 Gene

The flavin-containing monooxygenase 2 gene is one of five FMO genes found in mammals which encode a series of NADPH-dependent flavoenzymes that are capable of catalyzing the oxidation of numerous drugs and xenobiotics. FMO2 is expressed predominantly in the lung where it constitutes the major form of FMO. The human FMO2 gene (FMO2; 1q) differs from those of other mammals in that it possesses a CAG→TAG transition converting Gln472 to a stop codon. This lesion predicts a truncated polypeptide lacking 64 amino acids from its C-terminus. The FMO2 mRNA transcript is both abundant and stable. In vitro expression studies have demonstrated that the truncated protein product is translated, correctly targeted to the endoplasmic reticulum but has lost its catalytic activity. This mutation is not present in the

Fmo2 genes of gorilla and chimpanzee and must therefore have arisen in the human lineage in the last 5-7 Myrs. Interestingly, about 4% of individuals in African populations possess a CAG (Gln) triplet at codon 472. It is possible that this CAG allele represents a remnant of the original gene sequence in which case the lesion may have occurred during early human history. The alternative is that it is a reverse mutation brought about perhaps by gene conversion templated by a related FMO gene. The biochemical and toxicological significance of the loss of FMO2 activity in human lung is unclear and so it is premature to speculate as to whether its loss conferred a selective advantage upon bearers or whether the inactivating lesion could have become fixed by genetic drift alone.

4

Evolution of Split Genes

Intron Structure, Function and Evolution

Eukaryotic genes are not usually contiguous entities but arc instead interrupted by noncoding sequences termed introns. How were introns first acquired? Although there is at present no clear unambiguous answer to this question, there appears to be a continuous evolutionary line from the archaic self splicing group II introns of eubacteria and simple eukaryotes via protein-assisted self-splicing introns to the splicing exhibited by extant eukaryotic nuclear protein-encoding genes. It is therefore reasonable to speculate that group II introns could have been introduced by the invasion of eukaryotic cells by the bacterial endosymbiotic ancestor of the mitochondrion.

Whatever their origin, a close correlation exists between intron density and developmental complexity. Thus introns are largely, although not exclusively, confined to the eukaryotes. The archaebacteria, which form a third major taxon distinct from the eukaryotes and eubacteria, possess small introns in their tRNA and rRNA genes whilst eubacteria and a few eukaryotes possess introns that catalyse their own splicing. An average of one intron per kilobase (kb) of coding sequence is found in simple eukaryotes such as *Dictyostelium* and *Plasmodium*, 3-4 per kb in plants and fungi and 6 per kb in vertebrates. The average cellular gene in vertebrates contains about seven introns. Intron size also increases with phylogenetic complexity with intronic sequence accounting for only 10-20% of primary transcripts in the protista but as much as 95% in vertebrates.

As far as human genes are concerned, introns are the rule rather than the exception. Some human genes have nevertheless been found which lack introns. These include the sex determining (SRY; Yp11.3) gene, the POU domain transcription factor POU3F2 (6q16) gene, the thrombomodulin (THBD; 20p11.2) gene, the β2-adrenergic receptor (ADRB2; 5q32-q34) gene, the JUN proto-oncogene (1p31-p32), the recombination-activating genes RAG1 and RAG2 (11p13), the arylamine *N*-acetyltransferase (AAC1; 8p21.3-p23.1) gene, the 13 genes of the interferon-α (IFNA) gene cluster and the interferon-β gene (IFNB1) at 9p21, the heat shock 70 kDa protein genes (HSPA2, 14q22-q24; HSPA1A, 6p21.3), the pyruvate dehydrogenase E1α subunit (PDHA1; Xp22) gene, the formyl peptide receptor (FPR1; chromosome 19) gene, the serotonin receptor genes HTR1D (1p34.3-p36.3) and HTR1F, the casein kinase 2-α1 subunit (CSNK2A1; 20p13) gene, the ID2 gene (2p25) encoding an inhibitor of DNA binding, the purinergic receptor (P2RY1; chromosome 3) gene and the calmodulin-like genes (CALML3, 10p13-pter; CALML1, 7p13-pter). Intriguingly, >90% of the human genes that encode the G-protein-coupled receptor family are intronless, a finding which may reflect a retrotranspositional origin prior to gene duplication.

Introns in human gens vary enormously in size, from as little as 24 bp in the case of the parvalbumin (PVALB; 22q12-13) gene, to in excess of 600 kb in the case of the α1(V) collagen (COL5A1; 9q34.2-q34.3) gene. Minimum intron size may be determined by the need to prevent steric hindrance between splicing factors. Intron size is not usually well conserved between orthologous genes. Thus during the evolution of the higher primates, intron 8 of the lamin B (LMNB2; 19p13.3) gene increased very dramatically in size as a result of a repeat expansion, whilst intron 6 of the Ewing sarcoma breakpoint region 1 (EWSR1; 22q12) gene expanded progressively through successive retrotransposition and recombination events involving *Alu* sequences. Similarly, the paralogous murine phospholipase D1 and D2 genes differ in size as a result of a 20-fold expansion/contraction of intron size in one of the genes. There are, however, notable exceptions to the rule of lack of conservation; for example, the sizes of the 53 introns of the human and mouse α1(II) collagen (COL2A1; 12q12-q13.2) genes differ on average by only 13%.

In the context of intron size, one enigma is the pufferfish (*Fugu rubripes*) whose 400 Mb genome is 1/7 the size of the human genome, is relatively devoid of repetitive DNA, and contains comparatively short introns (75% are less than 120 bp in length). A dramatic example of the economy manifested by the *Fugu* genome is provided by the relative size of intron 7 of the Duchenne muscular dystrophy (DMD;

Xp21) gene in *Fugu* (2.4 kb) as compared to its human counterpart (109.6 kb). At least 40% of the human intron is made up of LINE elements, *Alu* sequences, THE-1 and related LTR sequences, interspersed repeat sequences, a *mariner* transposon and other repeats including microsatellites whose insertion served to double the size of the intron over the last 130 Myrs. This example is not unrepresentative of the size differences noted between orthologous human and *Fugu* introns; thus the neurofibromatosis type 1 (NF1; 17q11) gene spans only 27 kb in *Fugu* as compared to 335 kb in human. Whether or not the size of the *Fugu* genome represents the ancestral state of the early vertebrae genome is unclear but its size may well approach the minimum sustainable for a vertebrate. We can only speculate as to the possible reasons by selection has acted on a genome-wide basis to minimize intron size in this species. Perhaps the *Fugu* genome has been resistant to colonization by transposable elements and so has no endogenous source of reverse transcriptase to aid the process of retrotransposition.

Since a high proportion of human genes belong to gene families or super-families which have arisen through a process of duplication and divergence, it is hardly surprising that the locations of introns are often evolutionarily conserved. Such conservation may be evident in terms of the structures of orthologous genes, for example the phenylalanine hydroxylase (PAH; 12q22-q24) genes of *Drosophila* and human or the myoglobin gene of the abalone *Sulculus diversicolor* and the human gene encoding indoleamine 2,3 dioxygenase (IDO; 8p11-p12). It may also be apparent between paralogous genes, members of the same gene family within a given species, for example between the highly homologous human monoamine oxidase A and B genes (MAOA, MAOB; Xp11.23) or between the human gene encoding the α-subunit of the granulocyte-macrophage colony stimulating factor receptor (CSF2RA; Xp22.32) and other members of the cytokine receptor family. However it is also clear that some evolutionarily related genes can possess quite divergent intron distributions, for example those encoding the actin-regulatory proteins, gelsolin, (GSN; 9p33) the villins (VIL1, 2q35-36; VIL2, 6q22-q27) and the capping protein Cap G (CAPG; 2cen-q24) or the complement proteins C6 (chromosome 5), C7 (chromosome 5) and C9 (5p12-p14) or the fibrinogens (FGA, FGB, FGG; 4q31). A particularly dramatic example of post-duplication structural divergence is provided by the human microfibril-associated glycoprotein genes MFAP2 (1p35-p36) and MAGP2 (12p12-p13). These two genes comprise 9 and 10 exons respectively, but sequence and structural conservation as well as conservation of intron location is confined to exons 8 and 9 of the MFAP2 gene and exons 7 and 8 of the MAGP2 gene. These exons encode the first six of the seven precisely aligned cysteine residues at the center of both proteins. Divergent exon distribution can arise through intron insertion and deletion. Alternatively, it can also arise by recombination, an example of this is provided by the human prosaposin (PSAP; 10q22.1) gene. The PSAP gene is polycistronic, encoding a precursor for four saposins (A, B, C, and D). Saposins A, B and D are encoding by three exons, saposin C by only two. Analysis of intron locations has indicated that the PSAP gene evolved by two duplication events and at least one rearrangement. This rearrangement is thought to have involved a double crossover, between the first and second and between the second and third intron positions of the saposin B and C coding regions, after the introduction of introns.

Intron sequence themselves are not usually well conserved during evolution, a finding which is not surprising in view of the fact that most intronic sequence is likely to be nonfunctional. However, there are some notable exceptions: the single intron of the oligodendrocyte-myelin glycoprotein (OMG; 17q11-q12) gene exhibits 75% overall homology between human and mouse whilst the 53 introns of the COL2A1 (12q12-13.2) gene differ by 69% between the two species. Sequences in the third intron of the γ-actin (ACTG1; 17p11-qter) gene are highly conserved between human and *Xenopus*. Perhaps the most dramatic example of the conservation of intronic sequence is that evident in a comparison of 97 kb of the human and murine T-cell receptor α/δ (TCRA, TCRD; 14q11.2) gene locus which exhibits 66% overall sequence homology even though <6% corresponds to exonic sequence.

There are exceptions to the general rule that intron sequences evolve more rapidly than exons. For instance, the second exons of the human semenogelin genes (SEMG1, SEMG2; 20q12-q13.1) have evolved very rapidly when compared to their rat homologues, and more so even than the flanking introns. Similarly, the divergence between the red and green visual pigment (GCP, RCP; Xq28) genes in humans, chimpanzees and baboons is lower in intron 4 than in exons 4 and 5 of these genes. In this case, homogenization of intron 4 sequences is thought to have been brought about by gene conversion whilst selection has probably acted so as to confine gene conversion to the intron in order to retain the distinct functions of exons 4 and 5 between the two genes.

Evolutionary conservation may imply function but we are only just beginning to elucidate the function of conserved sequences within introns. The introns of some genes contain the coding sequences of other genes. Thus the OMG, EVI2A, EVI2B genes occur within intron 27b of the human neurofibromatosis type 1 (NF1; 17q11.2) gene and are transcribed in the opposite direction to the NF1 gene. An orthologue of the EVI2B gene is present in the corresponding *Fugu* intron but this intron reveals no trace of the OMG or EVI2A genes. This indicates that the EVI2B gene must have been inserted into the NF1 gene more than 450 Myrs ago whilst the OMG gene must have been a more recent acquisition.

Some intronic sequence motifs perform transcriptional regulatory functions. Positive regulatory (enhancer-like) elements have now been reported from the first introns of a considerable number of human genes, for example the genes encoding type X collagen (COL10A1; 6q21-q22.3), type I 3β-hydroxysteroid dehydrogenase (HSD3B2; 1p13.1) tissue inhibitor of metalloproteinase 1 (TIMP1; Xp11.23-p11.3), factor IX (F9; Xq27), heat shock protein 90 (HSPCB; 6p12), Bruton's tyrosine kinase (BTK; Xq21.3-q22) purine nucleoside phosphorylase (NP; 14q13.1), O6-methylguanine DNA methyltransferase (MGMT; 10q26), type I collagen (COL1A1; 17q21.3-q22), IgE receptor (FCER1B; 11q13), growth hormone (GH1; 17q22-q24), thymidylate synthase (TYMS; 18p11.32) and dystrophin (DMD; Xp21.2). Although enhancer sequences are most, commonly found within the first intron of genes, such sequence element are also occasionally found in other locations, for example the second intron of the human apolipoprotein B (APOB; 2p24) gene the third intron of the human oxytocin receptor (OXTR; 3p26) gene and intron 8 of the δ-aminolevulinate synthase 2 (ALAS2; Xp11.21). Negative regulatory elements (repressors) within introns have also been characterized, for example in the first introns of the human Bruton's tyrosine kinase (BTK; Xq21.3-q22) and acid maltase (GAA; 17q25.2-q25.3) genes. A highly conserved CCAAT element within intron 1 of the proliferating cell nuclear antigen (PCNA; 20p12) gene also appears to act as a negative regulatory element.

Other sequences within introns may affect nucleosome formation or play a regulatory role in the processing of the primary transcript either by modulating mRNA splicing or by influencing mRNA stability through RNA-DNA, RNA-RNA or RNA-protein interactions. For example, sequences within the second intron of the human β-globin (HBB; 11p15.5) gene are important in promoting efficient 3' end formation and appear to be essential to the stability of the cytoplasmic HBB mRNA. A further fortuitous function of introns may be to act as 'lighting conductors' for the retrotranspositional insertion of mobile elements thereby protecting the nearby coding sequences from inactivation. A glimpse of what surprises introns may have in store for us is to be found within the introns of the human U22 host gene. Seven fibrillarin-associated small nucleolar RNAs (U25-U31), with complementarity to different segments of rRNA, are encoded within different introns of this gene and are highly conserved between human and mouse. The spliced U22 host gene mRNA species has by contrast little coding potential, is short lived and is evolutionarily poorly conserved. The surprise therefore is that in the U22 host gene, it is the 'introns' rather than the 'exons' which appear to specify the functional products.

Evolution of Alternative Processing

Alternative Splicing

Alternative splicing allows the generation of different mRNAs (and therefore a diverse array of protein isoforms) from the same gene. It is therefore an important mechanism for the tissue-specific or developmental regulation of gene expression. The potential evolutionary advantages of alternative splicing have been discussed in detail by Smith et. al. (1989) and hence will now only be summarized briefly. Unlike gene duplication or rearrangement, alternative splicing does not change the gene structure or copy number and need not therefore be irreversible in genetic terms. Since existing splicing pathways need not necessarily be discarded in order to employ new ones, alternative splicing is likely to be particularly useful as a means to generate protein diversity during early development and in very long lived and terminally differentiated cells. The use of alternative splicing may also facilitate the efficient exploitation of intragenic duplications since if the transcript of the newly duplicated gene were to be alternatively spliced, the gene could continue to produce the old gene product as well as the new one. Although alternative splicing implies the existence of cell and developmental stage-specific splicing factors, it is still unclear whether it is a predecessor or a refinement of constitutive splicing. Smith et al. (1989) argued that the two processes could have evolved simultaneously. If, indeed, many genes have evolved by exon shuffling, then novel exon combinations may have been tested by alternative splicing. If these conferred a selective advantage, then they could have been fixed by mutation at the appropriate splice junction

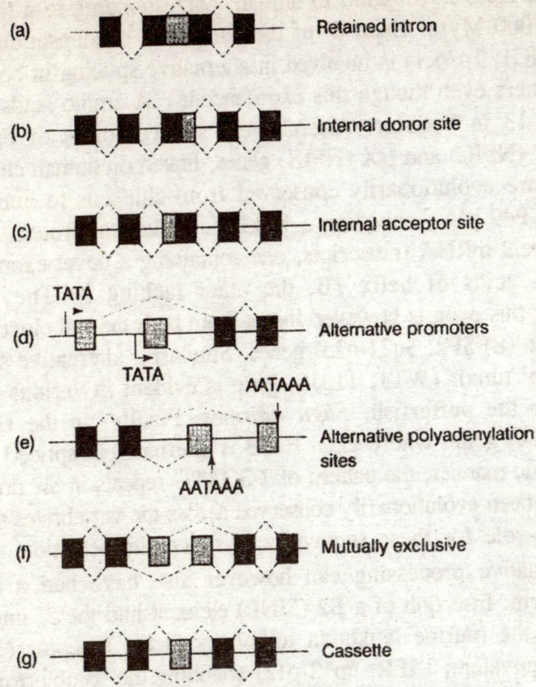

Fig. 4.1. Types of alternative splicing.

rendering them constitutive. Consistent with this view, most alternatively spliced exons appear to have originated by exon duplication.

There are now several examples of alternative splicing being evolutionarily conserved in orthologous genes. For instance, exon 9a of the proto-oncogene MTG8/ETO (CBFA2T1; 8q22) is alternatively spliced in both mouse and human. A developmental alternative splicing switch involving exon 16 of the erythroid protein 4.1 (EPB41; 1p) gene, occurs during mammalian erythropoiesis and this alternative splice also occurs in *Xenopus*. The reason for conservation of this switch over 350 Myrs of evolution probably lies in the fact that it controls the expression of a 21 amino acid peptide required for the protein's high affinity interactions with spectrin and actin that help to regulate erythrocyte membrane stability. The alternative splicing of exons 2 and 27 of the neural cell adhesion molecule L1 (NCAM1; 11q23-q24) gene has been conserved in human and the puffer fish (*Fugu rubripes*) demonstrating evolutionary conservation of the alternative splicing mechanism over some 430 Myrs. In the α-tropomyosin (TPM1, 15q22) gene, alternative splicing to produce tissue-specific isoforms has been conserved from *Drosophila* to human, corresponding to a timespan of at least 7000 Myrs. Exon 6' of the chromatin condensation regulator gene (CHC1; 1p36.1) is involved in alternative splicing in both humans and hamsters even though this exon encodes 31 amino acids in human but only 13 in hamster. Alternatively spliced forms of the nuclear factor I/C (NFIC) and I/X (NFIX) genes, linked on human chromosome 19p13.3 are evolutionarily conserved from chickens to humans. The chicken CP49 gene, encoding a 49 kDa cytoskeletal protein, generates two different mRNA transcripts, one containing a novel exon encoding 49 amino acids of helix 1B, the other lacking it. The transcript containing this exon is however absent from both the orthologous bovine and human (BFSP2; 3q21-q25) genes. Similarly, alternative splicing of the Wilms' tumor (WT1; 11p13) gene is evident in various mammals but not in the pufferfish, *Fugu rubripes*. Finally, in the fibronectin (FN1; 2q34) gene, where exon EIIIB is alternatively spliced in a cell-type-specific manner, the pattern of TGCATG repeats in the downstream intron has been evolutionarily conserved across the vertebrates suggesting a possible role for these sequences in splice site selection.

Alternative processing can however also have had a relatively recent origin. Insertion of a B2 (SINE) element into the 3' untranslated region of the murine leukemia inhibitory factor receptor (*lifr*) gene (human equivalent, LIFR; 5p12-p13) encoding the soluble form of the leukemia inhibitory factor receptor (LIFR) has, by potentiating alternative 3' mRNA processing and alternative splicing, given rise to a truncated mRNA species (relative to the mRNA encoding the membrane-anchored LIFR) which encodes soluble LIFR. In the rat, no such retrotranspositional event has occurred and the soluble form of LIFR is not found. The potential for alternative processing must therefore have arisen in the last 20-30 Myrs. Two species of mRNA encoding the metabotropic glutamate receptor subtype 5 (GRM5) gene occur in rat and human which differ in terms of the presence of a 96 bp insertion thought to result from alternative mRNA processing. Another example of the relatively recent evolution of alternative splicing has been noted in the lecithin: cholesterol acyltransferase (LCAT; 16q22) gene and is present in humans and the great apes but not in gibbons or Old World and New World monkeys. Known examples of single base-pair substitutions in splice sites of specific orthologous gene pairs that are responsible for the evolutionary emergence of alternative splicing.

One of the most dramatic inter-specific differences in alternative splicing involves the human *t* complex responder (TCP10; 6q27) gene and its murine orthologue. The human TCP10 transcript includes two exons not present in mouse transcripts, whilst mouse *tcp*10 transcripts include up to four exons that are not present in the human transcript. It is at present unclear whether a given exon has been incorporated into the transcript of one species or lost from the other. The recruitment and removal of exons is potentially explicable by single base-pair substitutions either eliminating splice junctions or converting intronic DNA sequence into novel splice sites. Islam et al. (1993) described this mode of evolution as 'punctuated equilibrium', a term that is used to refer to evolutionary situations in which phenotypic characters appear to leap from one equilibrium state to another in a short space of time. These authors suggested that this mode of evolution might have been facilitated by the existence of a gene family in which individual members possessed redundant functions. One family member would therefor have been free to pass through a nonfunctional state in which several base changes accumulated, resulting eventually a change in splicing phenotype. Whether selection or genetic drift could have been responsible was unclear. Unfortunately in this context, the authors did not see fit to discuss the intron phase of the added/removed exons.

Evidence for intergenic differences in alternative splicing during evolution has also come from the study of paralogous genes. Although the intron-exon distribution in the human annexin VI (ANX6; 5q32-q34) gene conforms exactly to that found in the human annexin I (ANX1; 9q11-q22) and II (ANX2; 15q21-q22) genes, exon 21 of the annexin VI gene (encoding 6 amino acids) is alternatively spliced (Smith et al., 1994). Similarly, the human chloride channel gene CLCN6 (lp36) contains an alternatively spliced 167 bp exon that is absent in the evolutionarily related CLCN1 (7q32-qter), CLCN5 (Xp) and CLCN7 (16p13) genes. The genetic basis of these examples of gene-specific alternative splicing has however not yet been elucidated. Known examples of single-base pair substitutions in paralogous genes responsible for the emergence of alternative splicing.

Alternative processing may become redundant after gene duplication as exemplified by the troponin I gene which encodes one of the three subunits of the troponin complex of the thin filaments of vertebrate striated muscle. Invertebrates and ascidians possess a single troponin I gene and this is alternatively spliced to generate proteins that differ in their N-terminal regions. By contrast, the three troponin I genes in the human genome (TNNI1, 1q31; TNNI2, 11p15; TNNI3, 19q13) and in the genomes of other vertebrates do not undergo alternative splicing although the 'extra' exon in the TNNI3 gene (as compared with the

TNNI1 and TNNI2 genes) is thought to correspond to the exon which is alternatively spliced in ascidians and invertebrates.

Alternative transcripts may also be generated by the differential utilization of polyadenylation sites. The differential utilization of two distinct polyadenylation sites in the plasminogen activator inhibitor 1 (PAI1; 7q22) gene (yielding two mRNA species of 2.6 kb and 3.6 kb) has been conserved between humans, orangutans and African green monkeys. However, only one PAI1 and mRNA species is apparent in lower primates and nonprimate mammals consistent with the acquisition of an extra polyadenylation site during primate evolution. The evolution of alternative processing may have come about through the differential utilization of polyadenylation sites via the newly described mechanisms of LINE element-mediated recombination. Finally, alternative transcripts may also arise through alternative promoter usage. There is very little information on inter-specific differences in promoter site selection.

Herbert and Rich (1999) have stressed the potential importance of RNA processing in evolutionary processes. Their perspective, which deserves close examination, was elegantly summarized in one of their introductory paragraphs:

"The outcome of one RNA processing event can affect the outcome of others resulting in regulatory 'networks' that influence both the information content and expression of the 'ribotype' (RNA pool), which in turn influences phenotype. Alternations in RNA processing generate different sets of ribotypes which are subject to natural selection on the basis of the phenotypes they produce. The evolutionary process requires the preservation of successful ribotypes in an inheritable (sic) form."

Aberrant Transcripts

There are numerous examples in the literature of aberrantly processed mRNA transcripts being produced in addition to correctly processed mRNAs in the cells of normal healthy individuals. It is possible that some of these 'aberrant transcripts' could be of functional significance but in the majority of cases they are likely to occur simply as a consequence of the absence of any selective pressure to avoid aberrant splicing. One example of this phenomenon is provided by the human glutathione peroxidase type 5 (GPX5) gene in which the majority of gene transcripts appear to be incorrectly spliced (Hall et al., 1998). If taken to its extreme, however, and the extent of aberrant splicing increases beyond a certain point, a gene may come to be effectively inactivated as has occurred in the case of the human chorionic somatomammotropin (CSHL1; 17q22-q24) 'pseudogene'.

mRNA Surveillance

The cell does however possess a mechanism that is capable of removing those aberrant transcripts that encode 'nonsense mRNAs' which would otherwise be expected to give rise to truncated proteins. Such mRNAs are often inherently unstable as a result of a process known as *mRNA surveillance* which involves nonsense-mediated mRNA decay. mRNA surveillance appears to be ubiquitous in eukaryotes with homologous proteins encoded by orthologous genes being found in organisms as widely separated as yeast (Upf1, Upf2, Upf3) and human (RENT1; 19p13). Presumably this mechanism has evolved in order to reduce the cost to the cell of producing non-productive transcripts.

Ectopic Transcripts

Extremely low levels of correctly spliced mRNA transcripts from tissue-specific genes have been demonstrated in supposedly 'non-expressing' cell types. Such *ectopic* or *illegitimate* transcripts have been found to occur at levels as low as one mRNA molecule per 5000-1000 cells, >1000-fold lower than the level of an average 'low abundance' mRNA. It remains unclear whether every cell is capable of generating ectopic transcripts or alternatively if the occasional cell in an otherwise non-expressing tissue is able to produce comparatively high levels of the transcript in question. Do ectopic transcripts have a biological role? Since the levels involved are often extremely low, it is hard to imagine that such a role involves significant protein synthesis. Perhaps what we observe simply represents a reasonable balance between the cost to the cell of producing nonproductive transcripts from many 'non-expressed' genes and the cost of turning off the transcription of many thousands of 'leaky' genes completely.

Exon Structure and Evolution

The average cellular gene in vertebrates contains about eight exons. A comprehensive analysis of 5061 exons from some 2705 intron-containing human genes has yielded a classification scheme for exons according to their transcriptional or translational boundaries. This scheme employs 12 different categories: (1) 5' terminal untranslated exons, (2) 3' terminal untranslated exons, (3) 5' terminal exons having a 5' untranslated sequence (UTR) followed by a coding sequence, (4) 3' terminal exons having a 3' untranslated sequence (UTR) followed by a coding sequence, (5) internal exons having a 5' portion of 5'

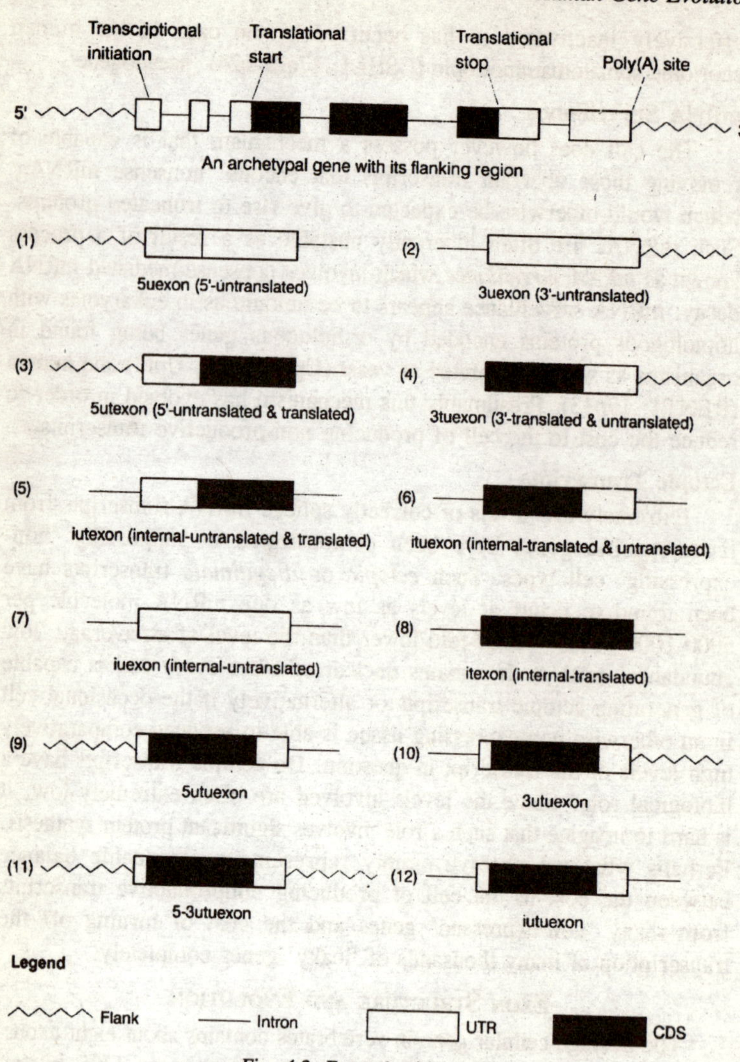

Fig. 4.2. Exon classification.

UTR followed by coding sequence, (6) internal exons having a coding sequence followed by a 3' portion of 3' UTR, (7) internal untranslated exons, (8) internal translated exons, (9) exons containing the complete coding sequence but which do not contain the transcriptional end, (10) exons containing the complete coding sequence but which do not contain the transcriptional start, (11) exons containing the complete coding sequence and both the transcriptional start and the transcriptional end, and (12) exon containing the complete coding sequence but neither the transcriptional start nor the transcriptional end. Although no doubt subject to ascertainment bias, the sample of exons studied by Zhang (1998) exhibited the following frequencies by type for the first 8 categories of exon: (1) 271, (2) 38, (3) 482, (4) 553, (5) 174, (6) 69, (7) 34, (8) 3440. Exons in categories (3) and (7) were in general found to be < 100 bp in length whereas exons in categories (2) and (4) were mostly 300-500 bp in length. There appears to be virtually no minimum size for an internal translated exon (category 8) in human genes; the smallest include the 4 bp exon 3 of the skeletal muscle troponin (TNNI1; 19q31) gene and two 3 bp exons (10 and 14) in the cardiac myosin binding protein C (MYBPC3; 11p11.2) gene. As far as the maximum size of an internal exon is concerned, the human F8C (3106 bp; Xq28), APOB (1572 bp; 2p23-p24) and MUC5B (10,690 bp 11p15.5) genes are currently league leaders.

Zhang (1998) found that the size of 5' UTR varied between human gens from 0 bp to 2077 bp with a mean of 136 bp. The size of the 3' UTR varies from –2 bp to 3427 bp with a mean of 589 bp. The –2 value, for the human α-D-galactosidase A (GLA; Xq) gene, is due to the fact that the coding sequence including the stop codon ends at nucleotide 11268 whilst the poly(A) addition site is located at 11266.

Introns Early or Introns Late?

Were introns used in the assembly of the first genes or were they added only later to previously contiguous coding sequences? The '*introns early*' theory or simply the '*exon theory of genes*' proposes that the genes encoding complex extant proteins emerged through the coalescence of primordial minigenes. These minigenes are held to have originally encoded protein modules and are now represented as exons whereas the non-coding linker DNA between the minigenes has survived as introns. Introns were then lost and novel exons made by fusing smaller exons together. By contrast, the '*introns late*' theory postulates that fully functional genes and introns inserted into them at different stages in their evolution. At its extremes, the debate between these two viewpoints is reminiscent of the old controversies between selectionists and neutralists. However, perhaps not altogether surprisingly, the debate has generated a significant amount of evidence supporting both theories.

One prediction of the introns-early hypothesis is that intron location should tend to demarcate structural or functional domains within proteins. This question has proved highly contentious. The extreme introns-late view is that no such correlation exists and their introns have been inserted quasi-randomly into the structure of genes. Consistent with this standpoint, several genes encoding evolutionarily ancient proteins (actins, alcohol dehydrogenase, carbonic anhydrase II, pyruvate kinase, globins) appear not to exhibit any correlation between intron position and structural features. However, more recently, a computer program designed to predict precisely the locations of module boundaries in proteins has helped to demonstrate that there is indeed a strong correlation between intron position and structural elements at least in the sample of 32 ancient proteins examined. The introns involved were invariably phase zero.

As we have noted already, the conservation of intron location is a common finding. That intron location is often conserved in evolutionarily ancient genes such as glyceraldehyde 3-phosphate dehydrogenase (GAPD; 12p13), carbamoylphosphate synthetase (CPS1; 2q35) and triose phosphate isomerase (TPI1; 12p13) is certainly compatible with the introns-early hypothesis. Most informative in terms of the introns-early/late debate, however, are discordant cases, examples of evolutionarily related genes from different taxa in which either intron position varies from between several nucleotides to several codons or where the intron is either present or absent. The introns-early theorists have explained these discrepancies by invoking *intron sliding* and deletion but despite the occasional convincing example [e.g. intron 8 of the histidyl-tRNA synthetase (HARS; chromosome 5) gene; intron 2 of the glucose-dependent insulinotropic peptide (GIP; 17q21.3-q22) gene which results in an 8 amino acid deletion of the prepropeptide of the rat protein as compared to human], the evidence for the widespread occurrence of intron sliding is still rather weak. The introns-late view is that intron sliding is inherently improbable because such a process would almost innevitably involve an intermediate stage that would alter the reading frame thereby leading to the loss or inactivation of the protein product. Therefore introns-late devotees have regarded discordant intron location, as found in the evolutionarily ancient α- and β-tubulin (TUBA1, 2q; TUBA2, 13q11; TUBB, 6p21-pter), aldehyde dehydrogenase (ALDH1, 9p21; ALDH2, 12q24; ALDH3, 17; ALDH5, 9; ALDH6, 15q26; ALDH9, 1; ALDH10, 17p11.2) and triose-phosphate isomerase (TPI1; 12p13) genes as evidence for the occurrence of multiple independent insertional or deletional events.

Several possible relationships may exist between the structural domains and the arrangement of exons in the gene. In many genes encoding globular proteins (e.g. the globins), the exons correspond

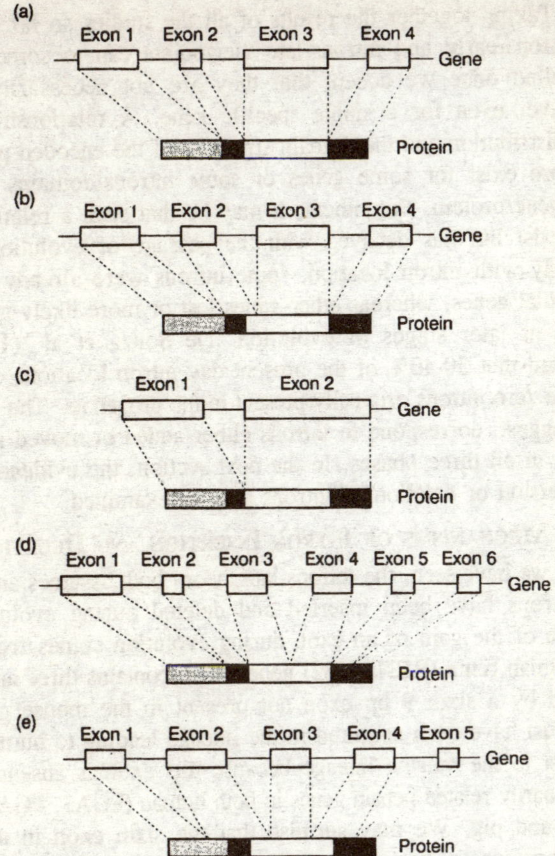

Fig. 4.3. Possible relationships between the arrangement of exons in a hypothetical gene and the structural domains of the protein it encodes, (a) each exon corresponds exactly to a structural domain, (b) the correspondence is only approximate, (c) an exon encodes two or more domains, (d) a single structural domain is encoded by two or more exons, (e) lack of correspondence between exons and domains.

closely to the structural domains thereby ensuring that when exon duplication has occurred, it has given rise to domain duplication. In the majority of cases, however, a more complex relationship exists between domains and exons. Thus it may be that the correspondence between exons and domains is only approximate. In some cases, an exon may encode two or more domains whilst in others, a single domain may be encoded by two or more exons. Finally, there may be no obvious correspondence between exons and domains.

Taking together the results of all the studies so far performed, the introns-early and introns-late viewpoints can to some extent be reconciled once we accept that they are not necessarily mutually exclusive, even for a single specific gene. A relationship between exon distribution and the domain structure of the encoded protein may therefore exist for some genes or some introns/domains within the same gene/protein. For others, it may be that such a relationship did once exist but has decayed with the passage of evolutionary time. Similarly with intron location, some introns were already present in primordial genes, whereas other genes lost or more likely gained their introns at later stages of evolution. De Souza et al. (1998) have suggested that 30-40% of the present day intron locations correspond to phase zero introns originally present in the progenote. The remainder, they suggest, correspond to introns either added or moved and appear equally in all three phases. In the next section, the evidence for both the insertion or deletion of introns will be examined.

Mechanisms of Intron Insertion and Deletion

As we have seen, the introns-late theory both assumes and requires that introns have been inserted and deleted during evolution. One example of the gain of an exon during evolution comes from the ten exon human renin (REN; 1q32) gene which contains three amino acids encoded by a sixth 9 bp exon not present in the mouse gene. This exon must have been acquired in the lineage leading to humans rather than lost in the murine lineage because this exon is absent from the evolutionarily related pepsin genes in both human (PGA3, PGA4, PGA5; 11q13) and pig. We may surmise that the sixth exon in the human REN gene was acquired by mutational change in intron 5 at some stage in the last 100 Myrs since the divergence of the rodent and human lineages. Another possible example of a late insertional event is provided by the human forkhead FKHL13 (FKHL13; 17q22-q25) gene which contains a single intron that has interrupted the forkhead DNA-binding domain.

The evolution of the human HLA-DRB6 (6p21.3) gene has been characterized not only by the deletion of an exon and the original promoter region but also by the de novo creation of an exon. This is thought to have occurred in association with the insertion of a retroviral LTR into intron 1 of the gene >23 Myrs ago, prior to the divergence of Old World monkeys from the human-ape lineage. Whether the exon/promoter deletion accompanied or followed the LTR insertion is unclear. Whichever, the gene is still capable of being transcribed.

That this is so is due to the creation of an open reading frame for a new exon by the insertion which serendipitously encoded a hydrophobic sequence that was able to function as a leader for the truncated HLA-DRB6 protein. The new exon also provided a functional donor splice site at its 3' end which potentiated in-register splicing with exon 2 of the HLA-DRB6 gene. Finally, since the LTR also provided a substitute promoter region, the downstream HLA-DRB6 gene could be transcribed.

Retrotransposition of an mRNA intermediate is one mechanism that would serve to erase completely the original exon-intron distribution of a gene. One potential example of this phenomenon is provided by the 68 kDa neurofilament protein (NEFL; 8p21) gene, a member of the intermediate filament multigene family that diverged over 600 Myrs ago. Other members of this family include desmin (DES; 2q35), vimentin (VIM; 10p13), glial fibrillary acidic protein (GFAP; 17q21), and the type I and II keratins. Whereas these latter genes possess seven or eight introns, six of which occur at homologous locations, the NEFL gene possesses only three none of which correspond in terms of their location to introns in the other known intermediate filament genes. Retrotransposition of a cDNA intermediate could account for the present day structure of the NEFL gene but three new introns would still have to have been acquired, presumably by insertion. In other cases of putative intron loss, the retrotransposition of semiprocessed mRNAs should also be considered. Such retrotransposed sequences will, by definition, contain fewer introns than their parent genes but the retention of some introns will often serve to obscure their origin.

The best characterized example of the insertion of an intron into a gene is from the sex-determining gene which in humans (SRY; Yp11.3) and other placental mammals is intronless. However, in dasyurid marsupials, an intron was inserted *de novo* into the *Sry* gene about 45 Myrs ago. The 825 bp intron lies within the coding sequence 550 bp from the start codon and was inserted in phase 1 (between the first and second bases of the codon). The intron, which contains a repetitive sequence element specific to marsupials, is correctly spliced out of the primary transcript. Since the SRY gene is essential for mammalian sex determination, it is very unlikely that intron insertion could have inactivated the gene even temporarily. We must therefore conclude that the inserted sequence probably contained functional splice junction motifs that ensured as its accurate removal from the marsupial *Sry* transcript.

Some gene families have evolved by multiple rounds of gene duplication accompanied by the gain or loss of both introns and exons,

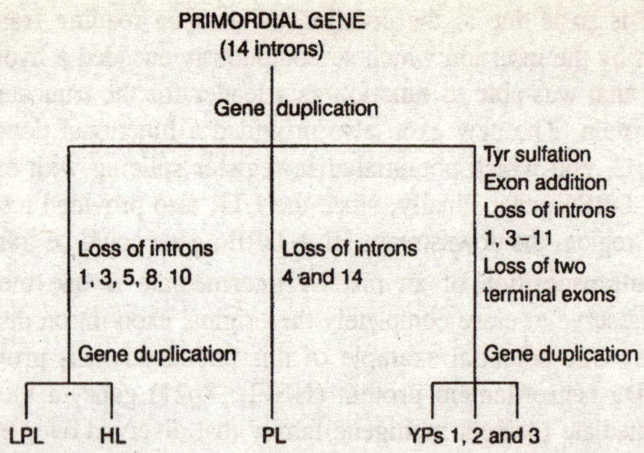

Fig. 4.4. A model for the evolution of the lipase gene family.

for example the human lipoprotein lipase (LPL; 8p22), hepatic lipase (LIPC; 15q21-q23) and pancreatic lipase (PNLIP; 10q24-p26) genes and the homologous yolk protein 1 gene from *Drosophila*. Another example is provided by the human thyroid peroxidase (TPO; 2pter-p24) and myeloperoxidase (MPO; 17q21.3-q23) genes.

The mammalian hepatocyte nuclear factor 1α (HNFα; TCF1; 12q24.3) gene possesses nine exons, whereas its avian and amphibian counterparts have ten. This difference is explicable in terms of the loss in mammals of intron 9 which in the chicken gene sub-divides the serine-rich transactivation domain. Interestingly, there is a sequence at the 3' end of intron 9 in the chicken gene which matches a conserved element used as a joining signal in immunoglobulin and T-cell receptor genes. Similarly, a perfectly conserved heptamer (CACAGTG) box, a recombinational signal sequence in immunoglobulin genes, is found at the junction of the fused exon 9-exon 10 in the rat HNFα gene. Horlein et al. (1993) speculated that V(D)J recombinase may have been involved in the excision of intron 9 in the mammalian gene.

The human surfeit 5 (SURF; 9p34.1) gene contains an intron in its 5' untranslated region which is not present in the mouse or rat *Surf5* genes. This additional intron is also present in apes, Old and New World monkeys and the prosimian *Galago*. Duhig et al. (1998) speculated that the intron was introduced after the divergence of primates and rodents but before the divergence of the human and prosimian lineages. A considerable number of examples are therefore now known of introns which have been either gained or lost during evolution as a result of the action of several distinct mechanisms. Such examples are certainly supportive of the introns-late theory.

EXON SHUFFLING

Exon Shuffling in the Evolution of Human Genes

Exon shuffling may be defined as the transfer of exons, encoding specific functional modules, between genes so that the module-associated functions are conferred upon the recipient proteins. Once of the first examples of exon shuffling to be recognized was from the human low density lipoprotein receptor (LDLR; 19p13.3) gene which contains with contiguous exons encoding epidermal growth factor (EGF)-like domains as well as seven repeats corresponding to LDL-binding domains. Other examples include the human hemopexin (HPX; 11p15.4-p15.5) gene which contains 10 exons each of which encodes a 45 amino acid repeat and the human factor XIII b subunit (F13B; 1q31-q32) gene which encodes a protein with ten 60 amino acid complement B-type ('sushi') domains each encode by a single exon. A large number of other human genes possess multiple copies of exons encoding specific domains. These encode proteins of the extracellular matrix (e.g. laminins, collagens), the serine proteases of coagulation a variety of receptors (e.g. integrins) and the immunoglobulins among many others. Exon shuffling is therefore a widely used evolutionary strategy although it appears to be confined to the genes of higher eukaryotes.

Until recently, exon shuffling was considered to result solely from intron-mediated recombination. However, an alternative mechanism, *LINE element-mediated recombination*, now appears to be capable of linking previously unlinked genomic DNA segments and may represent a novel means to bring about exon shuffling. Moran et al. (1999) have shown that LINE elements are capable of transducing exons from a downstream gene to new genomic locations thereby creating novel genetic combinations. This mechanism requires readthrough of the relatively weak LINE element polyadenylation signal by RNA polymerase to yield a processed mRNA transcript containing the LINE element fused to one or more exons from the neighbouring gene. Readthrough is made possible by the presence of more potent polyadenylation signals 3' to the LINE element and Moran et al. (1999) have shown that this can occur with high efficiency. The chimeric transcript then serves as a template for reverse transcriptase and if the resulting cDNA is subsequently integrated into the intron of a recipient gene, it may be recruited by that gene. This may help to explain why 3' terminal exons are often much longer than internal exons. Inefficient mRNA cleavage/polyadenylation might however lead to alternative splicing. A possible example of this process occurred about 10 Myrs ago during the evolution of the great apes: the co-transduction and integration of a LINE element and a fragment containing exon 9 of the cystic fibrosis transmembrane conductance regulator gene. Since some 6% of LINE element insertions occur within genes, it could be that this mechanism has played an important role in shuffling exons between genes, and may therefore have been central to the creation of the mosaic structures characteristics of so many genes in higher animals.

Phase Compatibility of Introns

Only a limited proportion of exons can be utilized for exon shuffling since the splice junctions of the shuffled exon must be phase compatible with their flanking exons in order to maintain the reading frame. Introns are classified as phase 0 if the intron lies between two codons, phase 1 if the intron lies between the first and second nucleotides of a codon, and phase 2 if it lies between the second and third nucleotides of a codon. The requirement for phase compatibility dictates that only symmetrical exons of class, 1-1 class 2-2 and class 0-0 [i.e. exons flanked by introns of the same phase (1, 2, or 0) are suitable for duplication or insertion. It is therefore not surprising that gene encoding mosaic proteins (proteins which comprise multiple domains and which are likely to have evolved by exon shuffling) contain a disproportionate number of class 1-1 exons encoding a wide variety of different modules, for example EGF-like, calcium-binding, fibronectin-like finger, kringle, complement B, LDL receptor, von Willebrand, thyroglobulin, C-type lectin etc. Class 1-1 introns predominate in the immunoglobulin, T-cell receptor and HLA genes, the Thy-1 glycoprotein (THY1; 11q22.3-q23) gene and the β2-microglobulin (B2M; 15q21-q22) gene, consistent with exon shuffling having been important in their evolution. Further examples of this phenomenon include the exons encoding the complement B-type domains in the human C4b-binding protein α gene (C4BPA; 1q32), the immunoglobulin-like and fibronectin type III modules in the human Axl oncogene (AXL; 19q13.1-13.2) gene and four of six thrombospondin modules of the human properdin gene (BF; 6p21.3). Genes with predominantly class 0-0 exons include those encoding the type III collagens and β-casein (CSN2; 4pter-q21) whilst class 2-2 exons are exclusively found in the glucagon (GCG; 2q36-q37) gene.

Using a large database of eukaryotic genes, Lone et al. (1995) demonstrated there to be an excess of *symmetrical* exons over expectation and estimated that at least 19% of exons had been involved in exon shuffling. Why the preponderance of class 1-1 exons? Patthy

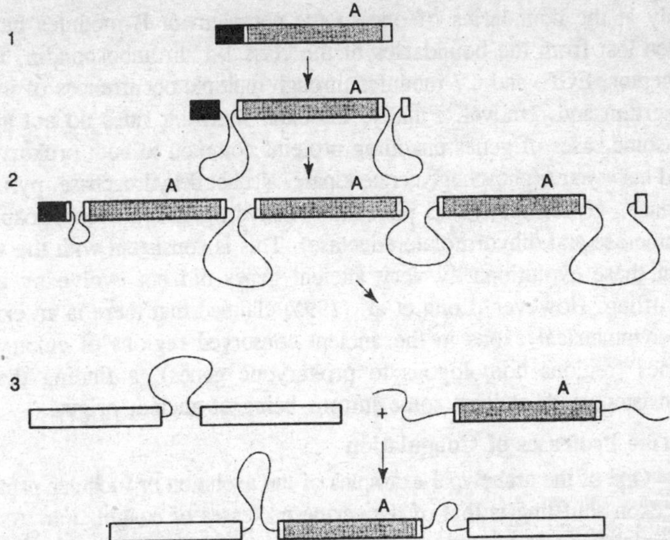

Fig. 4.5. Different stages in the conversion of a domain to a module.

(1994) suggested that since 92% of introns flanking signal peptide domains were phase 1, modularization of exported proteins with secretory signal peptide domains is likely to have employed class 1-1 introns. According to Patthy (1994), modularization proceeds in different stages: (i) insertion of introns of identical phase at boundaries of the protein fold, (ii) tandem duplication of the symmetrical module by intronic recombination, and (iii) module transfer to a novel location. Nonsymmetrical exons (those flanked by introns of different phase) are potentially much less versatile and may be expected to have been utilized much less often. Tomita et al. (1996) noted that when introns are inserted so as to disrupt codons, the site of insertion occurs much more frequently between the first and second bases than between the second and third bases. The reason for this remains unclear.

There are however various instances which do not conform to the above exon shuffling rules. In some cases, the original exon-intron organization of genes has become eroded with the passage of evolutionary time. Thus, although in the human perlecan (HSPG2; 1p35-p36) gene, the LLD receptor modules and the immunoglobulin-like modules are still flanked by phase 1 introns, the original introns have been lost from the regions of the gene encoding laminin A, laminin B and epidermal growth factor-like modules. Similarly, in the human complement 6 (C6, chromosome 5) gene, phase 1 introns are found only at the boundaries of one of the complement B modules having been lost from the boundaries of the class 1-1 thrombospondin, LDL receptor, EGF, and C7 modules through multiple occurrences of intron insertion and removal. Finally, the exon shuffling rules do not apply in some cases of genes encoding proteins common to both prokaryotes and eukaryotes (phosphoglycerate kinase, alcohol dehydrogenase, pyruvate kinase, glyceraldehyde-3-phosphate dehydrogenase, triosephosphate isomerase and dihydrofolatereductase). This is consistent with the view that these evolutionarily very ancient genes did not evolve by exon shuffling. However, Long et al. (1995) claimed that there is an excess of symmetrical exons in the ancient conserved regions of eukaryotic genes (regions homologous to prokaryotic genes), a finding that is consistent with at least some introns being of ancient origin.

Serine Proteases of Coagulation

One of the archetypal examples of the evolution of modular proteins by exon shuffling is that of the serine proteases of coagulation. As the completed primary sequences of hemostatic factors became available, it was noticed that certain domains of shared homology recurred many times in diverse proteins. The kringle domain is present in prothrombin as well as three proteins of fibrinolysis (tPA, urokinase and plasminogen) and may play a role in fibrin binding. A trypsin-like serine protease domain is common to all the proteolytic enzymes of coagulation. The EGF-like domain was first recognized among clotting proteins in factor X. Two copies of this motif are found in all the vitamin K-dependent proteins of coagulation (factor IX, factor X, factor VII, protein C) except protein S which has four and prothrombin which has none. The action of vitamin K in coaguloprotein synthesis is to promote carboxylation of N-terminal glutamic acid (Gla) residues by a liver microsomal enzyme system. The five proteins which share this N-terminal modification have a homologous propeptide domain that functions as a signal to the carboxylase.

The organization of coagulation factor genes reflects to a considerable extent their functional modular assembly as described above. It is apparent that since factors VII (F7; 13q34), X (F10; 13q34), IX (F9; Xq26.3-q27.1) and protein C (PROC; 2q13-q21) have virtually identical protein and gene structures, they must have emerged relatively recently by a process of gene duplication and divergence. Similarly, tissue plasminogen activator (PLAT; 8p12-q11.2), factor XI (F11; 4q35), factor XII (F12; 5q33-qter) and urokinase (PLAU; 10q24-qter) exhibit similar gene organization with respect to their protease domains but

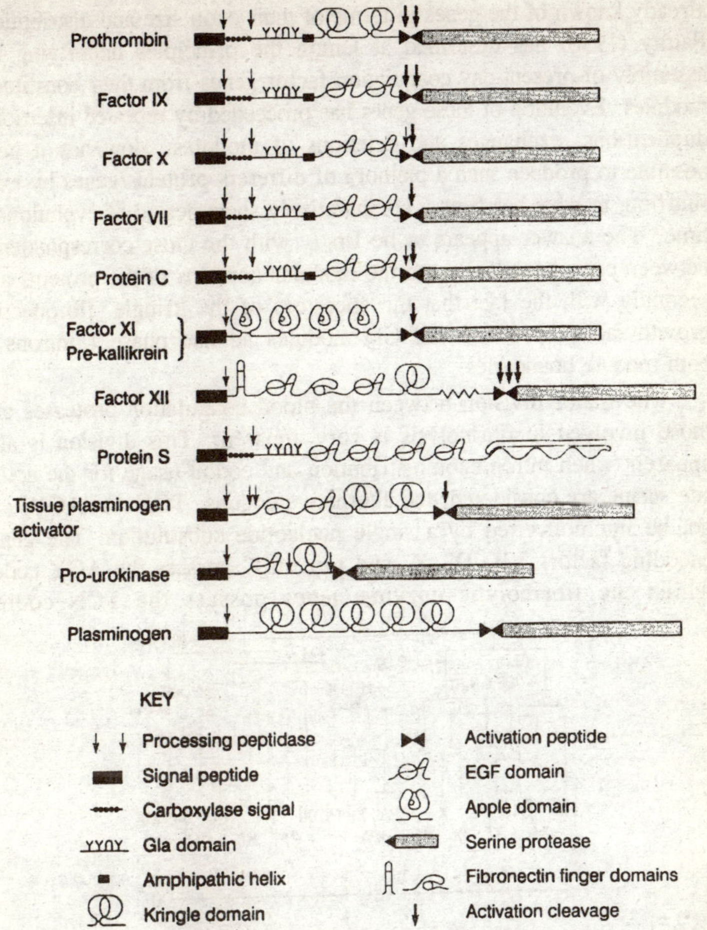

Fig. 4.6. Domains of hemostatic proteins.

differ from prothrombin (F2; 11p11-q12) and factors VII, IX, X and protein C. Kringle domains are encoded by a similar gene structure wherever they occur, as are the calcium-binding Gla domains. EGF-like domains are encoded by a single exon with type 1 splice junctions are the fibronectin type I (Fn I) domains.

The presence/absence of different modules (Gla and EGF-like domains, kringles) in the different coagulation factor proteins can be used to reconstruct their evolutionary past. The transfer of modules between proteins has arisen via exon shuffling and protein data can be used to construct evolutionary trees that are consistent with what is already known of the genes in terms of their exon size and distribution. Patthy (1990) has described at length the principles underlying the assembly of present-day coagulation factor genes from their constituent modules. Evolution of these genes has proceeded by repeated insertions, duplications, exchanges and deletions of modules. How has it been possible to produce such a plethora of different proteins/genes by exon shuffling in what has been a comparatively short period of evolutionary time? The answer appears to lie firstly with the close correspondence between exon boundaries and the modular domains of the proteins and secondly with the fact that the ancestors of the kringle, fibronectin, growth factor, protease and Gla modules all had phase 1 introns at both module boundaries.

The major division between the blood coagulation proteases and those involved in fibrinolysis is very apparent. This division is also apparent when intron/exon distribution and codon usage for the active site serine are considered: the alternative codons, TCN and AGY, can not be interconverted by a single nucleotide substitution. The genes encoding factors VII, IX, X, and protein C possess the AGY codon whilst the fibrinolytic enzyme genes possess the TCN codon.

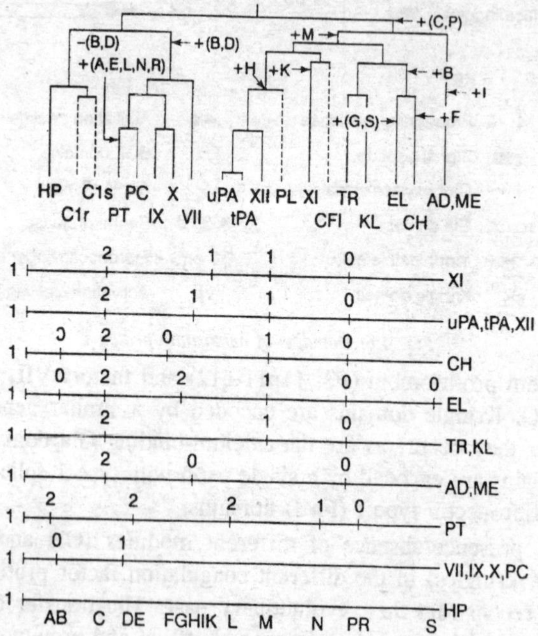

Fig. 4.7. Evolution of the serine protease family.

Phylogenetic trees for the protease and kringle domains are however not identical suggesting that these domains may have experienced different evolutionary histories. This implies that in multidomain proteins such as the serine proteases, each domain can represent an independent evolutionary unit.

Protein Folds, Primordial Exons and the Emergence of Exon Shuffling

Proteins are composed of combinations of secondary structural elements, α-helices and β-sheets, which are connected by loop regions at the surface of the molecule. Certain combinations of these elements (termed *motifs* or *folds*) are frequently used in proteins and are known as *domains*. Evolutionarily mobile domains are termed *modules*. Motifs can be as simple as in the case of the hexamer repeat unit that forms a left-handed parallel β-helix found in UDP-N-acetylglucosamine acyltransferase or more complex as in the 21-26 amino acid zinc finger motif. For practical purposes, a protein domain may be regarded as a structurally and/or functionally discrete portion of a protein that can fold independently into a stable tertiary structure and may be composed of single or multiple motifs. Since the zinc finger motif can fold independently, it may actually be regarded as a small domain in its own right. Most proteins contain two or more domains although some may just contain a single domain.

During evolutionary time, the amino acid sequences of proteins are eroded at a much faster rate than are the corresponding 3D structures. In other words, although it may be difficult to discern sequence conservation, structural conservation may still be evident. One example of this is provided by the α-defensin (DEFA1, DEFA4, DEFA5, DEFA6; 8p23) and β-defensin (DEFB1, DEFB2; 8p23) genes which, despite a complete lack of DNA sequence homology, and major differences between their encoded proteins in terms of their cysteine spacing and disulfide pairing, have nevertheless evolved from a common ancestor.

It is often quite difficult to ascertain whether structural similarities are *homologous* (i.e. based upon divergent evolution from a common ancestor) or *analogous* (i.e. based upon convergent evolution to a physically favoured secondary or tertiary structure). The α/β barrel domain is common among enzymes and examples of this domain are thought to share a common ancestor rather than to have evolved convergently to a stable fold. Such examples are a reflection of the evolution of extant protein structures from a small number of basic folds encoded by primordial exons.

Primordial exons encoding functional domains are now widely dispersed among many diverse proteins. Examples include the ankyrin repeat the spectrin repeat, the EGF-like domain, the WD-repeat, the ATPase domain and kringles. The EGF-like domain is present in ~1% of human proteins whilst the immunoglobulin domain may be even more prevalent. Other domains such as the Kunitz domain which is less prevalent, may have a more recent origin.

How many primordial exons were required to construct the huge array of extant proteins? Dorit et al. (1990) used the frequency with which exons in a 1200 exon database had been 'reused' between genes encoding different proteins to assess the size of the underlying exon pool; they estimated that the number of primordial exons was between 1000 and 7000. A similar study succeeded in distilling 1410 polypeptide chains down to 112 analogous fold families whose members exhibited an average of <18% sequence identity. Patthy (1991b) has pointed out, however, that this type of approach may tend to underestimate the true number of exons since it may not adequately have considered the pool of exons that participate in exon shuffling only rarely.

Mosaic proteins assembled from class 1-1 modules are confined predominantly to multicellular organisms (Patthy 1991a). Exon shuffling probably only came into its own with the radiation of the metazoa when the requirements of multicellularity fuelled the demand for novel and increasingly more complex proteins to potentiate intercellular communication. Perhaps the cellular machinery for exon shuffling only appeared at this time.

Pseudoexons

Pseudoexons may be defined as gene regions that are functional in a gene from one species and which have been conserved in non-functional form in another species. Good examples of pseudoexons are those found in the human αA-crystallin (CRYAA; 21q22.3), glycophorin B (GYPB; 4q28-q31), complement component receptor 2 (CR2; 1q32) and transketolase-related (TKT; Xq28) genes. In the CRYAA example, the pseudoexon is alternatively spliced into the mRNA in rodents and bats but was lost to other mammalian species about 30-40 Myrs ago. Exon III of the glycophorin B (GYPB 4q28-q31) gene is expressed in the chimpanzee and gorilla but has been silenced as a pseudoexons in humans through an exon 3 donor splice site mutation. This finding predicts a larger extracellular domain in the chimpanzee protein. Similarly, the human glycophorin E (GYPE; 4q28-q31) gene has lost two of its original exons. The recombinational use of these pseudoexons represents a major mechanism for allelic diversification in the glycophorins: depending upon the extent, location and type of recombination, the exchange or transfer of pseudoexons can result in the creation of novel inter-exon or intra-exon hybrid junctions that yield new mRNA splicing patterns.

Murine apolipoprotein E receptor 2 gene transcripts possess an extra exon as compared with transcripts from the human (LRP8; 1p34) gene. However, the human and marmoset LRP8 genes possess an intronic pseudoexon corresponding to the murine exon indicating that this exon has been lost in the primate lineage. The incorporation of the pseudoexon into the primate LRP8 mRNA transcript is prevented by two nucleotide substitutions in the adjacent 5' donor splice site. The pseudoexon contains a deletion which would have led to a frameshift but it is unclear whether the deletion occurred after inactivation of the exon or if the splice site substitutions served to allow the primate LRP8 gene to avoid incorporation of the mutation-containing exon.

The human insulin-like growth factor-II (IGF2; 11p15.5) gene is transcribed and processed into three different mRNAs under the control of three distinct promoters, two of which have counterparts in the mouse and rat . The 5'-most promoter, which is active in adult human tissues, controls the transcription of a cassette of three exons that comprises the 5' UTR of the human gene. The murine gene does not however contain a structural and functional homologue of exon 1 (and adjacent promoter) of the human gene. The absence of this cassette from murine Igf2 mRNA may account for the disappearance of Igf/2 transcripts from most murine tissues shortly after birth.

The presence of pseudoexons in a number of human genes has thus provided good evidence for the occurrence of exon inactivation (and hence also intron loss) during mammalian evolution. Such events are certainly compatible with the introns early theory. It is however unclear how common pseudoexons are in human genes since they will usually only be detected through the careful comparison of orthologous gene pairs, a procedure that will tend to underestimate their prevalence.

5

Gene Families

Origin of Human Genes

Some human genes appear to have originated comparatively recently. Indeed a very few have emerged in the relatively short space of time of that has elapsed since divergence from chimpanzee some 7 Myrs ago. Other human genes appear to have homologues in simple organisms implying a very ancient origin, in some cases even predating the divergence of prokaryotes and eukaryotes. In the following sections, representative examples of genes are cited in order to illustrate the enormous variability in the antiquity of extant human genes. To some extent of course, the origin of a given gene is a matter merely of semantic distinction. Thus Doolittle's (1992) truism stated: 'new proteins come from old proteins as a result of gene duplications followed by base substitutions'. This notwithstanding, it has still been possible to draw up a working classificatory scheme for proteins based on their 'apparent invention time'.

Genes with a Specifically Human Origin

There are relatively few known examples of human genes which originated in the last 7 Myrs after the divergence of the human and chimpanzee lineages. One such gene is the potentially functional immunoglobulin Vk gene located 1.5Mb telomeric to the human Ck gene which is putatively human-specific. Further studies of the immunoglobulin k locus (comprising one Ck, 5 Jk, and 76 Vk gene segments on the short arm of human chromosome 2), have provided evidence for the human-specific partial duplication of the locus. The duplicated portion of the immunoglobulin k locus, which contains 36 Vk genes and pseudogenes, is not found in the chimpanzee or gorilla.

Thus, the duplication event must have occurred after the branchpoint of humans from the great apes. Huber et al. (1994) estimated that this duplication event occurred 1 Myrs ago.

Another example of human specificity is provided by the salivary amylase (AMY1) genes clustered on human chromosome 1p21: three such genes are to be found in the human genome whereas only one is present in chimpanzee. The most parsimonious explanation is that amplification of the AMY1 gene sequences has occurred during the last 7 Myrs, the time since human and chimpanzee last shared a common ancestor. Similarly, the high-affinity immunoglobulin receptor genes (FCGR1A, 1q21; FCGR1B, 1p12; FCGR1C 1q21) were triplicated from a single primate ancestral Fcgr1 gene about 3 Myrs ago, after the divergence of chimpanzees from the human lineage. Finally, evidence for substantial deletions/translocations in the chimpanzee genome as compared to human is apparent from pulsed field gel electrophoretic studies of the region between the HLA-B and TNF genes on chimpanzee chromosome 5p21.3.

The gene coding regions of humans and chimpanzees are typically of the order of 99% homologous (e.g. BRCA1). As yet, however, there are no known differences between the human and chimpanzee genomes in terms of either the presence/absence of genes or gene copy number that could account for the considerable anatomical and behavioural differences between the two species. We may surmise that subtle differences in the expression of regulatory genes or alternatively differences in the responsiveness of those genes which serve as targets for the action of regulatory proteins, may help to explain why humans are not chimpanzees. Candidate genes could include not only those involved in developmental regulation but possibly also those situated adjacent to human-specific karyotypic rearrangements.

Human Genes which Originated after the Divergence of Old World Monkeys and New World Monkeys

The haptoglobin gene appears to have become amplified after the separation of Old World Monkeys from New World monkeys. The gorilla, chimpanzee, orangutan and rhesus monkeys have three haptoglobin genes, humans have two (HP, HPR; 16q22) whilst the spider monkey have only one. The apolipoprotien(a) (LPA; 6q27) gene probably also emerged during this time since its presence has only been detected in humans, apes, and Old World monkeys.

The number of rhesus blood group antigen genes varies between primate species, humans having two (RHD, RHCE; 1p34-p36.2), gorillas two, chimpanzees three, whilst orangutan, gibbons and all Old World and New World monkeys possess only a single RH-like gene. Duplication of the RH-like genes probably therefore occurred between 8 and 11 Myrs ago before divergence of the human and gorilla lineages.

Eosinophil-derived neurotoxin (EDN) and eosinophil cationic protein (ECP) are host defense proteins which are members of the ribonuclease family. They are encoded respectively by the RNASE2 and RNASE3 genes located on human chromosome 14q24-q31. Divergence of the nucleotide sequences of the two related genes was found to be consistent with a duplication event occurring some 25-40 Myrs ago after the divergence of Old World monkeys from New World monkeys. In agreement with this prediction, RNASE2 and RNASE3 orthologues are evident in chimpanzee, gorilla, orangutan and macaque but not in the marmoset, a New World monkey. Since their duplication, the two proteins have diverged very rapidly, with evolution promoting the acquisition of novel functions viz. increased cationicity/toxicity (ECP) and enhanced ribonuclease activity.

Human Genes which Originated during Primate Evolution

The ZNF91 Kruppel-associated box-containing zinc gene family on human chromosome 19p12-p13 is present in the great apes, the gibbons, Old World and New World monkeys but is not found in prosimians or rodents. The origin of this gene family is unclear. Presumably, the ZNF91 gene cluster arose by duplication/amplification at least 55 Myrs ago in the common ancestor of simians.

The three alcohol dehydrogenase genes (ADH1, ADH2, and ADH3), linked on human chromosome 4q22, also originated during the adaptive radiation of the primates; two successive duplications are estimated to have occurred 45±Myrs and 60±8 Myrs ago.

Other examples of human genes with a primate origin include the interferon-a gene (IFNA) and growth hormone (GH) gene clusters. The IFNA cluster, containing 13 members and located at chromosome 9p21, has emerged by a process of duplication and divergence over the last 26 Myrs. The GH gene cluster is also essentially a primate creation with several genes (GH1, GH2, CSH1, and CSH2; 17q23) having emerged over the last 25-50 Myrs by a process of duplication and divergence.

Some genes emerged in the last 100 Myrs during the adaptive radiation of the mammals. For example, the fucosyltransferase gene family (FUT1, FUT2, 19q13.1-qter; FUT3, FUT5, FUT6, 19p13.3;

FUT4, 11q21; FUT7, 9; FUT8, 14q23) is thought to have emerged during this period by a process of gene duplication, translocation and divergence.

α-Lactalbumin is a mammary gland-expressed calcium-binding protein which interacts with galactosyl transferase to promote lactose synthesis. Although this function is clearly confined to mammals, the α-lactalbumin gene (LALBA; 12q13) is thought to have arisen by duplication of the gene encoding lysozyme c which encodes a bacteriolytic enzyme present in the tissues and secretions of mammals, birds, reptiles and even insects. This gene duplication probably occurred 300-400 Myrs ago well before the evolutionary emergence of the mammals and the process of lactation.

Human Genes whose Origin Preceded the Divergence of Mammals

Other genes were specifically vertebrate creations, their emergence through duplication and divergence having perhaps been made possible by the genome duplications thought to have occurred prior to the adaptive radiation of this subphylum. A new wave of gene creation may have accompanied the emergence of the amphibians and early reptiles thereby equipping them for terrestrial life.

One example of a vertebrate creation appears to be the vitamin K-dependent serine proteases of blood coagulation (factors VII, IX, X, protein C and prothrombin). Prothrombin is present in bony fish (trout), cartilaginous fish (dog fish) and the hagfish, one of the modern representatives of the jawless Agnatha. There is no evidence, however, for the existence of thrombin or a thrombin-like protein in either protochordates or echinoderms. Whether the other four vitamin K-dependent factors of coagulation are present in fish is as yet unclear. If they are, the adaptive radiation of the vitamin K-dependent factors of hemostasis must have occurred during these space of some 50 Myrs ago. The subsequent evolution of the vitamin K-dependent factors of coagulation are present in fish is as yet unclear. If they are, the adaptive radiation of the vitamin K-dependent factors of hemostasis must have occurred during the space of some 50 Myrs between the divergence of the protochordates and the appearance of the *Agnatha*, some 450 Myrs ago.

Another specifically vertebrate invention was pulmonary surfactant which comprises a series of proteins which serve to reduce surface tension at the air-liquid interface in the lung. This was probably a prerequisite for air breathing. The human genome contains a number of clustered pulmonary surfactant genes among them SFTPA1 (10q21-q24) encoding pulmonary surfactant protein A. Orthologues of this gene have been found in airbreathing lung-fish and surfactant protein has been detected both in the swimbladder of goldfish and in the lungs of lungfish.

Transthyretin is a thyroid-binding protein which in human is synthesized mainly in the liver but also in the choroid plexus and retina. The transthyretin gene (TTR; 18q11-q12) is a vertebrate invention having made its first appearance in the stem reptiles some 300 Myrs ago. Transthyretin is expressed in the choroid plexus in reptiles. Liver synthesis of the protein only evolved later (and independently) in birds, eutherian mammals and in some marsupials.

Troponin I, together with troponins C and T comprise the three subunits of the troponin complex of the thin filaments of vertebrate striated muscle. Since invertebrates and ascidians possess a single troponin I gene at least two duplications must have taken place during the vertebrate lineage to generate the three troponin I genes evident in the human genome (TNNI1, 1q31; TNNI2, 11p15; TNNI3, 19q13).

The human ABO blood group gene (ABO; 9q34) has a common evolutionary origin with the chromosomally linked and now inactive α-1,3-galactosyltransferase (GGTA1; 9q33-q34) gene; the two genes emerged by duplication and divergence about 400 Myrs ago during the evolution of the early vertebrates.

Olfactory receptors are extremely important to mammals; dogs, rats and mice may have as may as 1000 genes encoding them. Although the human genome may contain fewer functional genes than other mammals, pseudogenes abounded. Indeed, it has been estimated that >0.1% of the human genome is composed of olfactory receptor genes and pseudogenes dispersed between at least 13 different chromosomes. [The classification of this gene family is still in its infancy but known members include OR1A1, 17p13; OR1D2, OR1D4, OR1D5, 17p13; OR1E1, OR1E2, 17p13; OR1F1, 16p13; OR1G1, 17p13; OR2D2, 11p15; OR3A1, OR3A2, OR3A3, 17p13; OR5D3, OR5D4, 11q12; OR5F1, 11q12; OR3A1, 11p15; OR10A1, 11p15]. The origin and emergence of the present-day size of the olfactory receptor gene family probably preceded the divergence of the mammals. However, at least in the primates, the olfactory receptor genes appear to have still been in a considerable state of flux with numerous translocations, duplications and deletions having occurred during the evolution of the great apes.

Human Genes whose Origin Preceded the Divergence of the Vertebrates

One example of a human gene whose origin preceded the advent of the vertebrates ~500 Myrs ago is the c-*myc* proto-oncogene. Present in all vertebrates, a homologue of the human gene (MYC; 8q24) is detectable in echinoderms but not in *Caenorhabditis* or *Drosophila*. Other examples include the insulin-like growth factor genes (IGF1 and IGF2; 12q22-q24 and 11p15.5 respectively) which emerged during the evolution of the protochordates more than 600 Myrs ago.

Human Genes whose Origin Preceded the Divergence of the Metazoa

Human genes with homologues in the fruitfly, *Drosophila*, must have originated prior to the divergence of the deuterostomes from the protostomes ~700 Myrs ago. A large number of *Drosophila* genes have been shown to have human homologues and *vice versa*. Examples (with human chromosomal locations where known) include proto-oncogenes *jun* (JUN; 1p31-p32) and *fos* (FOS; 14q24), *flightless I* (FLI1; 17p11.2), *minibrain* (MNBH; 21q22.2), *dorsal* (REL; 2p12-p13), *atonal* (ATOH1; 4q22), *archain* (ARCN1; 11q23), tumor suppressor gene lethal 2 giant larvae (LLGL1; 17p11-p12), transient receptor potential channel-related protein 1 (TRPC1) succinate dehydrogenase, iron-sulfur protein subunit (SDH1; 1p22.1-qter), *slowpoke* (SLO, chromosome 10), *enhancer of split* (TLE1; 19p13.3), *dodo* (PIN1), EGF receptor (EGFR; 7p12.1-p12.3), *dishevelled* (DVL1; 1p36), *son of sevenless* (SOS1, 2p21-p22; SOS2, 14q21), *awd* (NME1), *Ddx1* DEAD box polypeptide (DDX1; 2p24); *cut* (CUTL; 7q22) and the paired box domain family of transcription factors that play an important role in development (PAX1, 20p11.2; PAX2, 10q25; PAX3, 2q36; PAX4, 7q22-qter; PAX5, 9p13; PAX6, 11p13; PAX7, 1p36; PAX8, 2q12-q14; PAX9, 14q12-q14).

Homologs of the ETS1 proto-oncogene, located on human chromosome 11q23, have been found not only in *Drosophila* but throughout the metazoa in organisms as evolutionarily distant as sponges, anenomes, flatworms and nematodes. The origin of the ETS genes therefore appears to predate the divergence of the schizocoeles (arthropods, annelids etc) from the pseudocoeles (nematodes) more than 750 Myrs ago. STAT (Signal Transducers and Activators of Transcription) proteins represent an example of proteins present in *Drosophila* and *Dictyostelium* but not in yeast and so must have merged at some stage during the adaptive radiation of the metazoa. Homology is also evident between members of the human lipase gene family [lipoprotein lipase (LPL; 8p22), hepatic lipase (LIPC; 15q21-q23) and pancreatic lipase (PNLIP; 10q26)] and the yolk proteins of *Drosophila*.

Another example of a gene with an ancestry stretching back at least as far as 750 Myrs, is a human retina-expressed gene, CAGR1 (13q13) which is homologous to the *Caenorhabditis elegans* cell fate-determining gene, *mab*-21 Ahringer (1997) presented evidence that at least 50% of *Caenorhabditis* genes are likely to have counterparts in the human genome. Similarly, the human neurotrophic tyrosine kinase receptor gene has a homologue in the snail, *Lymnea stagnalis* although not apparently in *C. elegans*.

Ancient conserved regions, deemed to be regions of the greatest structural or functional importance on account of their evolutionary conservation are also evident in various other proteins with homologues in both human and *C. elegans*, for example adenylate cyclases, epidermal growth factor like domains, gelsolin, intermediate filament proteins, kinesins, neurotransmitter transporters and the ubiquitins.

Human Genes whose Origin Preceded the Divergence of Animals and Fungi

The 12,068 kb sequence of the 16 chromosomes of the yeast *Saccharomyces cerevisiae* genome has now been established. A considerable proportion of the organism's 5885 genes are significantly related to sequences in the human genome. Tugendreich et al. (1994) used the BLASTX P program to measure the extent of this homology: 29% of human cDNAs found in GenBank matched a yeast protein with a P value of less than 10^{-5}. More specifically, these authors showed that some important human disease genes manifest considerable sequence similarity to a yeast counterpart e.g. adrenoleukodystrophy (ALD, Xq28) with a yeast 70 kD peroxisomal membrane protein, the myotonic dystrophy protein (DMPK, 19q19) with a yeast cAMP-dependent protein kinase, the Wilms' tumor protein (WT1, 11p13) with a yeast zinc finger and Ret (RET, 10q11.2) protein underlying multiple endocrine neoplasia type 2A with a yeast cell division control protein. Similarly, human genes QM (Xq28) encoding a c-*jun*-associated transcription factor SEC13L1 (3p24-25) encoding a protein putatively involved in vesicle biosynthesis from the endoplasmic reticulum during protein transport DNECL (14q32) encoding cytoplasmic dynein, the MCM family (MCM2, 3q21; MCM3, 6p12; MCM4, 8q12-q13; MCM5, 22q13; MCM6, 2q21; MCM7, 7q21-q22) of DNA replication proteins, the PTEN (10q23.3) tumor suppressor gene and the G protein a-subunit gene family, have yeast homologues.

Since *S. cerevisiae* diverged from higher eukaryotes some 1000 Myrs ago, homologies between human and yeast genes are clearly very ancient. Such conservation at the amino acid level is likely to reflect the conservation of fairly basic biological functions. Thus, it comes as no surprise to find that the yeast genes encoding cytochrome c, histone deacetylase, the origin recognition complex the recombination protein Rad51 and the nucleotide excisions repair gene Rad23 have highly homologous human counterparts viz. CYC1 (8q24.3), HDAC1 (1p34), ORC1L (1p32), RECA (15q15.1), and RAD23A and RAD23B (19p13 and 3p25, respectively).

Sometimes the homology is regionally localized, for example the GTPase-activating protein-related domain of neurofibromin (encoded by the NF1 gene on human chromosome 17q11.2) exhibits extensive homology to the *S. cerevisiae* proteins IRA1 and IRA2. Such *ancient conserved regions* (ACRs) represent regions of the greatest structural or functional importance. Eukaryotic-specific ACRs are also evident in various other proteins with homologues in both human and yeast, for example hexokinases, β-transducins, protein kinase catalytic domains and the Src homology domain.

Human Genes whose Origin Preceded the Divergence of Plants and Animals

Some human genes have their counterparts in both plans and fungi implying that they originated before the divergence of the three kingdoms. Members of the *myb* gene family (MYB; 6q23.3-q24) are found in animals plants, fungi and even slime moulds. Myb proteins function as regulators of cell growth and differentiation by binding to DNA and regulating gene expression. Since slime moulds such as *Dictyostelium* are thought to have diverged before the main eukaryotic radiation which gave rise to animals, plants and fungi ~100 Myrs ago, the presence of Myb proteins in this organism serves to date the emergence of this ancient regulator of gene expression. Remarkably, despite this ancient origin, the Myb-related proteins of *Drosophila* and *Dictyostelium* are still able to recognize and interact with the same cognate DNA sequence as their vertebrate counterparts.

The High Mobility Group (HMG) proteins are another human gene family represented in both plants and fungi. Two members of this family, HMG1, and HMG2, are present on human chromosomes 13q12 and 4q31 respectively. Members of the HMG proteins superfamily are characterized by the possession of one or more HMG boxes, each of which comprises ~80 amino acid residues and is capable of interacting with DNA. Both the actins and the CCAAT-specific transcription factor, NF-Y have their counterparts in plants and fungi whilst the annexins and the cystatins are present in plants but not fungi.

Human Genes whose Origin Preceded the Divergence of Prokaryotes and Eukaryotes

Genes that are common to prokaryotes and eukaryotes must have arisen more than 2000 Myrs ago before the divergence of the two groups. A number of human genes fit into this category. Hexokinase is one of the best characterized examples with sequences available from bacteria, yeast, plants and vertebrates (three hexokinase genes HK1, HK2, and HK3 exist in humans on chromosomes 10q22, 2p12 and 5q35 respectively). A proposed multi-kingdom phylogeny of the hexokinase gene and depicts a series of duplication and fusion events that must have occurred during its evolution. The matrix metalloproteinases that play an important role in tissue remodelling and wound healing have homologues in plants, animals and bacteria.

Universal stress protein UspA of *E. coli* is homologous of the MADS-box transcription regulators of eukaryotes whilst the FUS6 family of eukaryotic proteins contains a putative DNA-binding domain related

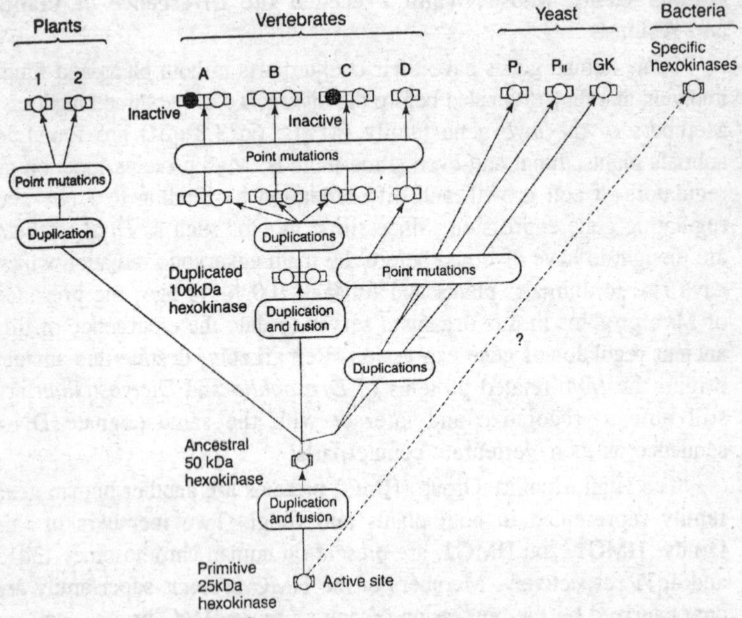

Fig. 5.1. A proposed phylogeny for the hexokinases.

to bacterial helix-turn-helix transcription regulators. Other human genes with homologues in *E. coli* are the Y-box family (YB1; 1p34) which encode nucleic acid-binding proteins which serve as transcriptional repressors in humans and the cold shock response in bacteria, the SIR2L gene family which are involved in cell cycle progression and chromosome stability and chromatin silencing, CYP51 (7q21), the most conserved member of the P450 monooxygenase family involved in the demethylation of sterol precursors, the MLH1 (3p21), MSH2 (2p21-p22), GTBP (2p16), PMS1 (2p31-q33) and PMS2 (7p22) mismatch repair genes, the RAD51A (15q15) DNA repair protein gene homologous to the RecA protein of *E. coli* and HES1 (21q22.3) with homology to the σ cross-reacting protein 27A of *E. coli*. A considerable degree of homology has also been noted between prokaryotic and eukaryotic RNA polymerases, prokaryotic σ factors and eukaryotic transcription factors TBP (TFIIB) and TFIIE, eukaryotic translation initiation factors eIF-1A and eIF-5A and their prokaryotic counterparts between prokaryotic and eukaryotic RNA-binding proteins and between prokaryotic and eukaryotic type II DNA topoisomerases and DNA polymerases. A considerable degree of homology has also been noted between prokaryotic and eukaryotic aminoacyl-tRNA synthetases.

The eukaryotic cell cytoskeleton contains a number of proteins, the most abundant being actins, tubulins, myosins and cytokeratins. The evolutionary origin of these proteins is intimately bound up with the origins of the eukaryotic cell itself. Thus γ-tubulin is thought to have evolved before the α- and β-tubulins diverged from each other and since these three tubulin isotypes are found in all eukaryotes, their origins must have preceded the eukaryotic divergence. Consistent with this postulated, sequences similarities exist between eukaryotic actins and tubulins and bacterial *ftsA* and *ftsZ* proteins.

There are many instances of genes present in *E coli* or yeast having multiple homologues in the human genome indicating that the gene family in question must have expanded during evolution. One example of this are the helicase genes of the RecQ family of which there are five known in the human genome (WRN, 8p; BLM, 15; RECQL, 12p12; RECQLA, 8q24.3; RECQL5, 17q25).

As one travels further back in evolutionary time, the similarity between gene sequences tends to decay to an extent that only specific portions may be recognizably homologous. These *ancient conserved regions* (ACRs) represent those regions of greatest structural or functional importance and often correspond to specific domains. ACRs common to both prokaryotes and eukaryotes have been noted in a diverse array of proteins, for example enolase, glyceraldehyde 3 -phosphate dehydrogenase, cytochrome c oxidase subunit I, aminoacyl-transfer RNA (II) synthetases, HSP70 and HSP90, phosphoglycerate kinase, pyruvate dehydrogenase E1α, pyruvate kinase, ribosomal proteins L3 and P0 and triosephosphate isomerase. Ceruloplasmin (CP; 3q23-q25) and coagulation factors V (F5; 1q23) and VIII (F8C; Xq28) manifest homologies to the small blue proteins of bacteria which have a role in electron transfer. A 60 amino acid domain found in cystathionine-β-synthase (CBS; 21q22.3) is also found in a bacterial ABC transporter protein and in putative protein found in archaebacteria.

Phylogenetic studies have suggested that the emergence of cytochrome oxidase (a key enzyme in aerobic metabolism), the aldehyde dehydrogenase (e.g., ALDH1, 9q21; ALDH3, 17; ALDH5, 9; ALDH6, 15q26; ALDH9, 1; ALDH10, 17p11.2), carbamoyl phosphate synthetase (a key enzyme of arginine and pyrimidine biosynthesis) (CPS1; 2q35), the protein synthesis elongation factors Tu (TUFM; 16p11) and G (EEF1G), tRNA splicing endonuclease and the glucose-6-phosphate isomerase gene family (GPI; 19q13) all predated the divergence of prokaryotes and eukaryotes. The RadA protein that catalyzes DNA pairing and strand exchange is present not only in all eukaryotes including human (RECA; 15q15.1) but also has homologues in both prokaryotes (RecA) and the archaea.

Phylogenetic studies have provided evidence for the extreme antiquity of the glutamine synthetase gene (Kumada et al., 1993). Glutamine synthetase is an essential enzyme of nitrogen metabolism with vital functions in both glutamine biosynthesis and ammonia assimilation. It is thought therefore to be indispensible to all living organisms. Comparison of the glutamine synthetase genes of extant organisms as diverse as bacteria and vertebrates allowed Kumada et al. (1993) to estimate the time of duplication of the ancestral glutamine synthetase gene to ~3500 Myrs ago, that is >1000 Myrs before the divergence of the prokaryotes and eukaryotes.

Emergence of Genes and Gene Families has Paralleled Organismal Evolution

Eukaryotic genomes have been largely constructed by a continual process of gene duplication and divergence utilizing a set of basic gene types many of which possess counterparts in the genomes of primitive organisms. Much of this gene diversification may have taken place in the 'Cambrian explosion' which is thought to have begun ~570 Myrs ago.

Genes which have emerged during primate evolution are relatively few in number and tend to involve the addition of new members to pre-existing multigene families (e.g. genes involved in host defence such as the immunoglobulins; the alcohol dehydrogenases whose expansion may have been associated with a move to a fruit diet). During mammalian evolution, certain genes emerged which were intimately associated with mammalian physiology such as lactation. The emergence of genes with novel functions (e.g. pulmonary surfactant) and the expansion of other gene families (e.g. coagulation factors and olfactory receptors) occurred in parallel with the emergence of the first terrestrial vertebrates. By contrast, genes whose origin preceded the divergence of animals and fungi, or animals and plants, encode proteins involved in fairly basic cellular processes. Finally, the truly primordial genes shared by both prokaryotes and eukaryotes encode protein products which are absolutely required for fundamental cellular processes such as DNA repair and replication, transcription and translation, as well as certain very basic enzymes of metabolism such as glutamine synthetase.

Iwabe et al. (1996) performed their own analysis of gene duplication during organismal evolution. They concluded that most gene duplications giving rise to novel functions predated the divergence of the vertebrate and arthropod lineages. However, genes encoding products that perform virtually identical functions but which differ in their tissue distribution (tissue-specific isoforms) underwent duplications independently in vertebrates and arthropods after divergence of their respective lineages. Finally, genes which encode proteins that are localized to cell compartments (compartmentalized isoforms) emerged by duplications which predated the separation of animals and fungi. Iwabe et al. (1996) concluded that there was a good correspondence between molecular evolution at the level of the gene, and tissue and organismal evolution.

Multigene Families

Gene Families

Dayhoff et al. (1978) defined a gene family as a group of related genes encoding proteins differing at fewer than half their amino acid positions. Owing to complex evolutionary histories, both the size of gene families and the relationship between their constituent members are very variable. In principle, the members of multigene families can evolve in three different way—by concerted evolution, divergent evolution or by a birth-and-death process. *Concerted evolution* may operate through either unequal crossing over or gene conversion both

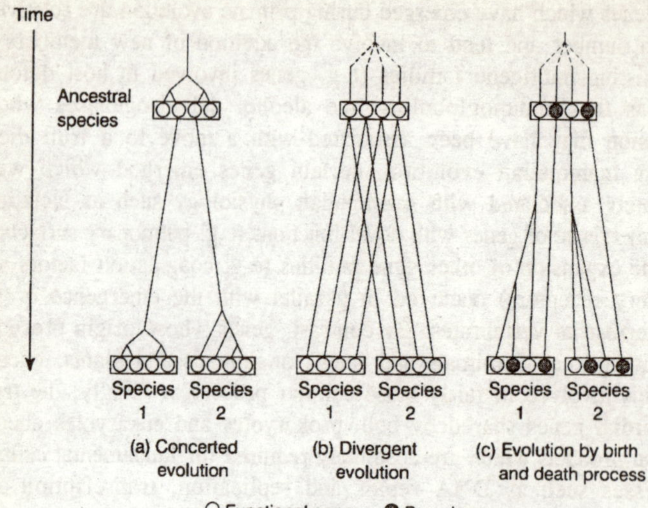

Fig. 5.2. Three different modes of evolution of multigene families.

of which serve to homogenize multigene family members. Examples of this process include the rRNA and histone genes. By contrast, some distinct groups of genes within multigene families appear to have been maintained over long periods of evolutionary time. An example of this *divergent evolution* is provided by the immunoglobulin V_H genes, the diversification of which may have conferred a selective advantage. In other multigene families, gene duplication serves to create new genes whilst other genes are inactivated by deleterious mutation or are eliminated by unequal crossing over ('*birth-and-death*'). Once again, the immunoglobulin V_H genes provide an example of this process. It can be seen therefore that some multigene families possess members that serve to illustrate the action of all three of the above processes. No one multigene family can be regarded as being archetypal. Each illustrates certain principles and so a variety of examples will be discussed.

Actin genes

A total of six actin genes are found in mammals although a considerably larger number of pseudogenes may be found. The actin genes are highly conserved evolutionarily and encode protein isoforms that differ from each other by only a few amino acids, mostly at their amino terminals. Those characterized in human are α-smooth muscle aortic actin (ACTA2; 10q22-q24), α-skeletal actin (ACTA1; 1q42.1), α-cardiac actin (ACTC; 15q14), β-nonmuscle cytoplasmic (ACTB; 7p12-

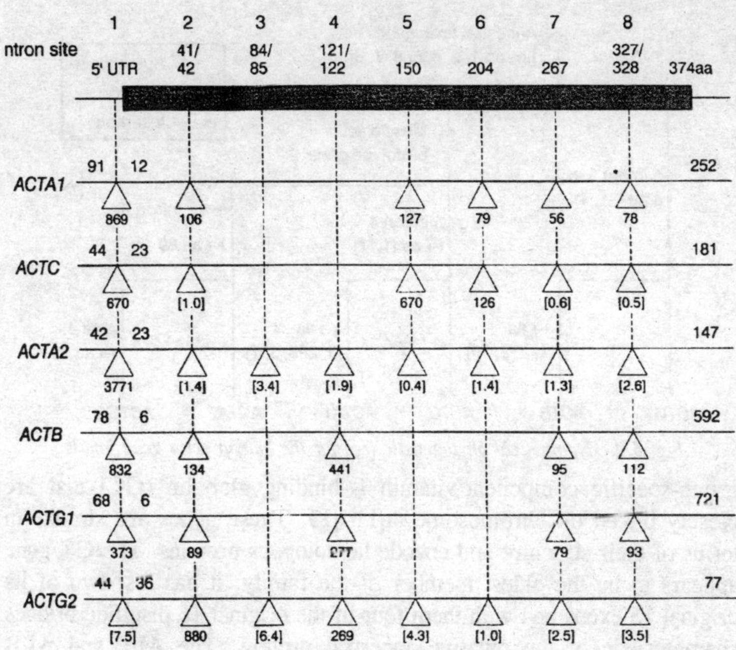

Fig. 5.3. Comparison of gene structures of the six human actin genes. Triangles indicate intron positions.

p15), γ-nonmuscle cytoplasmic (ACTG1; 17q25) and γ-smooth muscle enteric (ACTG2; 2p13.1) actins. Comparison of the structures of these six human actin genes indicates that comparable regions are identically interrupted by introns although the sizes of the introns differ. Comparison of nucleotide and amino acid sequences as well as gene structures also allowed Miwa et al. (1991) to propose a possible phylogenetic tree for the actin gene family. Since all four muscle actins differ from the cytoplasmic actins by substitutions at 19 amino acid residues, the muscle actins must have shared a common ancestral actin gene. Duplication and divergence then led to the emergence of the smooth muscle and striated muscle actin genes. Introns were gained or lost at different stages in the processes. The ancestral smooth muscle actin gene acquired an amino acid substitution at residue 89 before being duplicated; the smooth muscle γ-actin subsequently lost an amino acid residue at position 4.

Albumin genes

The human albumin gene family comprise four genes, encoding albumin (ALB), α-fetoprotein (AFP), α-albumin/afamin (AFM) and

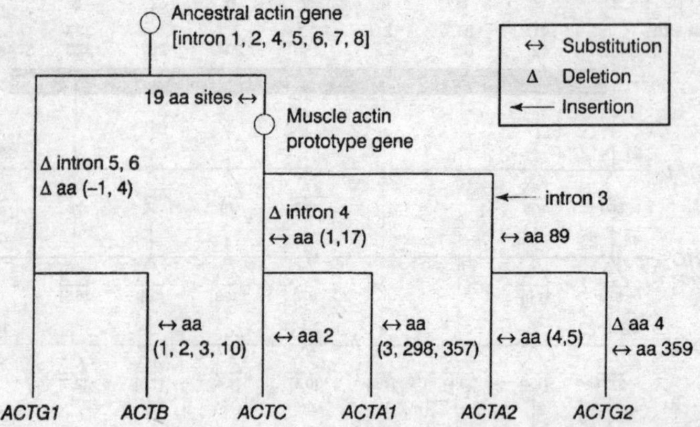

Fig. 5.4. Hypothetical phylogenetic tree for the human actin gene family.

group-specific component/vitamin D-binding globulin (GC) that are closely linked on chromosome 4q11-q13. These genes are similar in terms of their structure and encode homologous proteins. The GC gene appears to be the oldest member of the family; it has lost two of its original 15 exons and with them four of the original 18 disulfide bridges characteristic of the putative ancestral protein. The AFP and ALB genes have been evolving at a particularly rapid rate. However, the ALB gene still displays a high degree of evolutionary conservation in terms of its sequence which is perhaps a little surprising on account of its apparent nonessential nature evidenced by the absence of overt clinical signs in analbuminemic humans and rats.

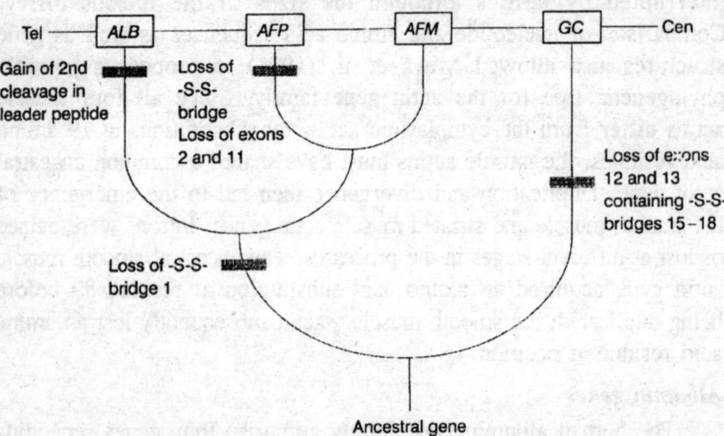

Fig. 5.5. Proposed model for the evolution of the human albumin gene family.

Albumin is synthesized in the adult liver whereas α-fetoprotein, being the fetal counterpart of albumin, is produced in the fetal liver and yolk sac. The mechanism of the developmental switch bringing about AFP gene repression and ALB gene activation is not yet understood.

The chromosomal region containing the albumin gene cluster has been involved in a number of pericentric inversions during the evolution of higher primates. As a result of one such inversion, the ALB and AFP genes were translocated to the short arm of chimpanzee chromosome 3 (analogous to human chromosome 4). Similar inversions in the gorilla and orangutan however left the ALB and AFP genes on the long arm of chromosome 3 in these species.

Apolipoprotein genes

The lipoproteins are the major carriers of cholesterol, triglycerides and other lipids in human plasma. They are encoded by a multigene family which is dispersed but not fully dispersed in the human genome: APOC1, APOC2, APOC4, APOE (19q13.2), APOC3, APOA1, APOA4 (11q23) and APOA2 (1q21-q23). The genes evolved from a primordial APOC1-like gene, which is thought to have existed ~680 Myrs ago, via a series of internal and complete gene duplications. Since both apoA1 and apoE genes have been found in fish, the duplication event from which they both arose must have occurred before the divergence of tetrapods and teleost fish. The structures of the genes show remarkable similarities: all (with the exception of APOA4) possess four exons with introns interrupting the coding sequence at very similar locations. The major difference between the genes is in the length of

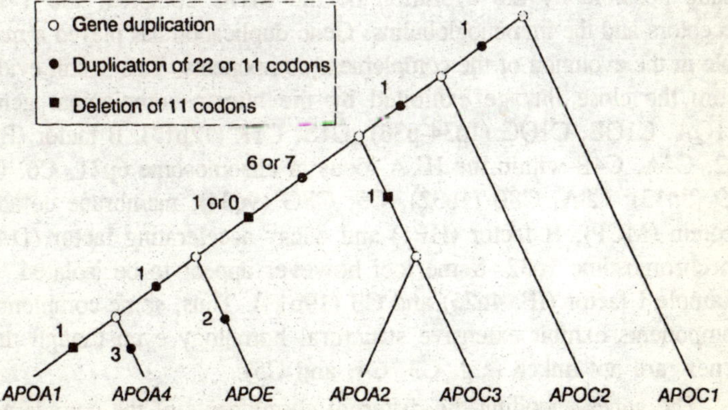

Fig. 5.6. A hypothetical scheme for the evolution of the apolipoprotein genes.

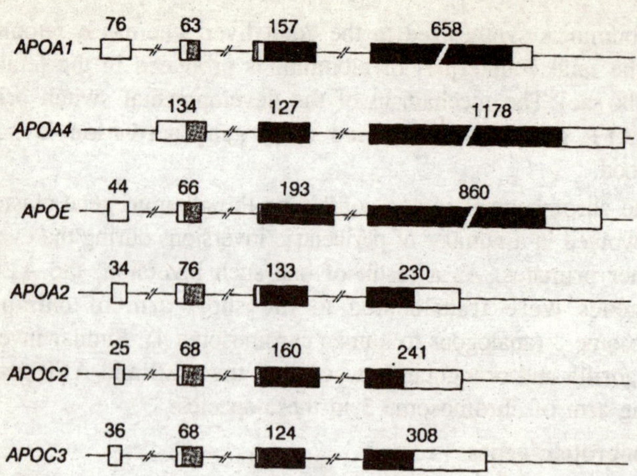

Fig. 5.7. Structural organization of the human apolipoprotein genes

the last exon which encodes a variable number of lipid-binding domains that contain multiple repeats of 22 amino acids each of which represents a tandem array of two 11-mers.

Two of the major known apolipoproteins do not fit easily into the above scheme: the 29 exon APOB (2p24) gene encodes a protein which may be distantly related to the other apolipoproteins whilst the APOD (3q26.2) gene encodes a protein with a high degree of homology to retinol-binding protein but little similarity to the other apolipoproteins.

Complement genes

The vertebrate complement system may be regarded as a primitive immune system which lacks the capability to recognize foreign antigens made possible by the evolution of the MHC complex, the T cell receptors and the immunoglobulins. Gene duplication has played a major role in the evolution of the complement components. This is still evident from the close linkage exhibited by the human complement genes: C1QA, C1QB, C1QG (1p34-p36), C1S, C1R (12p13), B factor (BF), C2, C4A, C4B within the HLA locus at chromosome 6p21, C6, C7, C9 (5p13), C8A, C8B (1p32), C5, C8G (9q34), membrane cofactor protein (MCP), H factor (HF1) and decay accelerating factor (DAF) on chromosome 1q32. Some loci however appear to be isolated, for example I factor (IF; 4q25) and C3 (19p13). Thus, some complement components exhibit extensive structural homology even though their genes are not linked (e.g. C3, C4, and C5).

The genes encoding the different components of the complement system belong to several evolutionarily unrelated gene families.

Phylogenetic analysis of the C3, C4 and C5 genes, together with the more distantly related α2-macroglobulin (A2M; 12p12-p13) gene, supports the view that C5 diverged first with C3 and C4 subsequently diverging before the separation of the jawed and jawless fishes. Similar analysis have been performed for the complement C1 components.

Exon shuffling has been extremely important in diversifying the structure of the complement genes thereby conferring novel functions upon the complement proteins. Thus for example, C1r, C1s, and C2 possess serine protease domains, C6, C7, C8α, C8β, and C9 possess thrombospondin and low density lipoprotein receptor domains, and C1r, C1s, C6, C7, C8a, C8b, and C9 possess homologies with epidermal growth factor.

Crystallin genes

The α-, β-, and γ-crystallins of the lens are ubiquitous in vertebrates. These proteins have ancient origins as evidenced by the relationship of the α-crystallins to small heat shock proteins and the relationship between the β- and γ-crystallins to proteins found in the bacterium *Myxococcus* and the primitive eukaryote *Physarum* respectively. In humans, there appear to be two active α-crystallin genes (CRYAA, 21q22.3; CRYAB, 11q22.3-q23.1), six active β-crystallin genes (CRYBB2, CRYBB3, CRYBA4, CRYBB1, 22q11.2-q12.1; CRYBA2, 2q34-q36; CRYBA1, 17q11.1-q12) and five active γ-crystallin genes (CRYGS, 3; CRYGA, CRYGB, CRYGC, CRYGD, 2q33-q35).

Vertebrate lens crystallins have undergone dramatic changes during evolution including gene duplication, gene inactivation, gene recruitment and changes in gene expression. The vertebrate lenses with the highest refractive index are found in fish and in the predominantly nocturnal rodents; in these lenses, γ-crystallin predominates. Most vertebrates have however reduced the refractive index of their lenses in order to aid accommodation and focusing at a distance; this process has been accompanied by the loss of γ-crystallins (for example, replaced by δ-crystallins in birds). Humans have reduced γ-crystallin levels by partial or complete inactivation of several members of the γ-crystallin gene family once found in the common ancestor of primates and rodents. Thus, of the human γ-crystallin genes, only CRYGC and CRYGD (2q33-q35) are still fully active. Several γ-crystallin genes have been inactivated by deletion or point mutation whilst the CRYGA and CRYGB (2q33-q35) genes have been down-regulated by promoter mutation and increased mRNA turnover respectively.

In addition to the ubiquitous crystallins mentioned above, there is a group of *taxon-specific* crystallins that are restricted to certain evolutionary lineages. These crystallins are pre-existing metabolic enzymes which have been recruited to a novel lens-specific role, not on account of their enzymatic activity but rather for structural reasons in that they are able to modify the refractive index of the lens. Thus, in most reptiles and birds, δ2-crystallin is the enzyme argininosuccinate lyase whilst δ1 crystallin has arisen by duplication of the δ2 crystallin gene but lacks its enzymatic activity. Various other examples of *gene recruitment* (also termed *gene sharing*) have been documented among the crystallins, for example lactate dehydrogenase B (crystallin ε, crocodiles and some birds), NADPH: quinone oxidoreductase (crystallin ζ; guinea pigs and camelids) and aldehyde dehydrogenase 1 (crystallin η; elephant shrews). Recruitment of these gene products to a new lens function has been achieved by the acquisition of lens expression for these genes through the emergence of novel promoter elements. In the case of ζ-crystallin, recruitment appears to have occurred independently in guinea pigs and camels; different regions of the same nonfunctional intronic sequence have been altered by single base-pair substitutions and micro-deletions and insertions to perform a role in lens-specific expression. By contrast, in birds, recruitment of the δ1-, δ2- and ε-crystallins appears to have come about through the modification of pre-existing promoters utilized for nonlens tissue expression.

Collagen genes

The collagens are multi-domain proteins which serve as structural molecules of the extracellular matrix. They have a very ancient origin as evidenced by their presence in sea urchins, annelids, *Drosophila* and even jellyfish and sponges. These proteins contain one or more domains with a triple helical conformation characterized by a repeating Gly-X-Y amino acid motif.

A total of 19 types of collagen (I to XIX) have so far been defined in humans; these comprise homo-or heteromeric assemblies of specific polypeptide chains. A minimum of 34 genes on 13 different human chromosomes are required to encode these chains. Based on their structures, the collagens may be divided into several distinct classes: fibrillar (I, II, III, V, XI), basement membrane (IV), fibril-associated with interrupted triple helices (IX, XII, XIV, XIX), filament-producing (VI), network forming (VIII, X) and anchoring fibril (VII). The class structure of the proteins and the homologies between the members of each class is reflected in the evolutionary relationships between the genes that encode them, their structure and often their chromosomal location.

Table 5.1. Types of human collagen and the genes that encode them

Collagen type	*Constituent chains*	*Gene symbol*	*Chromosomal localization*
I	α1(I)	COL1A1	17q21.3-q22
	α2(I)	COL1A2	7q22.1
II	α1(II)	COL2A1	12q13
III	α1(III)	COL3A1	2q31
IV	α1(IV)	COL4A1	13q34
	α2(IV)	COL4A2	13q34
	α3(IV)	COL4A3	2q36-q37
	α4(IV)	COL4A4	2q36-q37
	α5(IV)	COL4A5	Xq22
	α6(IV)	COL4A6	Xq22
V	α1(V)	COL5A1	9q34.2-q34.3
	α2(V)	COL5A2	2q31
VI	α1(VI	COL6A1	21q22.3
	α2(VI)	COL6A2	21q22.3
	α3(VI)	COL6A3	2q37
VII	α1(VII	COL7A1	3p21.3
VIII	α1(VIII)	COL8A1	3q12-q13
	α2(VIII)	COL8A2	1p32.3-p34.2
IX	α1(IX)	COL9A1	6q13
	α2(IX)	COL9A2	1p32.2-p33
	α3(IX)	COL9A3	20q13.3
X	α1(X)	COL10A1	6q21-q22.3
XI	α1(XI)	COL11A1	1p21
	α2(XI)	COL11A2	6p21.3
XII	α1(XII)	COL12A1	6q12-q13
XIII	α1(XIII)	COL13A1	10q22
XIV	α1(XIV)	COL14A1	8q23
XV	α1(XV)	COL15A1	9q21-q22
XVI	α1(XVI)	COL16A1	1q34
XVII	α1(XVII)	COL17A1	10q24.3
XVIII	α1(XVIII)	COL18A1	21q22.3
XIX	αl(XIX)	COL19A1	6q12-q14

The fibrillar collagens are proteins with a continuous triple helical domain containing uninterrupted Gly-X-Y motifs. Their genes are thought to have diverged some 800-900 Myrs ago but still share a very similar exon-intron organization comprising 52-54 exons of which 42 have specific lengths: 45 bp, 54 bp, 99 bp, 108 bp, and 162 bp representing multiples of the 9 bp encoding the Gly-X-Y motif. The latter three lengths are combinations of the first two suggesting that they may have been derived from an ancestral 54 bp exon by duplication and recombinations events. These events may have been mediated by recombination between introns. Partial gene duplication events are unlikely to have disrupted gene function because multiples of 9 bp would not have altered the reading frame of the proteins. The 66 exon COL5A1 and COL11A2 genes appear to have increased in size through an increase in the number of 54 bp exons. The COL3A1 and COL5A2 genes are both located at chromosome 2q31, indicative of a close evolutionary relationship.

Although the genes encoding the basement membrane collagens possess 46-52 exons, they differ from the fibrillar collagen genes in that they possess many fewer exons of lengths 45 bp and 54 bp. Further, the exons vary widely in size, do not always begin with a Gly codon

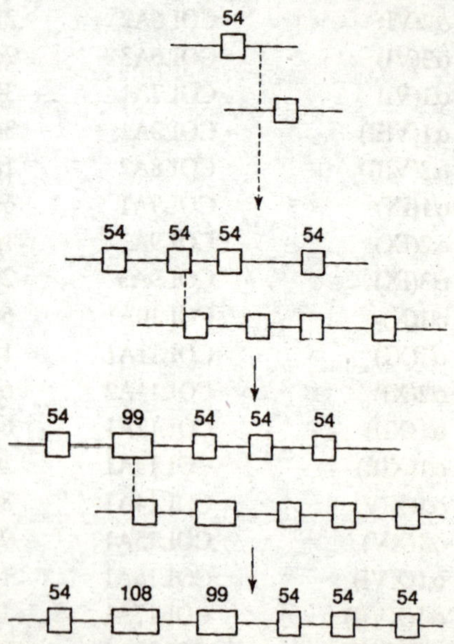

Fig. 5.8. A hypothetical scheme for the evolution of the exons in a region of the a2(I) collagen (COL1A2; 7q22) gene coding for the helical region of the polypeptide.

and often split codons. Their similar exon-intron organization testifies however to a common evolutionary origin. Moreover, the six type IV genes are arranged in syntenic pairs [COL4A1 and COL4A2 (13q34), COL4A3 and COL4A4 (2q36-q37), COL4A5 and COL4A6 (Xq22)] with head-to-head arrangement (sometimes sharing promoter elements), consistent with a model of gene duplication and divergence.

The fibril-associated collagens with interrupted triple helices are distinct from both the fibrillar and basement membrane collagens. The COL9A1, COL9A2, COL12A1, and COL19A1 genes are similar in structure indicative of a close evolutionary relationship: all contain some exons of length 54 bp. The COL9A1, COL12A1, and COL19A1 genes are also closely linked on chromosome 6q12-q14.

The genes encoding the filament-producing collagens, COL6A1 and COL6A2, possess a similar structure and are closely linked in head-to-tail orientation. Although they contain exons of lengths 45 bp and 54 bp, most are 63 bp in length. The type VII collagen gene, COL7A1, contains 118 exons some of which are of 45 bp or 54 bp in length but rather more are of 36 bp, the remainder invariably being multiples of 9 bp. By contrast to the extremely fragmented organization of the other collagen genes, the COL19A1 gene is remarkably compact, containing only three exons.

The evolution of the collagens is thus broadly consistent with a model of gene duplication and divergence coupled with intragenic exon duplications, deletions and fusions. The recurring sizes of many of the exons that encode the extant human collagens provides ample evidence for the exon amplification and contraction processes that must have fashioned them. Although the evolution of several families of related collagen genes has been studied in some detail, the evolution of the gene family in its entirety has not yet been pieced together. Since both the fibrillar and nonfibrillar collagens have a very ancient origin, this task is one of some very considerable complexity.

Genes for the fibroblast growth factors and their receptors

The fibroblast growth factors (FGF) constitute a family of polypeptide growth factors that have multiple functions in mitogenesis, angiogenesis and wound healing. They contain an extracellular portion with either 2 or 3 immunoglobulin-like domains, a transmembrane domain and cytoplasmic tyrosine kinase domain. The FGF family of proteins interacts with the membrane-associated FGF receptors (FGFR) and these interactions are important fora variety of developmental processes such as the formation of the mesoderm during gastrulation, the integration of growth, budding and patterning during early post-implantation, and the development of various tissues including the skeletal system.

A total of 14 FGF genes have been identified in the human genome: FGF1 (5q31), FGF2 (4q25-q27), FGF3 (11q13), FGF4 (11q13), FGF5 (4q21), FGF6 (12p13), FGF7 (15q15-q21), FGF8 (10q24), FGF9 (13q12), FGF10 (5p12-p13), FGF11 (17q21), FGF12 (3q28), FGF13 (Xq26) and FGF14 (13q34). Thus, with the exception of the closely linked and tandemly repeated FGF3 and FGF4 genes, the FGF genes are dispersed in the human genome. FGF genes are present in invertebrate genomes and probably expanded in number after the divergence of protostomes and deuterostomes. The mammalian FGF gene family has emerged by a process of phased duplicated and divergence. Some of the paralogous mammalian FGF genes may be derived from genome duplications that occurred early on in the evolution of the vertebrate genome, for example FGF4 and FGF6 on chromosomes 11 and 12, respectively, and FGF1 and FGF2 on chromosomes 4 and 5, respectively.

The generation of FGF diversity may have played a role in the various innovations of the vertebrate skeletal system and the novel FGF family members are likely to have co-evolved with FGRFs. FGFR genes are also found in invertebrate genomes and four are found in the

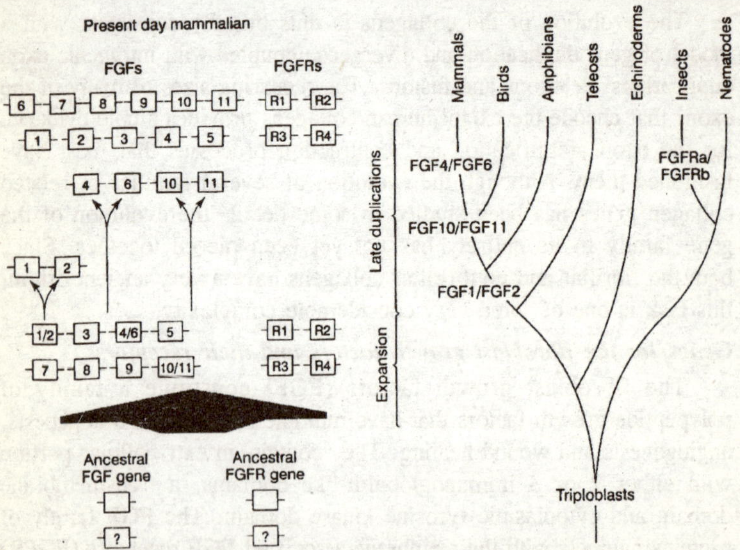

Fig. 5.9. A hypothetical scheme for the evolution of the fibroblast growth factors and their receptors.

human genome: FGFR1 (8p11), FGFR2 (10q26), FGFR3 (4p16) and FGFR4 (5q35-qter). However, by contrast with the FGF family expansion, the FGFR gene family appears to have undergone only a single phase of expansion.

GAGA receptor genes

The γ-aminobutyric acid (GABA) receptor is a pentameric ion channel complex which mediates fast inhibitory synaptic transmission in the central nervous system. The receptor exists as many different isoforms assembled from different combinations of subunit subtypes, α_{1-6}, β_{1-4}, γ_{1-4}, δ, ε, and ρ_{1-2}. The human GABA receptor genes are clustered at different chromosomal locations: GABRA2, GABRA4, GABRB1, GABRG1 (4p12-p13), GABRA1, GABRA6, GABRB2, GABRG2 (5q34-q35), GABRA5, BABRB3, GABRG3 (15q11-q13), GABRA3, BABRB4, GABRE (Xq28), GABRR1, GABRR2 (6q14-q21), and GABRD (1p). The organization of this gene family is consistent with an evolutionary model of intracluster gene duplication as well as the duplication of a primordial gene cluster. The tandem duplication of the ancestral a subunit gene appears to have occurred before the duplication and translocation of at least one of the original gene clusters.

G Protein α subunit genes

G proteins are ubiquitous in eukaryotes and serve to mediate signal transduction via appropriate receptors by coupling extracellular signals to intracellular effectors such as adenylyl cyclase, phospholipases and

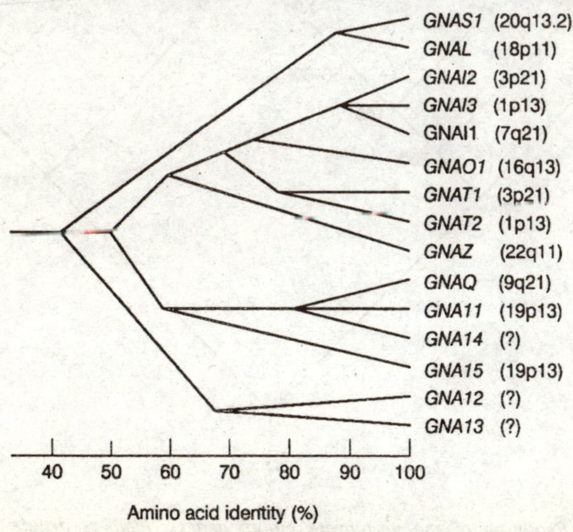

Fig. 5.10. Human G protein α subunit multigene family.

ion channels. More than a dozen G protein α subunit genes have been identified in the human genome and their evolutionary history appears to stretch back over 1500 Myrs. Whilst some members of the human gene family are chromosomally linked [e.g. GNAI2 and GNAT1 (3p21), GNAI3 and GNAT2 (1p13), GNA11 and GNA15 (19p13)], the remainder are scattered around the genome. This organization is potential explicable by a combination of successive genome duplications, tandem duplication, and duplication/translocation events.

Globin genes

Proteins homologous to the globins are ubiquitous in eukaryotes and even have counterparts among the prokaryotes. The myoglobin (MB; 22q11.2-q13) gene and the ancestor of the extant globin genes are thought to have arisen from a common ancestral gene encoding a heme-containing protein ~700 Myrs ago before the advent of the vertebrates. The α- and β-globin genes diverged from each other about 500 Myrs ago and at some stage became chromosomally separated. The α-globin cluster subsequently evolved by a process of successive duplication and divergence: the ζ/α gene divergence occurring about 400 Myrs ago and the θ/α gene divergence about 260 Myrs ago. The α1- and α2-globin genes arose from a further duplication event between 50 Myrs and 60 Myrs ago whilst the ψα1 pseudogene was inactivated ~45 Myrs ago. The 40 kb human α-globin gene cluster (16p13.3) therefore comprises the embryonically expressed ζ (HBZ) gene, the

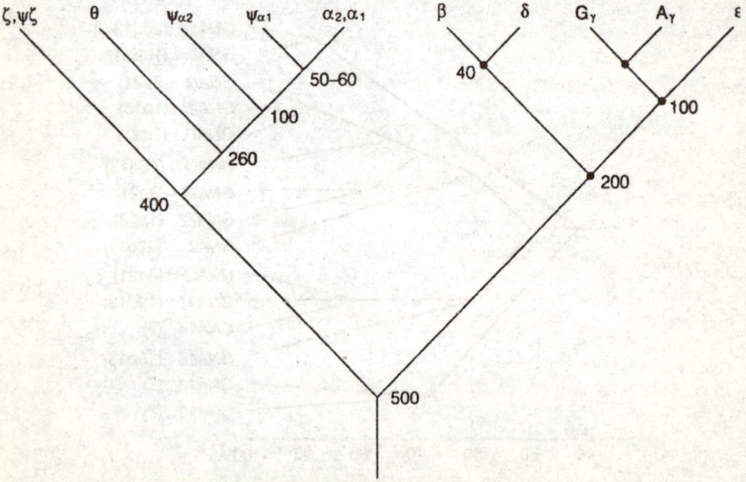

Fig. 5.11. Evolution of the vertebrate globin genes. Numbers denote approximate estimated times of divergence in Myrs before present.

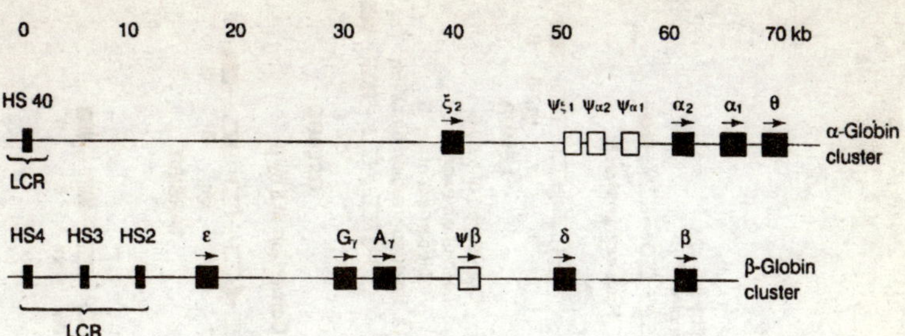

Fig. 5.12. Organization of the human α- and β-globin clusters.

post-natally expressed α2 (HBA2) and α1 (HBA1) genes, the θ-globin (HBQ1) gene (probably a transcribed pseudogene) and three conventional pseudogenes (HBZP, HBAP1, HBAP2). The genes in the α-globin gene cluster are therefore arranged in the order of their activation during ontogeny. The human HBA1 and HBA2 genes have remained virtually identical to each other as a consequence of crossing over and gene conversion.

The ancestral β-globin gene is thought to have duplicated about 200 Myrs ago to yield a δ/β ancestral gene and a ε/γ ancestral gene. The δ- and β-globin genes diverged form the δ/β ancestral gene about 40 Myrs ago. The ε- and γ-globin genes diverged from the ε/γ ancestral gene about 100 Myrs ago followed by the duplication of the γ-globin gene to form the Gγ an Aγ genes, an event which occurred before the divergence of the Old World and New World monkeys. The human β-globin gene cluster (11p15.5) therefore comprises the embryonically expressed ε (HBE1) gene, the fetal Gγ (HBG2) and Aγ (HBG1) genes, the post-natally expressed β (HBB) and δ (HBD) genes plus a single ψη pseudogene (HBBP1). Thus, as with the α-globin genes, the genes in the 75-kb β-globin gene cluster are arranged in the order of their activation during ontogeny.

The HBD gene is present in humans, apes and New World monkeys but has been inactivated in Old World monkeys. During evolution, gene conversion has operate on the HBD gene allowing it to acquire sequence characteristics of the HBB gene. Gene conversion events have however been even more frequent between the HBG2 and HBG1 genes. In addition to gene conversion, a number of other events have occurred during evolution which have altered the β-globin cluster in specific ways in different mammalian orders; these include insertion of repeat sequences, change in expression profile, gene duplication, gene fusion and gene loss or inactivation. Thus, the lagomorphs and rodents lack the η-globin locus whilst the artiodactyls lack the γ-

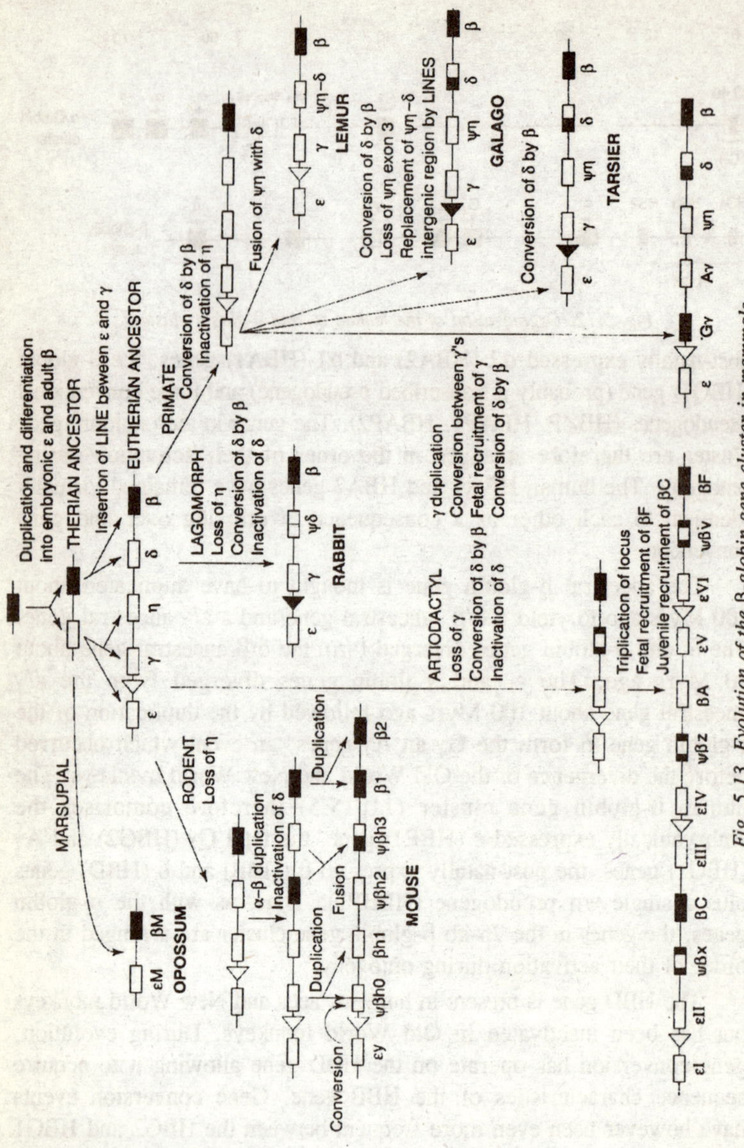

Fig. 5.13. Evolution of the β-globin gene cluster in mammals.

globin locus. As a consequence of some of these changes, the β-globin cluster in mammals varies from as little as 20 kb in lemurs to about 90 kb in goats.

The ε and γ globin genes were originally embryonically expressed and fetally inactive and this early expression pattern is still found in the galago and lemurs. In higher primates, the γ-globin gene was duplicated before the divergence of Old World and New World monkeys (perhaps by an unequal homologous crossing over event mediated by LINE elements) and became fetally expressed as a direct result of the accumulation of sequence changes in the 5' flanking region. In Old World monkeys, apes and humans, both the γ1- and γ2-globin genes are functional but the expression of the γ1 gene is three-fold higher than that of the γ2 gene. In New World monkeys, only one γ-globin gene is functional, usually γ2.

Growth hormone and somatomammotropin genes

The growth hormone and prolactin genes are thought to have emerged as a result of a duplication event some 470 Myrs ago. In nonprimates, with the exception of caprine ruminants, GH is encoded by a single gene whilst in primates, the gene has expanded to a gene cluster. Thus, the human gene encoding pituitary-expressed growth hormone (GH1) is located on chromosome 17q23 within a cluster of five related genes. The other loci present in the growth hormone gene cluster are two chorionic somatomammotropin genes (CSH1, and CSH2), a chorionic somatomammotropin pseudogene (CSHP1) and a second growth hormone gene (GH2). These genes are separated by intergenic regions of 6 kb to 13 kb in length, lie in the same transcriptional orientation, are placentally expressed and are under the control of downstream tissue-specific enhancer. The GH2 locus encodes a protein that differs from the GH1-derived growth hormone at 13 amino acid residues. All five genes share a very similar structure with five exons interrupted at identical positions by short introns.

On the basis of sequence data, it was initially thought that the evolution of the GH gene cluster had taken place comparatively recently over the last 15 Myrs by a process of gene duplication and divergence. The first event is thought to have been the duplication of a single ancestral GH gene to generate pre-GH and pre-CSH genes followed by the duplication of the newly created gene pair to yield GH1, CSH1, GH2, and CSH2. Finally, the CSH1 gene was duplicated to form two genes, one of which (CSHP1) became inactivated through mutation. Although this sequence of events is probably correct, the estimated timings now appear to be seriously inaccurate because a GH2 gene has subsequently been reported in macaque. The duplication event creating the GH1 and GH2 genes must therefore have occurred before the divergence of Old World monkeys and the anthropoid apes ~25 Myrs ago. No study has been performed on either prosimians or New

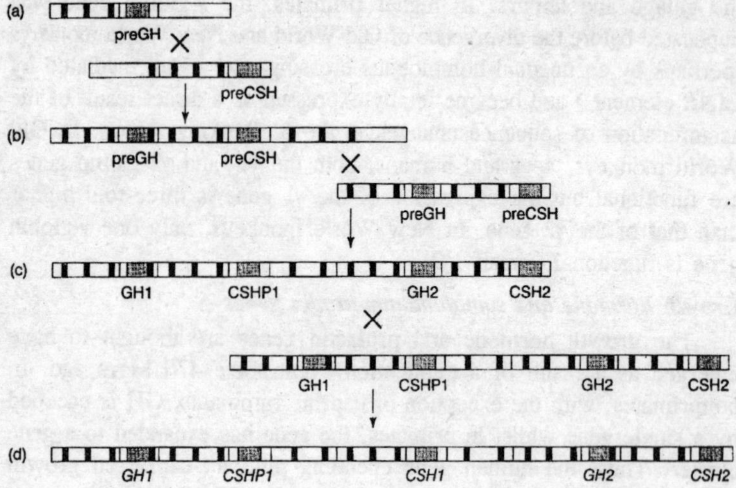

Fig. 5.14. Evolution of the human growth hormone gene cluster by serial unequal crossing over events. (a) Duplication to yield pre-growth hormone (GH) and pre-chorionic somatomammotropin (CSH) genes. (b) Unequal crossing over between pre-GH and pre-CSH genes to yield (c) the GH1, GH2, and CSH2 genes plus a CSHP1 locus which was to become inactivated. (d) Subsequently, the CSH1 gene emerged by duplication of the CSHP1 gene.

World monkeys and so the timing of this duplication event may have to be revised still further as new data emerge. The initially misleading conclusion as to the timing of the duplication events was probably due to a failure to consider the effect of gene conversion in minimizing sequence differences between the different GH loci.

The 70 kb human GH gene cluster contains some 48 *Alu* sequences some of which may have mediated the unequal recombination events responsible for the gene duplications. On the other hand, some *Alu* sequences have become duplicated along with their associated genes during the duplication process. One consequence of the relatively recent evolutionary changes in the GH gene cluster is that multiple sequence homologies and internal repetitions are still evident within it.

Glycophorin genes

The human glycophorins are encoded by a multigene family that has evolved from an ancestral gene that ceased to be functional at some stage during primate evolution. Glycophorins A and B are the major sialoglycoproteins of the human erythrocyte membrane and carry the MN and Ss blood group antigens respectively. They are encoded by the GYPA and GYPB genes which occur in a 330 kb cluster together

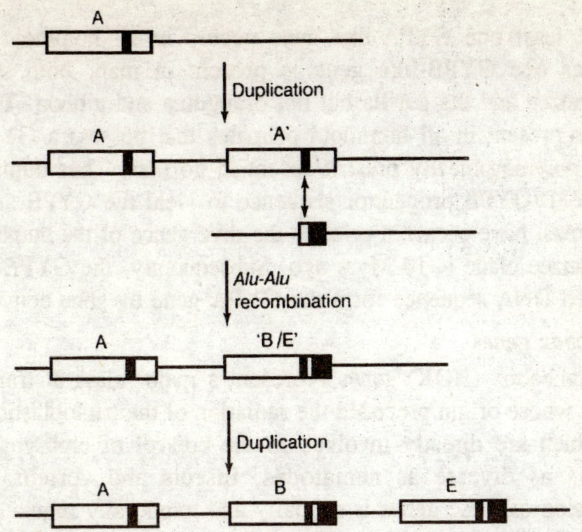

Fig. 5.15. Evolution of the human glycophorine genes, GYPA, GYPB and GYPE.

with the glycophorin E (GYPE) genes on chromosome 4q28-q31. This cluster is thought to have arisen by two successive duplications, the first creating the GYPA gene by duplication of an ancestral gene and the second (5-21 Myrs ago) generating the GYPB and GYPE genes. The GYPB gene differs from the GYPA gene by virtue of the presence, of, (i) a G→T transversion at the +1 position of the intron 3 donor splice site which serves to inactivate the expression of exon III and (ii) a 9 bp insertion at the 5' end of exon V. The GYPE gene differs from the GYPB gene in that exon IV has been inactivated by a splice site mutation, exon V contains a 24 bp insertion and the encoded protein has been shortened by 5 amino acids through the introduction of a premature Stop codon. Both the GYPB and GYPE genes lack exons 6 and 7 present in the GYPA gene that encode the cytoplasmic domain.

The ancestral gene from which the GYPA gene was derived is thought to have been inactivated but its remnants may still be present on chromosome 4, just downstream of the GPA gene. The ancestor of the human GYPB and GYPE genes underwent a rearrangement in its 3' region by recombination between *Alu* sequences and in so doing, acquired 3' sequence from another gene. This unequal crossing over event served to generate short last exons in both GYPB and GYPE genes; that of GYPB comprises a single amino acid whilst that of GYPE encodes an untranslated region.

At least one GYPA-like gene occurs in all hominoid primates whereas one GYPB-like gene is present in man, both species of chimpanzee and the gorilla but not orangutan and gibbon. The GYPE gene is present in all hominoid primates that possess a GYPB gene, but is polymorphically present/absent in gorillas. Thus duplication of the GYPB/GYPE progenitor sequence to yield the GYPB and GYPE genes must have occurred prior to the divergence of the human-gorilla-chimpanzee clade ~10 Myrs ago. Subsequently, the GYPE gene has acquired DNA sequence form the GYPA gene by gene conversion.

Homeobox genes

Homeobox (HOX) genes represent a major class of transcription factors whose origin preceded the radiation of the triploblastic metazoa and which are directly involved in the control of embryogenesis in animals as diverse as nematodes, insects and vertebrates. The expression of these genes is spatially and temporally regulated during embryonic development and plays an important role in establishing the system of positioning along the anterior-posterior axis of the embryo. HOX proteins share a common homeobox domain and influence gene transcription by binding to a 4 bp core sequence on the promoters of their target genes.

The HOX genes appear to have been duplicated on several occasions during evolution, perhaps even before the origin of angiosperms, fungi and the metazoa. Early in vertebrate evolution, an ancestral HOX gene cluster present in invertebrates and the cephalochordate *Amphioxus* was duplicated twice probably by whole genome duplication to give rise to the four linkage groups (HOX A, HOX B, HOX C, HOX D).

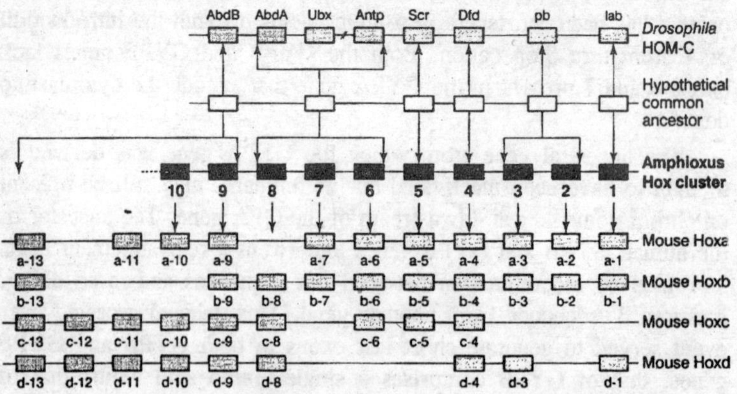

Fig. 5.16. Model for the evolution of the homeobox gene clusters of eukaryotes.

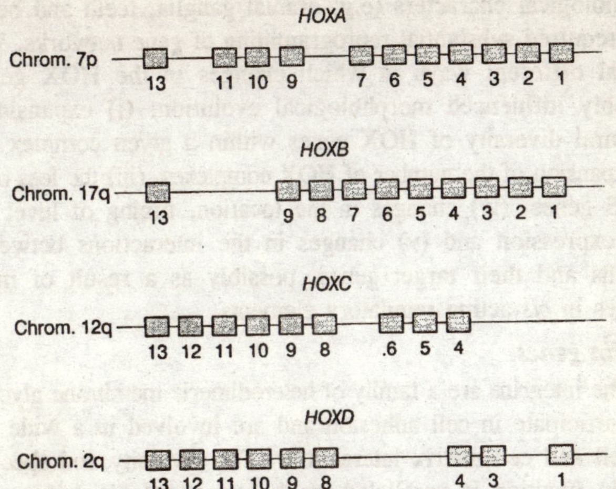

Fig. 5.17. Human HOX gene clusters.

From studies of HOX gene number, this duplication event is likely to have occurred after the divergence of ray-fined and lobe-finned fishes but before the radiation of the teleosts. It can be seen that cluster duplication must have been followed by the loss of some specific HOX genes and indeed this is evident from comparison of HOX gene number between clusters. The physical order of genes within the clusters has however been conserved during vertebrate evolution perhaps as a result of enhancer sharing between different HOX genes.

In human, most of the HOX genes are located in four clusters containing between them 39 HOX genes. The HOX A cluster is located at 7p14-p-15 and contains 11 genes: HOXA1, HOXA2, HOXA3, HOXA4, HOXA5, HOXA6, HOXA7, HOXA9, HOXA10, HOXA11 and HOXA13. The HOX B cluster is located at 17q21-q22 and contains 10 genes: HOXB1, HOXB2, HOXB3, HOXB4, HOXB5, HOXB6, HOXB7, HOXB8, IIOXB9, and HOXB13. The HOX C cluster is located at 12q13 and contains 9 genes HOXC4, HOXC5, HOXC6, HOXC8, HOXC9, HOXC10, HOXC11, HOXC12, and HOXC13. The HOX D cluster is located at 2q31-q32 and contains 9 genes: HOXD1, HOXD3, HOXD4, HOXD8, HOXD9, HOXD10, HOXD11, HOXD12, and HOXD13. In addition to the HOX gene clusters, other dispersed paralogous HOX genes are found in the human genome (e.g. MSX1, 4p16; HMX2, 10q25-q26) and may also be of functional importance.

The development of the vertebrate body plan both by elaboration of primitive chordate characters and the development of novel morphological characters (e.g. cranial ganglia, teeth and bone) must have required substantial reprogramming of gene networks. There are several different ways in which changes in the HOX genes have probably influenced morphological evolution: (i) expansion in the structural diversity of HOX genes within a given complex or class, (ii) expansion of the number of HOX complexes, (iii) the loss of specific HOXS genes, (iv) changes in the location, timing of level of HOX gene expression and (v) changes in the interactions between HOX proteins and their target genes, possibly as a result of mutational changes in *cis*-acting regulatory elements.

Integrin genes

The integrins are a family of heterodimeric membrane glycoproteins that participate in cell adhesion and are involved in a wide range of cell-cell and cell-matrix interactions. The diversity and specificity of integrin function is paralleled by the structural diversity potentiated by the existence of at least 16 different alpha chains and 8 different beta chains. Although, in principle, the alpha and beta chains could associate in a multitude of ways, in practice diversity is limited to certain combinations. These chains are encoded by a family of genes which are widely dispersed in the human genome. The alpha chain integrins can themselves be divided into two subgroups by virtue of the insertion of an I domain of about 180 amino acids in the extracellular region. The I-integrin alpha chain genes (ITGA1, ITGA2, ITGAD, ItGAM, ITGAL, and ITGAE) are thought to have arisen as a result of an early insertion into a non-I gene followed by gene duplication and divergence. The clustering of these genes on chromosomes 5 and 16 is thought to have resulted from relatively recent gene duplications. The non-I alpha chain genes (ITGA2, ITGA3, ITGA4, ITGA5, ITGA6, ITGA7, ITGA8, ITGA9) are largely confined to clusters on chromosomes 2, 12, and 7 locations which coincide closely with the homeobox (HOX) gene clusters. This is suggestive of the occurrence of genomic or chromosomal duplications involving both types of gene cluster. Some of the beta chain (ITGB) genes are also located on chromosomes 2, 12, and 17 indicating common ancestry with the non-I alpha chain genes.

Keratin genes

The keratins, cytoskeletal proteins of the epithelium, belong to two families: type I (acidic) and type II (basic). Since keratins are obligate heteropolymers, keratin intermediate filaments are composed of one type I and one type II polypeptide. This has led to consistent

Table 5.2. Types of human integrin chain and the genes that encode them

Integrin chain	*Gene symbol*	*Chromosomal location*
α1	ITGA1	5
α2	ITGA2	5q23-q31
α3	ITGA3	17
α4	ITGA4	2q31-q32
α5	ITGA5	12q11-q13
α6	ITGA6	2
α7	ITGA7	12q13
α8	ITGA8	?
α9	ITGA9	3p21.3
αIIB	ITGA2B	17q21.3
αE	ITGAE	?
αL	ITGAL	16p11.2
αM	ITGAM	16p11.2
αV	ITGAV	2q31-q32
αX	ITGAX	16p11-p13
αD	ITGAD	16p11.2
β1	ITGB1	10p11.2
β2	ITGB2	10q22.3
β3	ITGB3	17q21.3
β4	ITGB4	17q11-qter
β5	ITGB5	?
β6	ITGB6	2
β7	ITGB7	12q13.1
β8	ITGB8	7p

co-expression of type I-type II keratin pairs in different types of epithelial cells. The keratins are evolutionarily related to the family of intermediate filament proteins which includes vimentin and desmin. Type II keratins are not more homologous to the type I keratins than they are to other intermediate proteins indicating that the type I and type II keratins diverged from the common ancestor of all intermediate filaments at about the same time. This common ancestral gene probably had its origins among the lower eukaryotes. By contrast to the conserved central alpha-helical domains, the variable terminal glycine-rich domains of keratins 1 and 10 (comprising short 4–10 amino acid segments repeated 3–15 times) appear to have evolved by a series of tandem duplications brought about by unequal crossing over.

Table 5.3. Types of human keratin and the genes that encode them

Keratin type	*Gene*	*Chromosomal location*
Keratin 1	KRT1	12q11-q13
Keratin 2A	KRT2A	12q11-q13
Keratin 3	KRT3	12q12-q13
Keratin 4	KRT4	12p12-q11
Keratin 5	KRT5	12q
Keratin 6A	KRT6A	12q12-q21
Keratin 6B	KRT6B	12
Keratin 7	KRT7	12q12-q21
Keratin 8	KRT8	12
Keratin 9	KRT9	17q21
Keratin 10	KRT10	17q21-q23
Keratin 12	KRT12	17q11-q12
Keratin 13	KRT13	17q21-q23
Keratin 14	KRT14	17q12-q21
Keratin 15	KRT15	17q21-q23
Keratin 16	KRT16	17q12-q21
Keratin 17	KRT17	17q12-q21
Keratin 18	KRT18	12q11-q13
Keratin 19	KRT19	17q21-q23
Keratin, hair, acidic 1	KRTHA1	17q12-q21
Keratin, hair, acidic 2	KRTHA2	17q12-q21
Keratin, hair, acidic 3A	KRTHA3A	17q-12-21
Keratin, hair, acidic 3B	KRTHA3B	17q12-q21
Keratin, hair, acidic 4	KRTHA4	?
Keratin, hair, acidic 5	KRTHA5	17q12-q21
Keratin, hair, basic 1	KRTHB1	12q13
Keratin, hair, basic 2	KRTHB2	?
Keratin, hair, basic 3	KRTHB3	12q12-q13
Keratin, hair, basic 4	KRTHB4	?
Keratin, hair, basic 5	KRTHB5	12q12-q13
Keratin, hair, basic 6	KRTHB6	12q13

In the human genome, the genes encoding these keratin families have been found to be tightly clustered; at 17q12-q21 for the acidic keratins (e.g. KRT9, KRT10, KRT14, KRT15, KRT16, KRT17, and KRT19), and chromosome 12 for the basic keratins (e.g. KRT1, KRT2A, KRT5, KRT6A, KRT6B, and KRT8). There is one exception to this rule: KRT18 is a type I keratin gene but is located on chromosome 12q11-q13. There is some evidence for gene conversion between type I and type II keratin genes which could have facilitated their coevolution. Concerted gene duplications also provide evidence for coevolution of type I and type II keratin genes. Thus, in both families, the genes expressed in the embryo duplicated and diverged first, followed by the genes expressed in various differentiated cells. Further gene duplications gave rise to the hair keratin genes. This parallelism of gene duplication cannot be explained by gene conversion but as yet the underlying mechanism which apparently allows duplications in one family to influence duplications in the other is still unclear. Coevolution may have been driven by the obligate heteropolymer status of the proteins. The tight regulation of the coordinate expression of the type I and type II keratin genes implies that unbalanced production is likely to be deleterious. It may therefore follows that the duplication and functional divergence of a type I keratin gene might lead to a specific type II gene duplication being selectively favoured.

Genes of the major histocompatibility complex

The major histocompatibility complex (MHC) region on chromosome 6p21.3 spans 4 Mb DNA and contains the genes encoding the human leukocyte antigens (HLA), a large family of cell surface glycoproteins. The function of the MHC proteins, members of the immunoglobulin superfamily, is to present peptides to T cells. The MHC family is divided into class I and II whose members differ in terms of their structure, function, expression and polymorphism. Class I molecules (the classical transplantation antigens) are expressed on most cells whereas the expression of class II molecules is confined to antigen-presenting cells. Both class I and II molecules are heterodimers with four extracellular domains: $\alpha1$, $\alpha2$, and $\alpha3$ associated with $\beta2$–μιχρογλοβυλιν (class I) and $\alpha1$, $\alpha2$, $\beta1$ and $\beta2$ (class II). Thee molecules bind both self- and pathogen-derived peptides in a surface groove (*peptide-binding region*) and present these to cytotoxic T cells that lyse the pathogen-infected cell, thereby limiting host infection.

In humans, the class I loci are HLA-A, HLA-B, and HLA-C (classical or type Ia) and HLA-E, HLA-F, and HLA-G (non classical

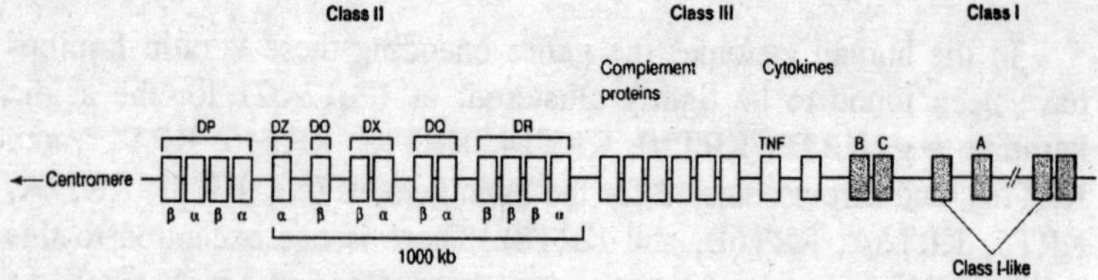

Fig. 5.18. Map of the human histocompatibility complex loci on 6p21.3.

or type Ib). The class I genes are interspersed with five MHC class I-related (MIC) loci (MICA, MICB, MICC, MICD, and MICE), five full-length pseudogenes (HLA-H, HLA-J, HLA-K, HLA-L, and HLA-X) plus several truncated pseudogenes and remnants. The 5 MIC genes are distantly related to the hemochromatosis (HFE; 6p21.3). The class II loci are clustered into three regions, HLA-DR, (HLA-DRA, HLA-DRB1, HLA-DRB2, HLA-DRB3, HLA-DRB4, and HLA-DBR5; NB. HLA-DRB3, HLA-DRB4, and HLA-DRB5 may be variably present or absent depending upon the haplotype), HLA-DQ (HLA-DQA1, HLA-DQA2, HLA-DQB1, HLA-DQB2, HLA-DQB3), and HLA-DP (HLA-DPA1, HLA-DPB1, HLA-DNA). The class I and II regions are separated by a gene-dense 1100 kb region containing a number of so-called type III genes including BF, C2, C4A, C4B, and TNF.

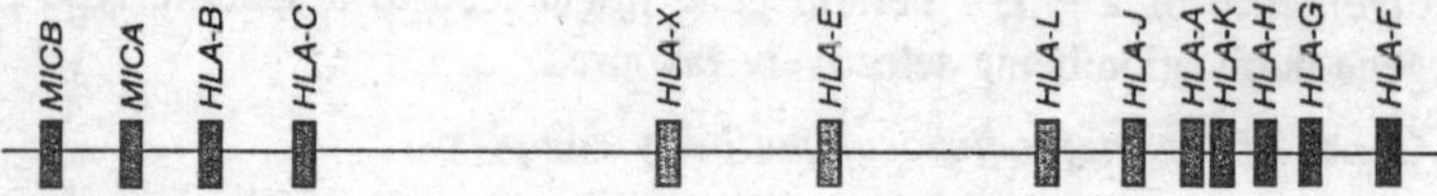

Fig. 5.19. Map of the class I region of the human histocompatibility complex.

The ancestral MHC molecule appears to have been assembled by combining its three constituent domains (peptide-binding domain, immunoglobulin-like domain and membrane-anchoring domain) by exon shuffling. Hughes and Nei (1993) estimated that the class II A and B genes diverged more than 500 Myrs ago. Class I genes then emerged by duplication and divergence during the primate radiation of the last 60 Myrs. The putative ancestral duplication(s) early in vertebrate evolution may well have provided an impetus to the emergence of the HLA complex in that the consequent redundancy could have created opportunities for the emergence of a variety of accessory and effector molecules.

So far, the MHC appears to be confined to vertebrates. In a phylogenetic analysis of class I genes form different mammalian orders, Hughes and Nei (1989b) found that the class Ib genes clustered with the class Ia genes. This is explicable either by postulating an independent

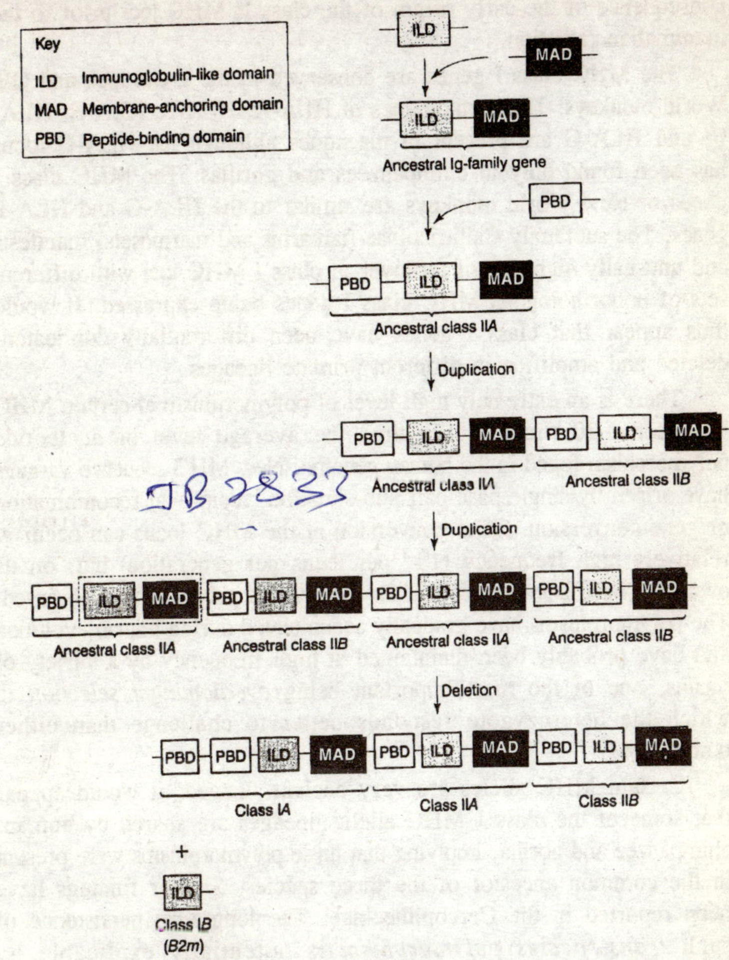

Fig. 5.20. Postulated origin of MHC class I and class II genes.

origin for the class Ib genes in different orders of mammals or by invoking the homogenizing influence of gene conversion between class I loci within each mammalian order. Class I genes evolved by a process of repeated gene duplication followed by a reduction in the level of expression of some genes, and the transcriptional silencing or deletion of others. This explains why orthologous relationships are found among class I loci of mammals of the same order but not among mammals of different orders. By contrast to the class I loci, orthologous relationships between mammalian class II loci are the rule rather than the exception, consequence of the early origin of the class II MHC loci prior to the mammalian radiation.

The MHC class I genes are conserved in the great apes and Old World monkeys. Thus orthologues of HLA-A, HLA-B, HLA-E, HLA-F, and HLA-G are present in macaques although the HLA-C locus has been found only in chimpanzees and gorillas. The MHC class I genes of New World monkeys are similar to the HLA-G and HLA-F genes. The subfamily Callitrichinae (tamarins and marmosets) manifests and unusually high rate of turnover of class I MHC loci with different sets of nonorthologous MHC class I genes being expressed. It would thus appear that class I genes have been differentially duplicated, deleted and amplified in different primate lineages.

There is an extremely high level of polymorphism at certain MHC loci, some 20 times higher than the average level of nucleotide polymorphism found in the human genome. New MHC sequence variants have arisen by single base-pair sub-situation, reciprocal recombination or gene conversion. Gene conversion at the MHC locus can occur at relatively high frequency (10^{-4} per locus per generation) but, on its own, it is insufficient to account for the level of polymorphism observed. The polymorphisms have gradually accumulated during human evolution and have probably been maintained at high frequency by a variety of means, one of the most important being *overdominant selection* in which the heterozygote responds better to challenge than either homozygote.

Certain MHC alleles are very ancient. Indeed, it would appear that some of the class I MHC allelic lineages are shared by human, chimpanzee and gorilla, implying that these polymorphisms were present in the common ancestor of the three species. Similar findings have been reported in the Cercopithecinae. The long-term persistence of such *trans-species polymorphism* is potentially explicable by overdominant selection because neutral alleles are unlikely to survive the process of speciation. Selection would be expected to act on MHC alleles that differed in their ability to bind and present foreign peptides. In a population exposed to pathogens, an individual heterozygous at many MHC loci would be able to present a larger number of foreign peptides than a homozygote and might therefore be resistant to a wider range of pathogens. Support for the overdominant selection hypothesis has come from studies that noted a significantly higher rate of non-synonymous than synonymous nucleotide substitution in the peptide-binding region (antigen recognition site), strong evidence for the action of positive selection. It should however be noted that negative or purifying selection may also operate on the HLA system so as to reduce diversity. One example of this is found in the peptide binding region of the HLA-E gene in New World monkeys.

The high frequency of polymorphism has probably also been maintained by other means such as temporal variation in selection pressure driven by changes in pathogens. Rare allele advantage (frequency-dependent selection) may play a role in that individuals with a rare MHC allele may respond better to challenge from new pathogen variants that have evolved in such a way as to evade the products of the more common MHC alleles. Population size and structure may also be important since diversity may be promoted by high effective population size or as a result of the agglomeration of smaller populations each bearing a few different alleles.

A high level of polymorphism can also be maintained by a 'genetic hitchhiker' effect. Thus the occurrence of at least 11 olfactory receptor genes (OR2C1) within the MHC complex may potentiate the selection of specific receptor alleles matched to a subset of MHC-determined odorants, such that those haplotypes that carry certain matched alleles would have a selective advantage in the population. If odorant discrimination is learned, this could have implications for the population genetics of species in which odorant-mediated kin recognition is important. Intriguingly, Wedekind et al. (1995) have claimed there to be evidence in humans for a preference for particular MHC-linked variations in sweat odors.

The common marmoset (*Callithrix jacchus*), a New World primate, possesses limited MHC class II variability as a result of the inactivation of the MHC-DP region and limited polymorphism at the MHC-DR and -DQ loci. This limited MHC class II repertoire could play a role in the apparently increased susceptibility of this species to viral, bacterial, protozoan and helminth infections.

Mucin genes

Many epithelial tissues such as trachea, mammary gland, pancreas, stomach, cervix and intestine produce high molecular weight glycoproteins known as mucins which are the major proteins of mucus. The mucins display only limited homologies with one another but do share the property of containing extensive tandemly repetitive regions. These regions can vary in length form 8 amino acid residues in MUC5AC to 169 residues in MUC6 and can be highly polymorphic. Several of the human mucin genes are located within a 400 kb cluster

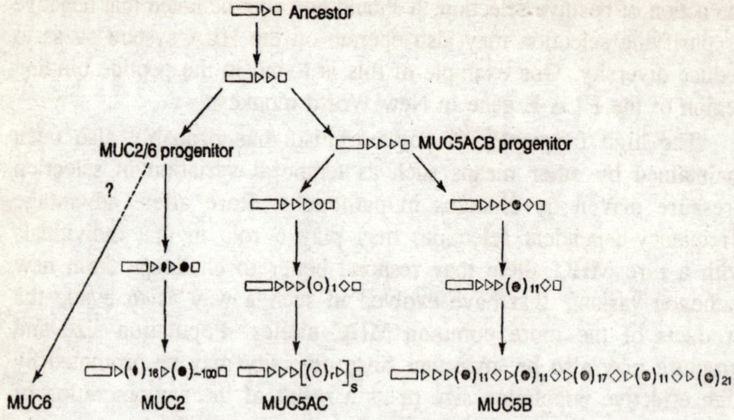

Fig. 5.21. Hypothetical scheme for the evolution of the MUC6, MUC2, MUC5AC, and MUC5B mucin genes from a common ancestor.

on chromosome 11p15 (MUC2, MUC5AC, MUC5B, MUC6), consistent with a series of successive gene duplications whereas others are solitary (MUC1, 1q21; MUC3, 7q22; MUC4, 3q29; MUC7, 4q13-q21; MUC8, 12q24). Although the human mucins display only a limited degree of homology with one another, some mucin genes possess exons of similar length and distribution consistent with their having evolved from a common ancestor.

As noted above, the mucin genes often contain internal tandemly repetitive domains. Thus the MuC2 gene contains two regions with a high degree of internal homology but no homology with each other. The first region comprises multiple 48 bp repeats interrupted by 21-24 bp segments whilst the second region is composed of 69 bp repeats arranged in a tandem array of up to 115 copies. The MUC5B gene contains an extremely large (10.7 kb) exon which encodes a 3570 amino acid protein that contains 19 subdomains which can be grouped into four larger composite units of 528 amino acids. ('super-repeats'). Similarly, the MUC4 gene contains an uninterrupted 18 kb exon encoding about 380 units of length 48 bp. Presumably these enormous exons have gradually become extended by serial internal duplication events that have not altered the splicing pattern, merely the length of the exon to be spliced.

Genes encoding RNA-binding proteins

RNA-binding proteins are involved in a wide range of biological functions including mRNA splicing, processing and translation. These proteins constitute a family insofar as they contain RNA-binding domains

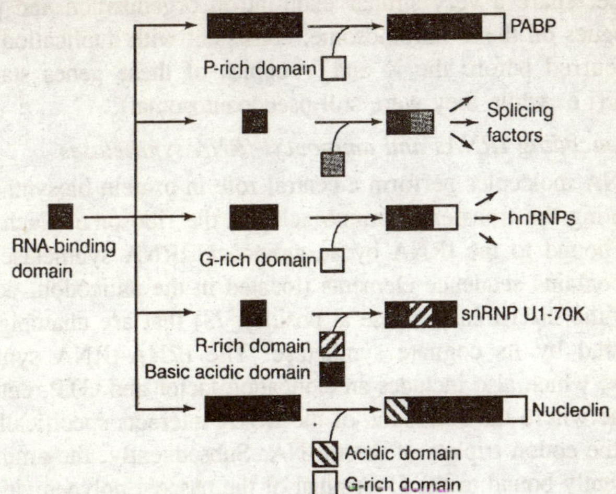

Fig. 5.22. Evolutionary scheme for various types of RNA-binding protein.

which share a common evolutionary origin that predates the divergence of prokaryotes and eukaryotes. Examples of human genes encoding RNA-binding proteins include the 70 kDa nuclear ribonucleoprotein (SNRP70; 19q13.3) gene, the heterogeneous nuclear ribonuclear riboprotein A1 (HNRPA1; 12q13.1) gene, the Arg/Ser-rich splicing factor (SFRS2; chromosome 17) gene, the poly(A) binding protein (PABPL1; 3q22-q25, 12q13-q14, 13q12-q13) genes, and the NCL (2q12-qter) gene which encodes nucleolin, a protein involved in the control of transcription of rRNA genes and in the nucleocytoplasmic transport of ribosomal components.

Sulfatase genes

Sulfatase catalyze the hydrolysis of sulfate ester bonds in a variety of substrates. Sulfatase genes have been described in lower cukaryotes and comprise an evolutionarily conserved multigene family in human: arylsulfatase A (ARSA; 22q13), arylsulfatase B (ARSB; 5q11-q14), arylsulfatase D (ARSD; Xp22.3), arylsulfatase E (ARSE; Xp22.3), arylsulfatase F (ARSF; Xp22.3), proximal to the pseudoautosomal boundary, steroid sulfatase (STS; Xp22.3) 4 Mb distant from the ARSD/ ARSE/ARSF cluster, iduronate-2-sulfatase (IDS; Xq27-28), galactose 6-sulfatase (GALNS; 16q24) and glucosamine 6-sulfatase (GNS; 12q14). The chromosomally dispersed sulfatase genes exhibit a relatively low degree of sequence identity consistent with these genes having emerged comparatively early during evolution. By contrast, the four genes located at XP22.3, within the pseudoautosomal region, are more similar in sequence, share a very similar exon-intron organization and possess homologues on the Y chromosome, consistent with duplication events that occurred before the X and Y copies of these genes started to diverge (i.e. while they were still pseudoautosomal).

Genes encoding tRNAs and aminoacyl-tRNA synthetases

tRNA molecules perform a central role in protein biosynthesis by potentiating the transfer of amino acids to the ribosome. Each amino acid is bound to the tRNA by an aminoacyl-tRNA synthetase. Each tRNA contains sequence elements (located in the anticodon, acceptor stem or the discriminator base at position 73) that are unambiguously recognized by its cognate synthetase. The tRNA-tRNA synthetase complex, which also includes an elongation factor and GTP, enters the ribosome where the anticodon of the tRNA interacts specifically with one of the codon triplets of the mRNA. Subsequently, the amino acid is covalently bound to the C-terminal of the nascent polypeptide chain. tRNA have similar sizes and tertiary structures but are subdivided into *isoaccepting groups* which, despite having different sequences, recognize the same amino acid.

It is unclear how many functional tRNA genes are present in the human genome. An early saturation hybridization study put the total number of tRNA genes at about 1300 per haploid human genome, an average of 65 copies for each tRNA species. Chromosomally allocated human tRNA genes include TRAN (alanine, 6p21-p22), TRE (glutamic acid, 1p36), TRG1 (glycine, chromosome 1), TRK1 (lysine, 17p13), TRL1 (leucine 1, 14q11-q12), TRL2 (leucine 2, 17p13), TRMI1 and TRMI2 (methionine, 6p23-q12), TRN (asparagine, 1p36), TRP1 and TRP2 (proline 1 and 2, 14q11-q12), TRP3 (proline 3, chromosome 5), TRQ1 (glutamine 1, 17p13), TRR1 (arginine 1, 17p13), TRP3 (arginine 3, 6p21-p22), TRR4 (arginine 4, 6p21-p22), TRT1 (threonine 1, chromosome 5), and TRT2 (threonine 2, 14q11-q12). Thus tRNAs from different isoaccepting groups often cluster together, for example 6p22 (TRAN, TRMI1, TRMI2, TRR3, TRR4), 14q11-q12 (TRL1, TRP1, TRP2, TRT2) and 17p13 (TRK1, TRL2, TRQ1, TRR1). Other tRNA genes are dispersed and, since these genes are often flanked by short 8-12 bp direct terminal repeats, they may have arisen by RNA-mediated transposition.

A human gene (TRSP) encoding an opal suppressor phosphoserine tRNA has been characterized on chromosome 19q13. This tRNA is potentially capable of suppressing nonsense mutations since it is able to recognize a different codon than that corresponding to the amino acid it carries. This ability is conferred upon it by dint of a base change in its anticodon which allows it to translate a UGA Stop codon and insert a serine at this position.

Phylogenetic studies of tRNA genes have shown that members of isoaccepting groups do not invariably cluster together. It would appear that during evolution, tRNAs have been recruited from one isoaccepting group to another by mutations in the anticodon.

The aminoacyl-tRNA synthetases constitute a large gene family in the human genome including glutaminyl/prolyl (EPRS; 1q32-q42), lysyl (KARS, 16q23-q24), alanyl (AARS; 16q22), arginyl (RARS; 5pter-q11), valyl (VARS1; 9), histidyl (HARS; chromosome 5), asparginyl (NARS; chromosome 18), threonyl (TARS; 5p13-cen), methionyl (MARS; chromosome 12), isoleucyl (IARS; 9q21), glycyl (GARS; 7p15), tryptophanyl (WARS; 14q32), cysteinyl (CARS; 11p15.5) and leucyl (LARS; 5cen-q11). Whilst many of the members of this gene family are chromosomally widely dispersed, similar chromosomal locations for some members are suggestive of linkage. The aminoacyl-tRNA synthetases may be grouped into two distinct classes, each with ten members, based upon sequence data and structure. These groups are thought to have arisen from two progenitor aminoacyl-tRNA synthetases by gene duplication and divergence. Despite their ancient origin and their presence in both prokaryotes and eukaryotes, sequence similarity between aminoacyl-tRNA synthetases from eukaryotes and prokaryotes may be as low as 15%.

Interestingly, there is a relationship between synthetase class and the nucleotide that is conserved at position 73. Eight of the ten tRNAs that are aminoacylated by class I synthetases have an alanine residue at this position whereas those that are aminoacylated by class II synthetases manifest greater nucleotide diversity. It is clear that the genes for both tRNAs and aminoacyl-tRNA synthetases have a very long evolutionary history. Their origins and the means by which they may have coevolved are as yet unclear although recent work has suggested that the tRNA synthetases may have been preceded by their tRNAs.

Ubiquitin genes

The ubiquitin genes are extremely highly conserved 76 amino acid proteins which are required for ATP-dependent nonlysosomal intracellular protein degradation of defective proteins and proteins with a rapid turnover. Although found in all eukaryotes, they have not so far been found in prokaryotes.

The human genome contains multiple ubiquitin-related sequences most of which are processed pseudogenes. There are however at least four functional ubiquitin genes which together have provided an interesting challenge to carefully crafted general definitions of the gene. These genes belong to one of the two distinct classes. The first is exemplified by the ubiquitin C (UBC; 12q24) gene which contains 9 direct repeats of the ubiquitin amino acid sequence with no spacer regions and no introns. The UBC locus has a single termination codon and yields a 2500 nucleotide mRNA which upon translation generates a polyubiquitin precursor molecule which is post-translationally cleaved to active ubiquitin. Unequal crossing over events at the UBC locus have been responsible for polymorphism in ubiquitin repeat unit number. A second human polyubiquitin gene (UBB, 17p11-p12) with 3 repeat units yields a 1000 nucleotide mRNA. The second class comprises ubiquitin fusion genes which encode a single ubiquitin fused in-frame to a ribosomal protein of either 52 (UBA52; 19p13) or 76-80 (UBA80) amino acids.

The number of nucleotide differences between ubiquitin repeats within a given species is very low as compared to those between species. Thus, repeats within a locus share a more recent common ancestor than any two repeats in different specie. This finding is explicable in terms of concerted evolution, probably mediated by unequal crossing over or gene conversion. In human, concerted evolution is evident both within and between ubiquitin loci but appears to occur at a higher rate for some repeats than others.

An overview of the evolution of multigene families in eukaryotes

During eukaryotic evolution, gene families have been created by sequential, phased rounds of gene duplication with specific genes becoming duplicated as a result of whole genome duplications, subchromosomal regional duplications, and through the rather more discrete duplication of individual gene loci. Since the number of genes in a gene family varies quite widely, we may surmise that some sequences are more predisposed to duplicate than others. As a consequence of their common origin, multigene family members usually have a similar structure both in terms of their nucleotide sequences and exon-intron organization (although the gain or loss of some introns in some members can serve to obscure their common ancestry). Whereas some genes have retained their syntenic relationship with each other after duplication and have remained chromosomally linked for relatively long periods of evolutionary time, others have been translocated to another chromosomal location. In some gene families, new rounds of gene duplication have then led to the formation of gene clusters. Once diversification of gene clusters, sub-clusters and individual genes had occurred through the acquisition of comparatively subtle mutational changes, the duplication of entire gene clusters as well as portions of clusters would then have led to the emergence of a sub-family structure within the gene family.

Diversification of individual genes within multigene families has occurred in a host of different ways in different lineages. Amino acid substitutions may appear to constitute relatively subtle structural changes but these changes can be quite dramatic in functional terms if, for example, substrate specificity is altered. Diversification can also proceed by internal duplication and deletion (e.g. gain or loss of individual exons), the insertion or removal of individual amino acid residues, repeat expansion and by the more complex processes of gene conversion, fusion and recombination. Family members have sometimes been inactivated or lost so that in somc cases, the immediate ancestors of extant genes may now no longer exist (molecular 'missing links'). In others, promoter changes have led to the emergence of differences in gene expression either in terms of tissue specificity or level of expression or in responsiveness to inductive stimuli whether of environmental (e.g. temperature), systemic (e.g. hormonal) or intra-cellular origin (e.g. transcription factors). Some multigene family members have evolved more quickly than others whether as a result of selection pressure or the stochastic processes of neutralist evolution. What is common to all the gene families cited above is the principle that gene duplication has created redundancy which has then allowed evolutionary experimentation through diversification, ultimately leading to the recruitment of a new generation of genes encoding proteins with novel properties.

Highly Repetitive Multigene Families

Histone genes. Histones are basic nuclear proteins which make up the nucleosome within the chromatin fibre. Pairs of H2A, H2B, H3 and H4 form the octamer with H1 being responsible for linking the nucleosomes and potentiating the formation of higher order chromosome structure. In mammals, the histones may be sub-divided into three types: (i) main-type replication-dependence histones, (ii) replication-independent 'replacement' histones, and (iii) tissue-specific histones. The genes encoding the replication-dependent and tissue-specific histones lack introns, give rise to non-polyadenylated mRNAs, contain

3' elements with dyadic symmetry essential for mRNA processing, and are chromosomally clustered. By contrast, the genes encoding the replacement histones can contain introns, give rise to polyadenylated mRNAs and are solitary rather than clustered.

Three main clusters of histone genes are apparent in the human genome and contain between them about 60 genes. Two clusters, located at 6921.3, are separated by ~2 Mb and contain all replication-dependent H1 histone (H1F1, H1F2, H1F3, H1F4, H1F5) genes and surrounding core histone (H2A, H2B, H3, H4) genes. The other cluster at 1q21 is smaller consisting of at least four core histone genes. Various solitary replacement histone genes have been located on different chromosomes, for example H1° (H1F0; 22q13), H2A.X (H2AX; 11q23), H2A.Z (H2AZ; 4q24) and H3.3B (H3F3B; 17q25). Finally, testis-specific histone genes H3F3A and H1FT have been localized to chromosomes 1q42 and 6p21, respectively.

Histones are very ancient proteins as evidenced by homologies between prokaryotic and eukaryotic histones H2A/B, H3, and H4. Interestingly, homology also exists between the core histones and the CCAAT-binding factor suggesting that transcriptional regulation and nucleosomal packing may have been intimately related for a very considerable period of evolutionary time.

In lower eukaryotes, histone genes usually occur in long tandemly repetitive arrays but in mammals, although the genes are clustered, they are less ordered and do not occur as tandem repeats. A map of part of the major histone cluster on human chromosome 6 and it is evident that the arrangement of the histone genes is fairly random; the only regularity is the arrangement of the divergently orientated H2A and H2B gene pairs. For both main-type and replacement histone types, chromosomal localization, gene number and gene organization appear to be fairly well conserved between human and mouse. This indicates that the separation of these two groups of histone genes must have occurred at an early stage in mammalian evolution. Chromosomal clustering may be a prerequisite for coordinate regulation. An unexpectedly high degree of within-species homogeneity of tandemly repeated histone genes is explicable in terms of unequal crossing over and/or gene conversion.

The residual sequence similarity (pairwise 15-20%) exhibited by the four core histones together with their structural similarity argues strongly for descent from a common ancestor, albeit a very ancient one. A phylogenetic analysis of the core histones has suggested that the histones that form dimers (viz. H2A/H2B and H3/H4) have very similar trees and appear to have coevolved. H3 and H4 have evolved ~10 fold more slowly than H2A and H2B and are very highly conserved. The H2A.Z variant arose early and appears to be more highly conserved than the main-type 2A histone whilst H3.3 variants have arisen independently on many occasions. By contrast, H2A.X arose comparatively recently during the evolution of the vertebrates.

Ribosomal RNA genes

The ribosomal RNA (rRNA) genes occurs as 300-400 copies in the human genome. They are organized in tandemly repeated blocks which are located on the acrocentric chromosomes: RNR1 (13p12), RNR2 (14p12), RNR3 (15p12), RNR4 (21p12), and RNR5 (22p12). The 44 kb rDNA repeat unit contains a 13.3 kb RNA polymerase I-transcribed portion and a 31 kb nontranscribed spacers region. The 13 kb transcript is then processed to generate the 28S, 18S and 5S mature rRNA molecules with respective lengths of 1.8 kb, 0.15 kb, and 5.8 kb.

The 28S rRNA contains both conserved and variable regions. The former are constant in size and sequence whereas the latter are variable and exhibit species-specific differences between primates. The variable regions potentiate hairpin loop formation and are essential for rRNA secondary structure.

The number of rRNA genes on a given chromosome varies in the population, while in any one individual, rRNA gene number varies between clusters. The rRNA clusters appear to have evolved in concerted fashion. Thus rRNA genes on non-homologous chromosomes are far more similar to one another than would be expected if these genes had evolved independently. Several mechanisms have been proposed to account for this relative homogeneity of rDNA including unequal crossing over between rDNA sequences on non-homologous chromosomes. Gene conversion may then act so as to homogenize rDNA repeats within each cluster. On the basis of linkage disequilibrium data, Seperack et al. (1988) have suggested that gene conversion and unequal crossing over may also occur within a chromosome (i.e. by sister chromatid exchange)

5S ribosomal RNA genes

The human 5S rRNA (RN5S1) genes occur in clusters of 2.3 kb and 1.6 kb tandem repeat units which differ from each other by virtue of a deletion in the 3' flanking region. There appear to be 300-400 copies of 5S rRNA genes per haploid human genome. Between 100 and 150 genes are derived from the 2.3 kb cluster (probably located at chromosome 1q42), 5-10 from the 1.6 kb cluster whereas the remaining 200-300 genes/gene variants are not found in repeat structures and may be dispersed around the genome. In addition to these genes, a large number of 5S rRNA pseudogenes exist which brings the total number of 5S rRNA homologous sequences per haploid genome to around 2000.

Small nuclear RNA genes

The human U2 snRNA (RNU2) genes are clustered in tandem arrays of between 6 and 30 copies at 17q21-q22. The primate U2 snRNA arrays have evolved in concerted fashion with each repeat being essentially homogeneous within a species although somewhat different between species, an observation consistent with the action of gene conversion. By contrast to the situation in higher primates, the U2 snRNA genes of both the mouse and the prosimian, *Galago crassicaudatus*, are dispersed rather than clustered suggesting that the arrays characteristic of higher primates may have resulted from amplification of a common ancestral gene. Once established in the simian lineage, however, the U2 tandem repeat array has remained at the same chromosomal locus through multiple speciation events over a period of >35 Myrs.

Some 30 copies of the human U1 snRNA (RNU1) genes are also clustered at 1p36.3. This site is distinct from the cluster of U1 snRNA pseudogenes at 1q12-q22 that outnumber their functional counterparts by some 15-30 fold. This contrasts with the situation found for the U2 snRNA genes, where the active genes outnumber the pseudogenes.

Gene Superfamilies

The superfamily is used to describe a group of gene families whose individual members share a common evolutionary origin, possess common features within a family but differ with respect to certain other features between families. A selection of some of the most important human gene superfamilies will be presented.

Cadherin genes

The cadherins are membrane-associated glycoproteins which act as calcium-dependent cell adhesion molecules but which may also be involved in signal transduction. Cadherins may be classified into four groups (classical, desmosomal, protocadherins, cadherin-related proteins) which are structurally similar, each containing an extracellular domain consisting of between 4 and 30 + repeats of an 110 amino acid cadherin-specific motif. Although classical cadherins appear to be confined to the vertebrates, the superfamily has ancient origins with members present in organisms as diverse as nematodes and humans. The classical and desmosomal cadherins are thought to have evolved more recently by a process of duplication and divergence from the primordial proto-cadherin-like proteins.

A cluster of human classical cadherin genes is present at chromosome 16q22.1 (CDH1, CDH5, CDH3) with CDH13 and CDH15 located at 16q24 and CDH11 not yet regionally localized on chromosome 16. Other genes encoding classical cadherins are present on 5p13-p14(CDH12) and 18q11 (CDH2) whilst a protocadherin (PCDH7) gene has been mapped to 4p15.

Cytochrome p450 genes

The cytochrome P450 enzymes comprise a large liver-expressed family of heme-containing electron transport molecules which are involved in the oxidative metabolism of a wide range of substrates including steroids, drugs and xenobiotics. This family can be subdivided into sub-families on the basis of structural and functional criteria and many are extremely polymorphic, for example CYP2D6. With nearly 40 CYP P450 genes identified (and perhaps between 20 and 150 yet to be identified), this superfamily is one of the large superfamilies represented in the human genome: CYP1A1 and CYP1A2 (15q22), CYP1B1 (2p21), CYP2A6, CYP2A7, CYP2A13 (19q13.1), CYP2B6 and CYP2B7 (19q13.2), CYP2C8, CYP2C9, CYP2C10, CYP2C18, CYP2C19 (10q24), CYP2D6 (22q13.1), CYP2E (10q24-qter), CYP2F1 (19q13.2), CYP2J2 (1p31), CYP3A4 (7q22.1), CYP4A11 (1), CYP4B1 (1p12-p34), CYP7A1 (8q11-q12), CYP11A (15q23-q24), CYP11B1 and CYP11B2 (8q21), CYP17 (10q24.3), CYP19 (15q21), CYP21 (6p21.3), CYP24 (20q13), CYP26A1 (10q23-q24), CYP27A1 (2q33-qter), CYP27B1 (12q13.3-q14) and CYP51 (7q21). Several gene clusters are apparent, for example on 10q24 and 19q13, and this is likely to reflect a history of gene duplication and divergence. A common origin may however also be reflected by the possession of a similar exon/intron arrangement as in the case of the CYP17 and CYP21 genes or the CYP11, CYP24, and CYP27 genes. The phylogeny of the cytochrome P450 superfamily has been explored to some extent and various attempts have been made to date the various gene duplication events which have given rise to extant genes. Much of the increase in CYP P450 gene number took place around 400 Myrs ago. This was the time when tetrapods first began to colonize the land and feed upon the plants that had become established in the late Silurian and early Devonian periods. Since these terrestrial plants probably contained toxic compounds, expansion of the CYP P450 gene family to provide a large set of detoxifying enzymes was probably an adaptive response to this chemical challenge.

The CYP51 gene encodes sterol 14-α-demethylase which plays an important role in sterol biosynthesis in fungi and plants as well as animals. As such, CYP51 is the only P450 family member which is recognizable across all eukaryotic phyla. The CYP51 gene also appears to have its counterparts in prokaryotes and the gene family may have originated before the divergence of the eukaryotic and prokaryotic kingdoms. When members of the various mammalian and fungal P450 gene families were aligned and compared, more than 80 intron locations were identified. Since it is unlikely that all of these introns were present in the primordial eukaryotic P450 gene, it may concluded that P450 gene structures have evolved very considerably over the last 2 billion years either by intron insertion or intron sliding.

Cystatin genes

The cystatin superfamily comprises a number of proteins, many of which are cysteine protease inhibitors, but which have evolved to take on a variety of different physiological functions. The family has emerged through a process of duplication and divergence from a primordial gene which is thought to have existed more than 1200 Myrs ago. Human cystatin superfamily genes belong to family 1 [cystatins A and B (CSTA, 3cen-q21; CSTB, 21q22.3)], family 2 cystatins C, S, SA and SN (CST3, CST4, CST2, and CST1; 20p11.2), family 3 kininogen (KNG, 3q21-qter) or family 4 histidine-rich glycoprotein and α_2HS-glycoprotein (HRG, 3q27; AHSG, 3q27-q29). The evolutionary relationship of these genes remains to be unravelled.

G-protein-coupled receptor genes

The G-protein-coupled receptor (GPCR) superfamily can be separated into five different groups on the basis of their natural ligands: (i) peptides and peptide hormones, for example endothelin receptors (EDNRB, 13q22; EDNRA, chromosome 4), adrenocorticotropic hormone receptor (MC2R; 18p11.2), angiotensin receptor (AGTR1; 3q21-q25), thyrotropin receptor (TSHR; 14q31), (ii) neurotransmitters, for example dopamine receptors (DRD1, 5q35; DRD2, 11q23; DRD3 3q13; DRD4, DRD5, 4p15-p16), (iii) other regulatory factors, for example thrombin receptor (F2R; 5q13), (iv) sensory stimuli, for example opsins, and (v) unknown ligands. The GPCRs are evolutionarily and phylogenetically related, most of them possessing a common motif of 7 transmembrane domains. A multitude of human genes encoding G-protein-coupled receptor with unknown ligands are also known and these are widely scattered around the genome, for example GPR1 (15121.6), GPR2 (17q21), GPR3 (1p34-p36), GPR4 (19q13), GPR5 (3p21), GPR6 (6q21-q22), GPR7 (10q11-q21), GPR8 (20q13), GPR9 (8p11-p12), GPR10 (10q25-q26), GPR12 (13q12), GPR13 (3p21-pter), GPR15 (3q11-q13), GPR18 (13q32), GPR20 (8q24).

Heat shock genes

The heat shock proteins are evolutionarily ubiquitous ('universal') proteins that function as molecular chaperones under both physiological and stress conditions. These proteins recognize and stabilize partially folded proteins during the processes of translation, translocation across membranes and multimer assembly. *Escherichia coli* possesses one heat shock protein gene (*dnaK*), yeast possesses nine as does *Drosophila*. This superfamily can be subdivided into three major classes based upon the molecular weight and degree of homology of the proteins.

The most highly conserved class is the HSP70 family represented in human by at least eleven genes: HSPA1A, HSPA1B and HSPA1L linked on chromosome 6p21, HSPA2 (14q24), HSPA3 (chromosome 21), HSPA4 and HSPA9 linked on chromosome 5q31, HSPA5 (9q34), HSPA6 and HSPA7 linked on chromosome 1q, and HSPA8 (11q23-q25). Some members of the HSP70 family are ubiquitously expressed, others are tissue-specific, some are constitutively expressed, others are expressed only in response to stress. The HSP70 proteins also differ in their subcellular localization, for example cytosol, nucleus, nucleolus, endoplasmic reticulum, mitochondrion. The HSP70 genes are extremely ancient having arisen before the diversification of cellular life into bacteria, archaebacteria and eukaryotes. Indeed, sequence data from these proteins have been used to argue for the origin of eukaryotes via a fusion between archaebacteria and gram-negative bacteria. Not surprisingly in view of their universality, they exhibit extreme evolutionary conservation. Chromosomal localization has also been conserved evolutionarily as evidenced by the linkage of three HSP70 genes to the HLA/MHC complex in both humans and rodents.

The 90 kDa heat shock proteins comprise a second family of evolutionarily highly conserved proteins which have counterparts in bacteria as well as in all eukaryotes. Phylogenetic analysis has suggested that the first of the HSP90 gene duplications occurred very early on in the evolution of the eukaryotes. Human homologues of this gene family are dispersed in the genome and include HSPCA (1q21-q22) and HSPCB (6p12). The human genome also contains a group of heat shock proteins of molecular weight 15-30 kDa which are highly conserved and evolutionarily related to the α-crystallins. Interestingly, one of the human genes encoding heat shock protein 27 (HSPB2) is closely linked to the α-B-crystallin gene on chromosome 11q22-q23.

Insulin and insulin-like growth factor genes

The insulin gene family has ancient origins among the primitive eukaryotes. Human representative of this superfamily include insulin, (INS; 11p15.5), insulin receptor (INSR; 19p13), insulin-like growth factor I (INF1; 12q22-q24), insulin-like growth factor II (IGF2; 11p15.5) and the relaxins (RLN1 and RLN2; 9pter-q12). The origin of the IGF1 and IGF2 genes, which encode important regulators of growth and development, antedates the emergence of the first vertebrates, since they appeared early on in the evolution of the protochordates more than 600 Myrs ago. It is likely that in primitive organisms, insulin-like peptides functioned so as to promoter the uptake and utilization of nutrients, and as a consequence, stimulated growth. With increasing complexity, nutrition and growth became uncoupled and the insulin family diversified into proteins and receptors with differing capacities for regulating metabolism and growth.

Interferon genes

The interferon superfamily of viral defence proteins comprises two main classes of gene (type I and type II). The type II interferons have only one member, interferon-γ, encoded by a gene (IFNG) on chromosome 12q14. By contrast, type I interferons comprise several sub-families of genes most of which are clustered in a 400 kb region of chromosome 9p21. This cluster contains 13 interferon-α genes (IFNA1, IFNA2, IFNA4, IFNA5, IFNA6, IFNA7, IFNA8, IFNA10, IFNA13, IFNA14, IFNA16, IFNA17 and IFNA21), a single interferon Ω gene (IFNW1), a single interferon-β gene (IFNB1) and several pseudogenes. The genes are arranged in tandem and most of the functional genes are oriented with their 3' ends pointing in a telomeric direction. There are two exceptions, IFNA1 and IFNA8, which together with four pseudogenes orient towards the centromere. This is consistent with the occurrence at some stage of an inverted duplication within the gene cluster with its breakpoint between IFNP12 and IFNP11.

The evolutionary relationships of the various IFNA family members have been determined by phylogenetic analysis and are consistent with the emergence of this gene family by a process of gene duplication

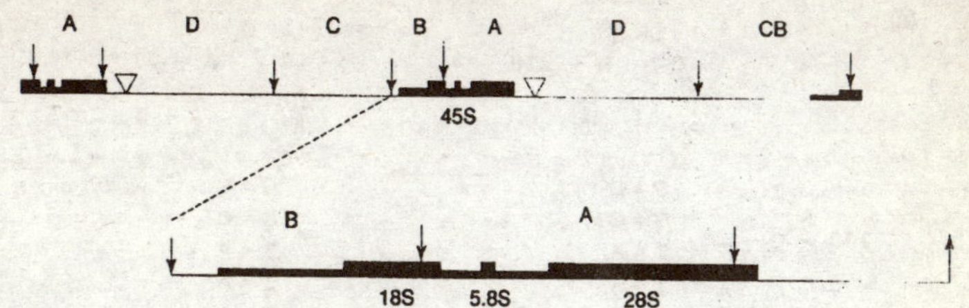

Fig. 5.23. Organization of the human ribosomal gene clusters.

and divergence. The proximal genes, IFNA1, IFNA2, IFNA5, IFNA6, IFNA13, appear more closely related to each other than they are the distal genes, IFNA4, IFNA7, IFNA10, IFNA16, IFNA17, and IFNA21, whilst IFNA8 is equally divergent from both groups. Since these two groups are located at opposite ends of the gene cluster, this may reflect an early multi-gene duplication. The gene cluster appears to have evolved by gene duplication as a result of unequal crossing over involving tandem units of IFNA and IFNW1 genes but divergence between gene sequences as well as subsequent gene deletions and duplications have served to erase the evolutionary history of parts of the gene cluster. The IFNA14 gene, located midway between the two groups, is similar to the distal group in its 5' half and the proximal group in its 3' half. This is explicable either in terms of a unequal crossing over event between distal and proximal interferon genes or a gene conversion event which has corrected the 5' half of the gene against a donor proximal gene. Several of the human IFNA genes differ from the other family members by one or more relatively subtle mutations. Golding and Glickman have shown in their insightful and prescient (1985) study that these changes are explicable in terms of their templation by the local DNA sequence environment. Thus, the in-frame deletion of a GAT codon in the IFNA2 gene may have been templated by a 5 bp inverted repeat 9 bp 5' to the observed deletion.

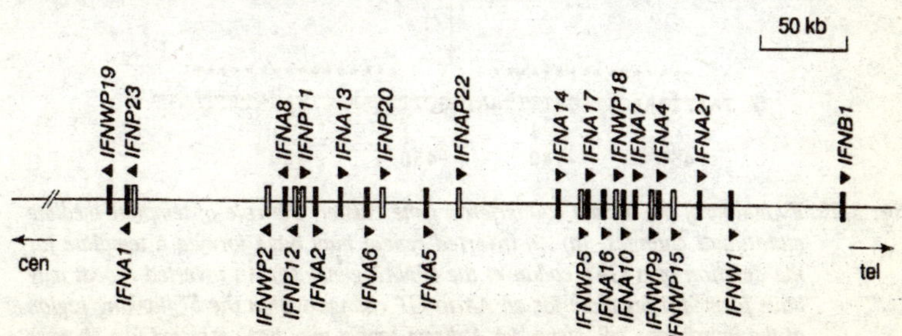

Fig. 5.24. Map of the human type-I interferon gene cluster on chromosome 9p22.

(a)

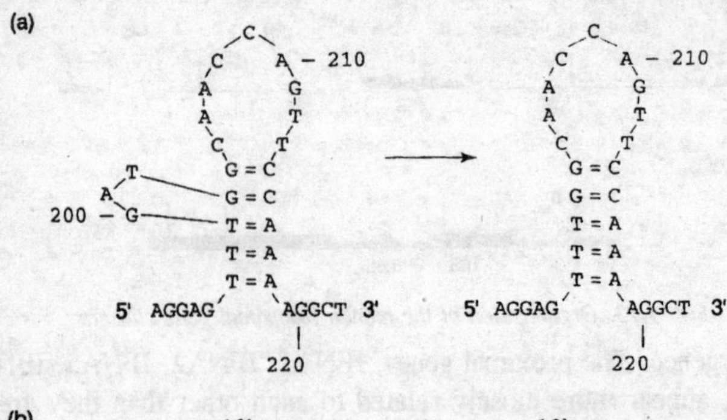

(b)

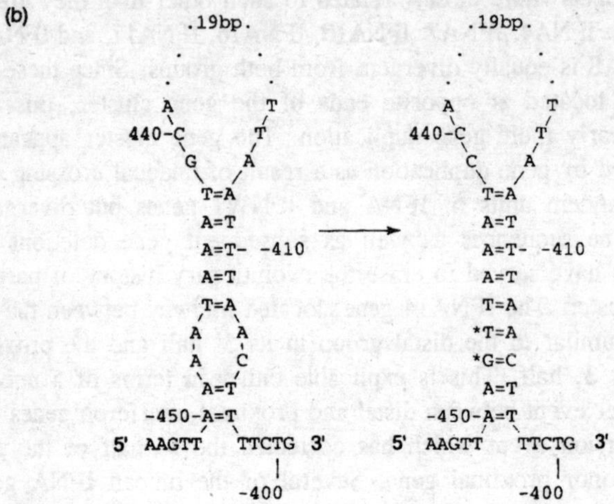

(c)

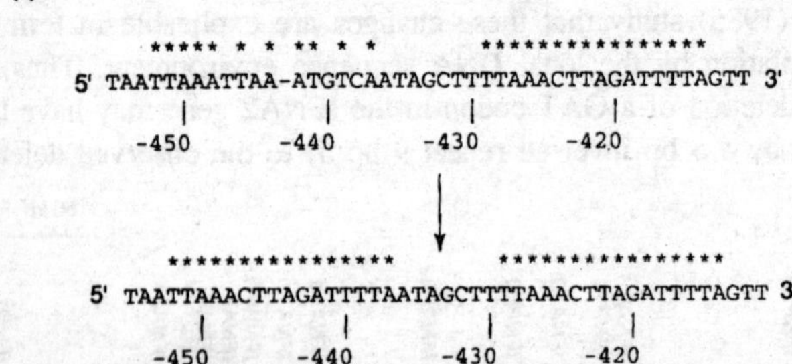

Fig. 5.25. Evolution of the human α-interferon gene cluster; the role of template-mediate mutational changes. (a) An inverted repeat may have formed a template for the deletion of a GAT codon in the IFNA2 gene. (b) An inverted repeat may have formed a template for an AA to GT change within the 5' flanking region of the interferon "α9" gene. (c) A direct repeat may have directed five changes in the 5' flanking region of the IFNA10 gene.

Similarly, a 9 bp inverted repeat (separated by 25 bp DNA) may have templated in AA to GT change in the 5' flanking region of the interferon "α9" gene whilst a 16 bp direct repeat could have templated five distinct sequence changes (two transitions, two transversions and a guanine insertion) as the product of one mutational event.

The type I interferon genes not only lack introns but also exhibit sequence homology indicating that they share a common ancestry. Since the spacer regions between primate IFNA genes still retain some sequence similarity, we may surmise that some of the gene duplication events have been relatively recent. Indeed, Miyata and Hayaschida (1982) estimated that they arise within the last 26 Myrs. The duplicational expansion of the IFNA gene cluster may have been driven by the need to produce large quantities of interferon rapidly, by selection for a novel temporal or spatial pattern of expression, or by selection for a specialized novel function, perhaps associated with a variant receptor with lower affinity for its ligand.

The absence of introns in the IFNA and IFNB genes probably reflects the organization of the ancestral gene which could conceivably have been a retrotransposed copy of an intron-containing interferon gene such as that encoding interferon-γ (IFNG; 12q14) which contains three introns. The IFNA and IFNB genes also exhibit an axis of internal symmetry, presumably the result of an internal duplication which must have occurred prior to the divergence of the two gene families.

Nuclear receptor genes

Nuclear receptors are ligand-activated transcription factors that regulate the expression of their target genes by binding to specific cis-acting sequences in their promoter regions. This superfamily may be divided into a minimum of three distinct groups, one containing the receptors for steroid hormones (glucocorticoids, androgens, estrogens, progesterone, etc), a second containing receptors for vitamin D, thyroid hormone and retinoic acid and a third containing various 'orphan' receptors which putatively interact with ligands that still remain to be identified. Nuclear receptors exhibit a modular organization and contain at least four domains: an A/B domain involved in transactivation, a highly conserved zinc finger-containing domain (c) involved in DNA binding, a hinge (D) domain and a carboxy terminal (E) domain that is required for ligand binding, dimerization and transcriptional regulation.

Human genes belonging to this superfamily include the androgen receptor (AR; Xq11-q12), estrogen receptor (ESR1; 6q25), glucocorticoid receptor (GRL; 5q31), mineralocorticoid receptor (MLR; 4q31), progesterone receptor (PGR; 11q22), retinoic acid receptors α (RARA; 17q21) β (RARB; 3p24) and γ (RARG; 12q13), thyroid hormone receptors α (THRA; 17q21) and β (THRB; 3p24) and vitamin D receptor (VDR; 12q12-q14). Genes encoding the 'orphan' receptors include hepatocyte nuclear factor 4 (HNF4A; 20q12-q13), the COUP transcription factors (TFCOUP1; 5q14; TFCOUP2; 15q26), the retinoid X receptors (RXRA, 9q34; RXRB, 6p21.3; RXRG, 1q22-q23), and the peroxisome proliferator-activated receptors α (PPARA; 22q12-q13), γ (PPARG) and δ (PPARD; 6p21). It can be seen that the retinoic acid and thyroid hormone receptor genes are arranged in two syntenic groups.

The ancestral nuclear receptor gene may have originated very early by fusion of DNA-binding and steroid-binding domains. Indeed, Escriva et al. (1997) have proposed that the ancestral nuclear receptor was an orphan receptor which subsequently acquired its ligand-binding potential. Since numerous vertebrate nuclear receptors have *Drosophila* homologues, the nuclear receptor superfamily must have diversified before the divergence of arthropods and vertebrates more than 500

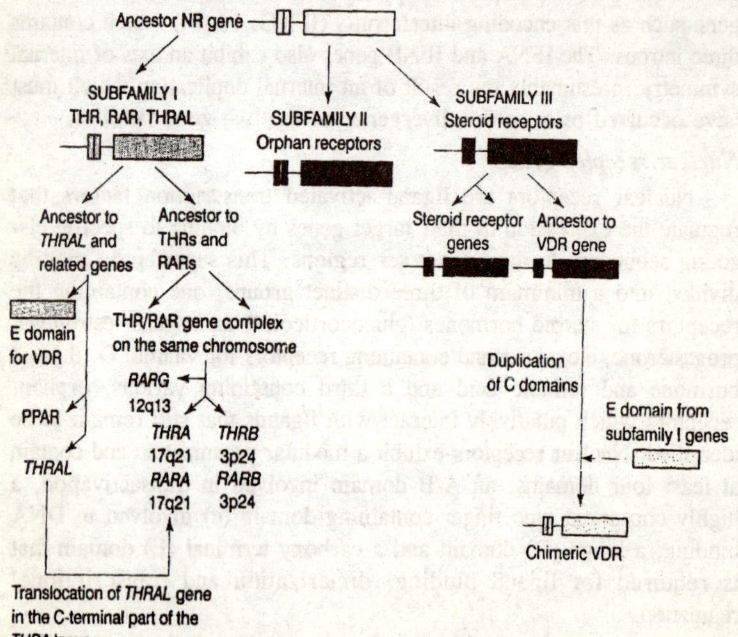

Fig. 5.26. Evolutionary schema for the human nuclear receptor genes.

Myrs ago. This diversification may have proceeded along the line of the schema laid out. Two waves of gene duplication occurred, one very early on giving rise to the different receptor groups and a second in the vertebrate lineage. This was accompanied by domain shuffling between genes. Sequence from the thyroid hormone receptor α-related gene, THRAL, appears to have been translocated to the THRA locus thereby creating the final THRA exon. This must have occurred early on the in mammalian evolution since this organization is present in rat and human but not in chicken. An E domain from subfamily I has been acquired by the VDR gene which distinguishes it form the other genes in the steroid hormone receptor family. There appears however to be no relationship between the phylogenetic position of a liganded receptor and the chemical nature of its ligand.

Protein kinase C genes

The member of the mammalian protein kinase C superfamily are involved in a number of signalling systems underlying a range of cellular processes. The characterized members of the human superfamily can be divided into three subfamilies, class I: α (PRKCA; 17q22-q23), β (PRKCB1; 16p11) and γ (PRKCG; 19q13), class II : δ (PRKCD; 3p) and θ (PRKCQ; 10p15), class III: ε (PRKCE), class IV: τ (PRKCI; Xq21.3), and ζ (PRKCZ). These genes have an ancient origin, their homologues having been found in both nematodes and yeast. The deduced amino acid sequences are highly conserved among mammals (>95% homology between humans and rodents) although homology is lower (45-65%) between nematodes and mammals. The expression pattern of the different proteins can however vary markedly. Structurally, each subfamily is characterized by a different arrangement of regulatory domains which serve to determine signalling specificity.

Serine proteases genes

The serine protease superfamily provide an archetypal example of evolution by gene duplications and divergence coupled with exon shuffling. Thus, mast cell chymase (CMA1; 14q11.2), the digestive pancreatic proteases [trypsin (PRSS1 and PRSS2; 7q35), chymotrypsin (CTRB1; 16q23) and elastase (ELA1; 12q13)] and the hepatic proteases of coagulation (e.g. thrombin; F2; 11p11-q12) have all evolved from the same common ancestral precursor gene. This ancestral gene appears to predate the divergence of eukaryotes and prokaryotes and may itself have been the product of an internal duplication. Enormous structural and functional diversity has been generated both by the addition of non-proteolytic domains and by changes in the enzymes active sites which account for differences in their specificity.

The vitamin K-dependent serine proteases of coagulation exhibit substantial sequence and structural homology and their evolution. Human representative of this serine protease family include factor VII (F7; 13q34), factor IX (F9; Xq27), factor X (F10; 13q34), prothrombin (F2; 11p11-q12), protein C (PROC; 2q13-q14) and protein S (PROS1; 3p11). Various serine proteases have also been recruited to functions in fibrinolysis e.g. plasminogen (PLG; 6q26-q27), tissue-type plasminogen activator (PLAT; 8p11-q12) and urokinase (PLAU; 10q24-qter) and, phylogenetically, these are closely related to the contact factors: factor XI (F11; 4q34), factor XII (F12; 5q33-qter) and plasma kallikrein (KLK3; 4q34-q35).

The division between the proteases of coagulation and fibrinolysis is apparent at the level of exon/intron organization but also in terms of codon usage for the active site serine residue. The genes encoding the vitamin K-dependent factors of coagulation possess an AGY codon whereas the fibrinolytic enzyme genes exhibit a TCN codon also found in the serine protease genes of eubacteria and invertebrates. These alternative codons cannot be interconverted by a single nucleotide substitution. Brenner (1988) proposed that the AGY codon could have been derived from an active cysteine residue encoded by TGY in a cysteine protease that existed billions of years ago. Irwin (1988) has however argued that the AGY codon evolved from a TCN codon on at least two separate occasions, once on the lineage leading to the vitamin K-dependent factors of coagulation and once on the lineage leading to plasminogen and apolipoprotein(a).

Hypervariability of the active site region is apparent in some serine proteases indicating that rapid evolution of the reactive center may have driven the functional divergence of these enzymes. Since a similar phenomenon is apparent for the serine protease inhibitors, it may be that the proteolytic enzymes and their inhibitors have coevolved.

Serpin genes

The serine protease inhibitors (Serpins) are a superfamily of proteins with over 100 members in mammals, and counterparts in invertebrates, plants and even some viruses. Serpins interact with the substrate-binding sites of their cognate proteases via an exposed binding site of canonical conformation. The protease and its inhibitor rapidly form a tightly bound 1:1 stoichiometric complex with the reactive center of the serpin acting as a bait for the appropriate serine protease. Mammalian serpins share a common ancestor some 500 Myrs ago and include α1-proteinase inhibitor (PI; 14q32), α1-antichymotrypsin (AACT; 14q32), heparin cofactor II (HCF2; 22q11), antithrombin (AT3; 1q23), α2-antiplasmin (PLI; 17p13), pigment epithelium-derived factor (PEDF; 17p13), C1-inhibitor (C1I; 11q11-q13), protein C inhibitor (PCI; 14q32), and plasminogen activator inhibitors 1 and 2 (PAI1, 7q21-q22; PAI2, 18q21.3). Other less well characterized mammalian serpins include cytoplasmic antiproteinase 2 (PI8; 18q21.3), maspin (PI5; 18q21.3), nexin (PI7; 2q33-q35), kallistatin (PI4; 14), neuroserpin (PI12; 3q26). Thus, serpin gene clusters are found at 14q32, 17p13 and 18q21.3 in the human genome.

Not all serpins act as protease inhibitors. Thus, angiotensinogen (AGT; 1q41) and corticosteroid-binding globulin (CBG; 14q32) act as blood pressure regulatory hormones whilst ovalbumin functions as an avian egg storage protein. Although there is no ovalbumin gene in the human genome, there are several genes encoding ovalbumin-like serpins (ov-serpins) which are clustered either on 18q21.3 (PI10, SCCA1, SCCA2) or 6p25 (PI2, PI6, PI9). The ov-serpins on chromosome 6 share more amino acid sequence identity with one another than they do with their chromosome 18 counterparts with the exception of PI8. By contrast, most of the chromosome 18 ov-serpins share greater sequence identity with a chromosome 6 ov-serpins than with each other. To account for these observations, Bartuski et al. (1997) proposed that the 6p25 loci arose first and then gave rise to the chromosome 18 cluster.

Dendrograms derived from comparisons of serpin nucleotide and amino acid sequences and of intron positions differ significantly. The exon/intron organization of extant serpin genes is therefore unlikely to be explicable simply in terms of the loss of introns from a large primordial gene. Wright (1993) proposed that the serpin gene family arose from an early recombination event which fused the amino and carboxyl domains. The subsequent insertion of sequence (possibly intronic) served to create β-sheet A and stabilized the new structure. Few of the introns demarcate regions of secondary or tertiary structure and further insertions, deletions and migrations of introns must have occurred.

Substrate specificity manifested by the serpin is determined, at least in part, by the P1 residue of the bait loop which contains Met or Val for elastase, Lys for trypsin, Leu for chymotrypsin and Arg for thrombin. After gene duplication, divergence in terms of substrate specificity has been very rapid and has resulted from 'accelerated evolution' of the reactive center region. This rapid evolution is thought to have taken place by a combination of mechanisms, from genetic drift to gene conversion to positive Darwinian selection.

Zinc finger genes

The term 'zinc finger' refers to a 28 amino acid sequence motif which binds zinc ions thereby stabilizing the structure of small DNA-binding domain. It has been estimated that between 300 and 700 human genes encode zinc finger-containing proteins. Most of these proteins belong to the *Kruppel*-type family and act as transcription factors. Zinc finger proteins often contain multiple zinc finger motifs, the number varying from between 2 and 37.

To date, some 250 human zinc finger genes have been allocated a symbol and some have been chromosomally localized. Inspection of the available mapping data reveals at lest 9 distinct clusters of zinc finger genes in the human genome: 3p21-p22 (ZNF52, ZNF64, ZNF35, ZNF166, ZNF167, ZNF168), 6p21 (ZNF76, ZNF165, ZNF173, ZNF204), 8q24 (ZNF7, ZNF16, ZNF34), 10p11 (ZNF11A, ZNF25, ZNF33A, ZNF37A), 10q11 (ZNF11B, ZNF22, ZNF33B, ZNF37B), 11q23 (ZNF123, ZNF15, ZNF128, ZNF129, ZNF145), 19p12-p13 (ZNF43, ZNF56, ZNF58, ZNF66, ZNF67, ZNF85, ZNF90, ZNF91, ZNF92, ZNF208), 19q13 (ZNF42, ZNF45, ZNF83, ZNF93, ZNF132, ZNF134, ZNF135, ZNF136, ZNF137, ZNF146, ZNF154, ZNF155, ZNF160, ZNF175), 22q11 (ZNF69, ZNF70, ZNF71, ZNF74). This superfamily has presumably evolved through cycles of gene transposition and duplication. Duplications in the ZNF91 family are known to have occurred some 55 Myrs ago in the common ancestor of the simians.

The human ZNF45 and ZNF93 genes are very similar to their murine orthologues although the human genes encode more zinc finger repeats than their murine counterparts, consistent with the occurrence of intragenic deletions/duplications. Similarly, the human transcription factor MOK2 (MOK2; 19q13.2-q13.3) contains 10 zinc finger motifs in comparison to 7 in the murine homologue.

Olfactory receptor genes

The olfactory system is thought to be capable of distinguishing several thousand odorant molecules. This is potentiated by olfactory receptors (ORs) which are responsible for the recognition and G protein-mediated transduction of specific odorant signals. With possibly 1000 members in the human genome, the OR gene superfamily constitutes by far the largest family encoding G protein-coupled receptors. OR genes are intronless and occur in clusters that are present at more than 25 chromosomal locations in the human genome. However, more than 70% of human OR-homologous sequences are probably pseudogenes.

Human OR gene clusters appear to be disproportionately located in subtelomeric regions and have been subject to frequent duplications and inter-chromosomal rearrangements some of which appear to have been mediated by recombination between repetitive sequence elements. OR genes within the clusters belong to at least four different subfamilies which display as much sequence variability within clusters as between clusters. The classification of human OR genes is still in its infancy but chromosomally localized members include OR1A1, 17p13; OR1D2, OR1D4, OR1D5, 17p13; OR1E1, OR1E2, 17p13; OR1F1, 16p13; OR1G1, 17p13; OR2D2, 11p15; OR3A1, OR3A2, OR3A3, 17p13; OR5D3, OR5D4, 11q12; OR5F1, 11q12; OR6A1, 11p15; OR10A1, 11p15.

The olfactory system is combinatorial in that one OR can recognize multiple odorant molecules, and one odorant molecule may be recognized by multiple ORs, whilst different odorants are recognized by different combinations of ORs. Each nasal olfactory sensory neuron expresses only one allele of a single OR gene. In the olfactory epithelium, different sets of ORs are expressed in distinct spatial zones; neurons expressing a given OR gene are located in the same zone but in that zone they are interspersed with neurons expressing other ORs. However, OR genes expressed in the same zone map to numerous different loci whereas a single OR locus may contain genes expressed in different zones. Since some of the OR pseudogenes have been found to be expressed, it is possible that some of those neurons that only express a single receptor type may be nonfunctional.

The origin and emergence of the olfactory receptor gene family probably preceded the divergence of the mammals. In the primates, the olfactory receptor genes appear to have been in a considerable state of flux with numerous translocations, duplications and deletions occurring during the evolution of the great apes. The origin of ORs is unclear but their expression during spermatogenesis and their presence in mature sperm cells suggests that their original function could have been in sperm physiology.

Genes which Undergo Programmed Rearrangement

The immunoglobulin and T-cell receptor genes are members of the immunoglobulin superfamily and share the common property of being assembled from multiple gene coding segments in the lymphocyte. In the germline, the genes encoding the variable portions of these receptors are usually split into variable (V), J (joining) and D (diversity) segments that are joined together by a somatic process of site-specific recombination that involves the recombination-activating proteins RAG1 and RAG2 to generate exons that encode the antigen-binding portion of the polypeptide.

Immunoglobulin genes

A diverse repertoire of human antibodies is generated by the combinatorial somatic rearrangement of a relatively small number of gene segments, variable (V_H), diversity (D) and joining (J_H) segments for the heavy chain V region, and variable (V_L) and joining (J_L) segments for the light chain V region. The J_H and J_L segments are spliced to the constant region genes of the heavy (C_H) and light (C_L) chains following mRNA transcription.

The human immunoglobulin genes are mainly distributed between three chromosomal locations: 14q32 for the heavy chain loci, 2p12 for the k light chain loci and 22q11 for the λ light chain loci. The 1100 kb human heavy chain locus contains 51 functional V_H gene (IGHV) segments, ~30 D segments (IGHDY), 6 functional J_H segments (IGHJ) and 9 functional C_H genes: μ (IGHM), δ (IGHD), γ3 (IGHG3), γ1 (IGHG1), α1 (IGHA1), γ2 (IGHG2), γ4 (IGHG4), ε1 (IGHE) and α2 (IGHA2). The human C_H gene locus clearly exhibits dyadic symmetry as a result of a duplication event that included the Cγ-Cγ-Cε-Cα gene cluster in the ancestor of the great apes. The symmetry is imperfect in extant species because the Cε gene was lost through independent deletion events in humans, chimpanzees, gorilla and the white-handed gibbon (*Hylobates lar*) whereas in the orangutan, the Cα gene was deleted as well as the Cε gene.

The k light chain locus contains 76 V_k genes (IGKV) and pseudogenes, 5 functional J_k segments (IGKJ) and single C_k gene (IGKC). This locus consists of two copies of a DNA region arranged with opposite polarity, the result of an intra-locus duplication which has occurred since the divergence of human and chimpanzee. The λ light chain locus contains 36 functional V_λ genes (IGLV), 33 V_λ pseudogenes, 34, V_λ 'relics' containing large deletions or insertions, 6 J_λ segments (IGLJ) and C_λ genes (IGLC). In addition to the above, 8 V_H segments (IGHV2) and a cluster of D segments (IGHDY2) have been identified on chromosome 15q11.2 with a further 16 V_H segments (IGHV3) on chromosome 16p11.2. These 'orphon' V_H sequences are thought to have been translocated to chromosomes 15 and 16 some 20 Myrs ago and ~40% may be functional. A small cluster of orphon Vk sequences has also been found on chromosome 22, and flanking direct and inverted repeats have been invoked to account for their transposition.

It has been apparent for some time that the heavy and light immunoglobulin chains are homologous and that these proteins must have been assembled by extensive duplication of a short ancestral 100 amino acid polypeptide chain. This domain, the immunoglobulin fold, has now been found in a very wide array of different proteins throughout the animal kingdom. The immunoglobulins of the immune system appear however to be a vertebrate creation.

The immunoglobulin genes are thought to have originated very early in the evolution of the jawed vertebrates. Phylogenetic analyses of the human V_H heavy chain genes, C_H genes, V_λ light chain genes and pseudogenes and Vk light chain genes are consistent with a history of multiple, successive duplication events, both extensive and of more limited extent, during the evolution of all branches of the immunoglobulin gene family. This mode of evolution is reminiscent of a 'birth and death process' rather than any mechanism of concerted evolution. Some immunoglobulin genes have been present in vertebrate genomes for >400 Myrs, others are more recent creations that have emerged during the mammalian radiation. The expansion of the immunoglobulin gene family was almost certainly an adaptive response to the wide range of novel pathogens that challenged early tetrapods in their new terrestrial environment.

The evolution of the immunoglobulin gens is subject to at least two distinct types of constraint: existing gene segments must be conserved in order to meet recurring immunological challenges but diversity must be continually created to meet new immunological challenges. The duplication of gene segments provides a means to achieve this ensuring that old structures/functions are maintained at the same time as new ones are being created.

T-cell receptor genes

The T-cell receptor plays an important role in antigen recognition during the immune response. The four T-cell receptor chains (α, β, γ, and δ), members of the immunoglobulin superfamily, dimerize to form two types of receptor (α/β), γ/δ). In the human genome, these chains are encoded by four genes; TCRA and TCRD (14q11.2), TCRB (7q35), and TCRG (7p14-p15). The location of the TCRD gene is highly unusual in that it lies within the TCRA gene: the multiple gene units are organized as Vα-Vδ-Dδ-Jδ-Cδ-Jα-Cα. Each locus comprises a large number of tandemly arrayed variable (V) genes, diversity (D) segments (TCRB and TCRD), a number of clustered joining (J) elements and one or two constant (C) region genes. These gene segments undergo somatic rearrangement during lymphocyte differentiation to yield either V/J or V/D/J genes. As with the immunoglobulin genes, the origin of the T-cell receptor genes probably occurred very early in the evolution of the jawed vertebrates. The tandem arrays of V genes seen in extant vertebrate genomes have arisen subsequently by serial duplication, generating the sizeable repertoire of T-cell receptor V genes essential for creating diversity of antigen recognition specificity.

The TCRB gene has been the best characterized of the four loci with the sequencing of 685 kb from the region. This region contains 46 variable gene segments, 19 pseudogenes and two clusters of D, J, and C segments. A further 22 additional sequences, termed 'relics' by the authors, were also identified. These sequences exhibited limited homology to V gene segments and represent partial pseudogenes extensively altered by insertions and deletions. Some 30% of the TCRB locus is composed of genome-wide interspersed repeats but these sequences do not appear to have facilitated the duplication of locus-specific repeats. Some of these repeats within the TCRB locus have been shown to harbor trypsinogen genes (PRSS1, PRSS2). These genes must have been conserved for at least 350 Myrs because a trypsinogen gene cluster is present at this location in both mouse and chicken. Higher primates contain similar numbers of V gene segments at the TCRB locus: human (47), gorilla (45), orangutan (57) macaque (57) with the greater number of V genes in orangutan and macaques being due to amplification of the TCRBV7, 9 and 23 subfamilies.

Convergent Evolution

Most members of the gene and protein families have been subject to *divergent evolution*, the process by which homologous proteins or domains develop from a common origin but gradually acquire their own unique identity in terms of structure and function. By contrast, *convergent evolution* refers to the similarity between two protein structures, amino acid sequences or nucleotide sequences due to their independent evolution from different origins rather than through their possession of a common ancestor. Evolutionary convergence therefore implies 'adaptive change in which lesser related entities come to appear more related than they are'. The term convergence has however been used in many different contexts with very different meanings. Doolittle (1994) distinguished between *functional* convergence, *mechanistic* convergence and *structural* convergence.

Functional convergence is exemplified by various pairs of enzymes that have evolved independently to catalyse the same biochemical reactions, for example superoxide dismutases, aldolases, alcohol dehydrogenases and topoisomerases. Myoglobin from the abalone *Sulculus diversicolor* exhibits functional convergence with vertebrate myoglobin although it is not in any way homologous to it. Instead, the *Sulculus* myoglobin gene is evolutionarily related to the human indoleamine, 2,3-dioxygenase (IDO; 8p11-p12) gene. The original *Sulculus* myoglobin gene may have been lost and a modified *Ido* gene could have evolved as a substitute.

Perhaps the best example of mechanistic convergence is provided by the serine proteases subtilisin and chymotrypsin which, although unrelated evolutionarily, have independently evolved similar enzymatic mechanisms. Chymotrypsin has a catalytic triad with a histidine at residue 57, an aspartate at residue 102 and a serine at residue 195. The bacterial protein subtilisin possesses a completely different structure but also has a catalytic triad comprising an aspartate at residue 32, a histidine at residue 64 and a serine at residue 221. The human genome contains representatives of both families. Thus, the chymotrypsinogen B (CTRB1, 16q23), chymotrypsin-like protease (CTRL; 16q22), trypsin 1 (PRSS1; 7q35), trypsin 2 (PRSS2; 7q35), elastase 1 (ELA1; 12q13), plasminogen (PLG; 6q26), and prothrombin (F2; 11p11-q12) genes are members of the chymotrypsin family of serine proteases. Human genes encoding subtilisin-like serine proteases include the proprotein convertases (PCSK1, 5q15-q21; PCSK2, 20p11; PCSK4, chromosome 19; PCSK5, 9q21.3), furin (PACE; 15q25-q26) and paired basic amino acid cleaving enzyme 4 (PACE4; 15q26). Another example of mechanistic convergence is provided by the cold shock domain protein family and the family of RNA-binding proteins that contain an RNA-binding domain. Both of these domains contain conserved ribonucleoprotein motifs on similar single stranded nucleic acid-binding surfaces.

Structural convergence, on the other hand, many reflect the tendency of specific amino acid sequences to fold into certain favoured conformations. Thus, structurally dissimilar families of transport proteins have been found to exhibit similar structural units consisting of six tightly packed α-helices which may comprise all or part of a transmembrane channel.

In none of the above examples is there any evidence for *sequence convergence*. For two sequences to be shown to display convergence in this strict sense, they would not only have to be shown to be evolutionarily unrelated but chance would also have to be excluded as a reason for their similarity. Indeed, sequence convergence would imply the occurrence of adaptive amino acid replacements that have been positively selected during evolution. Although sequence convergence has been invoked in many different situations, there are still no convincing examples of unrelated sequences which adhere to these criteria such that they warrant the use of the term *sequence convergence*.

COEVOLUTION

Coevolution can be said to be operating in cases where the process of evolutionary change experienced at one locus is influenced by changes that have occurred at another locus. At its simplest, coevolution can occur when two genes are intimately associated as in the case of shared bidirectional promoter elements. More complex situations involve unlinked genes. Fryxell (1996) proposed that ' the acquisition of a novel function by a duplicated gene could be facilitated by pre-existing heterogeneity in proteins that interact directly with the product of the duplicated gene'. Thus, the duplication and functional divergence of one gene might serve to create an altered genetic environment that could promote the divergence of duplicate copies of genes encoding proteins that interact with the protein products of the first gene.

In a study of the interspecies diversity manifested by a series of 48 ligand-receptor pairs, Murphy (1993) demonstrated there to be a linear relationship between receptor divergence and ligand divergence. Interestingly, the inter-species differences in receptor structure were nonrandomly distributed and largely confined to the extracellular domains that interact with the ligand. Since this study employed ligand-receptor pairs from diverse systems (host defence proteins, neurotransmitters, hormones, growth factors and cell adhesion proteins), it is reasonable to suppose that the coevolution of genes encoding interacting proteins is not an uncommon phenomenon.

The coevolution of families of receptors and their ligands is perhaps best exemplified by the insulin-neve growth factor family and their receptors. These ligand-receptor pairs include insulin (INS; 11p15.5) and insulin receptor (INSR; 19p13), insulin-like growth factor 1 (IGF1; 12q22-q23) and its receptor (IGF1R; 15q25-qter), brain-derived neurotrophic factor (BDNF; 11p13) and neurotrophin 5 (NTF5; 19q13.3), and their cognate receptor neurotrophic tyrosine kinase receptor type 2 (NTRK2; 9q22.1), neurotrophin 3 (NTF3; 12p13), and neurotrophic tyrosine kinase receptor type 3 (NTRK3; 15q25), nerve growth factor (NGFB; 1p13.1) and neurotrohpic tyrosine kinase receptor type 1 (NTRK1; 1q21-q22). The ligand-encoding gene share a common ancestry as do the genes encoding their receptors. As new trophic factors emerged by duplication and divergence, so their cognate receptors evolved by a similár parallel process. Other examples of the coevolution of ligands and their receptors include interleukin 8 (IL8; 4q13-21) and its receptors (IL8RA, IL8RB; 2q25), interleukin 4 (IL4; 5q23-q31), and interleukin 4 receptor (IL4R; 16p11-p12), the gonadotropins and their receptors and the nuclear receptors and their ligands.

Coevolution occurs between genes encoding different subunits of heterodimeric proteins, for example the *myc* proto-oncogene (MYC; 8q24) and its dimerization partner *max* (MAX; 14q23), the genes encoding the type I and II keratins and the genes encoding the α-and β-integrin chains. Coevolution can also occur between agonists and antagonists, for example interleukins 1α (IL1A; 2q13) and 1β (IL1B; 2q13-q21), and the interleukin 1 receptor antagonist (IL1RN; 2q14.2) which binds to the IL1 receptor and blocks IL1α and IL1β binding without inducing a signal of its own; all three genes share a common ancestry. The cytokines and their receptors have both arisen by a process of gene duplication and divergence and although the relative timing of these duplicative events is still unclear, evidence is emerging for ligand-receptor coevolution.

One prediction of Fryxell's (1996) hypothesis is that duplication/divergence events in functionally related gene families may be temporally correlated. This prediction appears to be bourne out at least for the α- and β-integrin chain genes and the fibroblast growth factor/fibroblast growth factor receptor genes.

Whilst numerous examples of 'ligand promiscuity' have been recognized e.g. the type I interferon genes clustered on 9p21 'receptor promiscuity' appears to be much rather. Examples of receptor promiscuity include the interleukin 8 receptors (IL8RA, IL8RB; 2q35) and the interferon receptors, α, β and ω. 1 (IFNAR1) and 2 (IFNAR2) and interferon receptor γ2 (IFNGR2) whose genes are closely linked to each other on 21q22.1

The coevolution of ligand-receptor pairs by parallel pathways of gene duplication and functional divergence has probably been facilitated in two quite distinct ways. Firstly, the close linkage of multiple genes encoding either ligands or their receptors will have served to promote gene duplication. Secondly, whose genome duplications early on in vertebrate evolution may have, by simultaneously increasing both ligand and ligand receptor diversity, provided the raw material for selection to recruit novel receptor-ligand interactions thereby potentiating the dramatic increase in the biochemical and physiological complexity characteristic of the vertebrates.

6

Regulating Factors

The eukaryotic transcriptional apparatus is much more complex than that of prokaryotes and this complexity is bound up with the fact that many eukaryotic genes are silenced by being tightly packed in chromatin. Transcriptional repression was a necessary pre-condition for the development of large genomes since increasing gene number required gene expression to be restricted both spatially and temporally. That nucleosomal packing and transcriptional repression have, in eukaryotes, long been associated with each other is evidenced by the discovery that the core histone fold is not confined to the histones but has also turned up in a number of transcriptional coactivators and repressors. This ancient association is also evident in the High Mobility Group (HMG) superfamily which may be divided into two distinct subfamilies comprising transcription factors and chromatin structure regulatory proteins, respectively. The evolution of cellular processes probably involved several distinct stages: metabolism and translation preceded transcription whereas transcription must have preceded transcriptional repression by chromosomal packing at the dawn of the eukaryotic era. Genome expansion was potentiated by chromosomal packing whilst gene regulation became ever more complex as new transcription factors emerged. The last universal ancestor probably possessed basic molecular components of metabolism and translation while having a prokaryote-like genome organization together with a transcriptional system reminiscent of the archaea.

Promoters and Enhancers

The basic mechanism of transcriptional activation has been conserved across the spectrum of eukaryotes from yeast to human. It is therefore not surprising tnat two sequence elements, the TATAAA and CCAAT boxes which serve to coordinate the basal transcriptional machinery, are found in all eukaryotes. Indeed, these motifs occur in the promoters of a wide variety of RNA polymerase II-transcribed genes. Similarly, the initiator element, a functional analogue of the TATA box, appears to be ubiquitous in eukaryotes. Other sequence motifs also occur, indifferent combinations and permutations, thereby giving eukaryotic promoters their modular character.

Transcriptional activators bind to promoters sequences in different combinations and permutations, some factors being coordinated by *cis*-acting DNA sequence motifs, others by protein-protein interactions on the promoter. The effect of any given activator is therefore likely to depend on the other activator and repressor proteins present which may bind either competitively or cooperatively. Such a process of 'combinatorial control' has, by maximizing flexibility potentiated the rapid evolution of quite elaborate gene regulatory networks. Those seeking a reference source to cis-acting DNA sequence motifs may consult the Transcription Regulatory Regions Database or the Eukaryotic Promoter Database both of which contain information on regulatory regions of eukaryotic genes and include data on transcription factor binding sites, promoters, enhancers and silences.

If we take the simple case of two genes recently duplicated at the genomic DNA level, then we might reasonably expect that their promoter regions would initially be very similar. As a consequence, the expression profiles of these genes might also be expected to be very similar if not identical. However, gene duplication generates redundancy allowing the promoters to be freed from the constraints of selection thereby enabling them to acquire mutations. Such mutations can lead to the inactivation of the gene by abolishing its expression or alternatively can, in rather more subtle ways, serve to change its expression pattern.

In principle, promoters can change either by the slow steady acquisition of single base-pair substitutions in pre-existing sequence motifs or, more dramatically and abruptly, by *promoter shuffling*, the gain or loss of individual regulatory elements exchanged between genes in cassette fashion. Promoter shuffling may have occurred in cases of homologous genes that contain dissimilar upstream regulatory elements but also in cases of nonhomologous genes containing similar upstream regulatory elements. For promoter shuffling to be viable, the transposed sequence element must be capable of exerting an influence on the transcription of the gene that has just acquired it. That this proposition is a reasonable one is evidenced by the result of studies reported by Kermekchiev et al. (1991). These authors tested 27 combinations of different promoters and enhancers and found that the relative *in vitro* efficiency of the enhancers was roughly the same irrespective of the promoter used. Another way in which promoter sequences can change abruptly is through gene conversion as descried for the human growth hormone (GHI; 17q22-q24) and γ-globin gene promoters.

In subsequent sections the similarities and differences between the promoters of extant mammalian genes will be explored. In addition, some of the mechanisms by which evolution has recruited specific sequences to a promoter function and fine-tuned their interactions with transcription factors will be described.

Evolutionary Conservation of Cis-acting Elements and 'Phylogenetic Footprinting'

Sequence conservation in promoter regions is usually held to be an indicator of functional importance and may therefore be used as a rough guide to the location of cis-acting regulatory elements that might bind trans-acting factors. Such *phylogenetic footprinting* has proven successful in locating *trans*-acting factor binding sites in a considerable number of human genes including the ε-globin (HBE1; 11p15.5) gene, the γ1-globin (HBG1; 11p15.5) gene, the β-globin (HBB; 11p15.5) gene locus control region, the cytochrome c oxidase subunit Vb (COX5B; 2cen-q13) gene, the Duchenne muscular dystrophy (DMD; Xp21) gene, the hormone-sensitive lipase (LIPE; 19q13) gene and the cystic fibrosis transmembrane conductance regulator (CFTR; 7q31.3) gene among others. It was also instrumental in identifying the CArG sequence motif [CC(A/T rich)$_6$GG] important for the regulation of the cardiac α-actin (ACTC; 15q14) gene.

Phylogenetic footprinting has been used to identify regulatory elements in the promoter regions of the rapidly evolving SRY (Yp11.3) genes of 10 mammalian species. A total of 10 putative regulatory elements were identified by these means. Interestingly, differences were apparent between the SRY promoters not only in terms of the presence/absence of specific motifs and motif copy number, but also in the spacing between motifs, their relative location and orientation (sense or antisense strand). Conserved elements in the SRY gene promoters within each taxonomic group (primates, bovids, rodents) did however tend to occupy orthologous positions.

The study of Vuillaumier et al. (1997) identified eleven DNase I-hypersensitive phylogenetic footprints in a 3.5 kb region of the CFTR

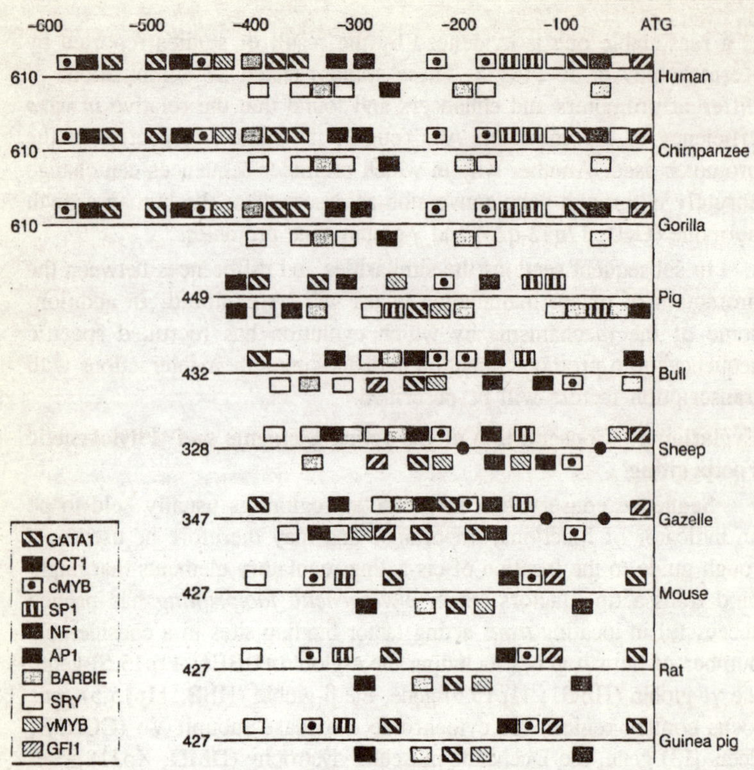

Fig. 6.1. Comparison of the promoter sequences from the SRY genes of ten mammals.

gene (7q31.3) promoter when eight mammals from four different orders were compared. Two of these footprints, which corresponded to the cAMP response element and the PMA-responsive element, were conserved in all species examined, a finding which probably reflects the vital importance of transcriptional control through the cAMP and diacylglycerol pathways respectively. By contrast, a 300 bp purine pyrimidine stretch, thought to represent a negative element of basal transcription, was found to be present only in the *Cftr* genes or rodents. Studies of the globin genes have also shown that some motifs may be of functional importance in one species but not be conserved between species, for example two Spl-binding sites in the human HBB gene locus control region.

A variation on this theme is *differential phylogenetic footprinting* which utilizes the promoter sequences of two extant species representing the most closely related lineages between which a difference in developmental expression pattern can be detected. An example of its successful use was in the study of the galago and human γ-globin (HBG1) gene promoters. Differential phylogenetic footprinting has also been used to study the promoters of the apolipoprotein AI (APOA1; 11q23.3) genes of human and African green monkey in order to assess the functional significance of species-specific differences in expression. Sorci-Thomas and Kearns (1995) identified seven sites within the proximal promoter region of the APOA1 gene which differed between human and African green monkey. These authors then tested the functional significance of these sequence changes by mutating the human promoter to the simian sequence and testing promoter strength by reporter gene expression assay. Substitutions at three of the sites (–189, –144, and –48 relative to the transcriptional initiation site) were found individually to increase the activity of the wild-type human promoter to ~60-65% of that of the African green monkey. In addition, two double mutations (–144/–48 and –189/–144) restored promoter activity to the same level as found in the monkey. Thus, we may therefore infer that several substitutions in the APOA1 gene must have occurred during primate evolution which together served to determine the specific level of APOA1 gene transcription in the different species.

Even if promoter elements are conserved, their relative location may not be. Thus the presence of four perhaps five DNase I hypersensitive sites in the β-globin gene locus control region is conserved between placental mammals, marsupials and monotremes but the spacing between the elements differs between species.

Nonhomologous Genes Containing Similar Regulatory Elements

Some *cis*-acting elements are common to a wide variety of different genes and many be evolutionarily very ancient. Thus, the TGTGACGTCTTTCAGA cAMP-responsive element in the promoter of the human vasoactive intestinal polypeptide (VIP; 6q26-q27) gene is similar not only to elements found in the promoters of nonorthologous murine and avian genes but also to sequences described in yeast and adenoviral promoters and to the *E. coli* consensus sequence recognized by the cAMP receptor protein.

Even if we confine ourselves to the same species, completely unrelated genes often also possess similar upstream regulatory elements. Thus, the promoters of the human E-selectin (SELE; 1q22-q25) and β-interferon (IFNB1; 9p22) genes both contain NFkB, ATF-2 and HMG I(Y) binding sites. Other examples of promoter similarity include the serum response elements present in the regulatory regions of the human apolipoprotein E (APOE; 19q13.2) *c-fos* (FOS; 14q24.3) and β-actin (ACTB; 7p12-p15) genes or the AGGCGGCCCTTT motif in the apolipoprotein B (APOB; 2p24), apolipoprotein CIII (APOC3; 11q23-qter) and α1-antitrypsin (PI; 14q32.1) genes. It is as yet unclear if these elements have evolved by the slow fine tuning of existing sequences or if they have instead been introduced by promoter shuffling.

Paralogous Genes Containing Dissimilar Regulatory Elements

Evolutionarily related genes in the same organism often differ in their expression profiles. Thus, the various paralogous members of the family of murine cytochrome P450-16 (testosterone 16-α-hydroxylase) genes are evolutionarily related yet are regulated quite differently. Similarly, the chromosomally unlinked human annexin I (ANX1; 9q11-q22), VI (ANX6; 5q32-q34) and VII (ANX7; 10q21.1-q21.2) genes are evolutionarily related and encode proteins that are structurally and biochemically very similar. However, these genes possess very different set of promoter and enhancer elements which are presumably responsible for their distinct patterns of tissue expression. Can we relate differences in expression profile of paralogous genes to changes in upstream regulatory elements?

A number of studies of paralogous gene promoters have been performed and differences in promoter sequence between members of the same multigene family have indeed been found that may account for differences in transcriptional efficiency. Some paralogous promoters differ with respect to the presence or absence of specific regulatory elements. Others differ in terms of more subtle single base-pair substitutions introduced into specific *cis*-acting motifs which serve to increase or decrease the binding affinity of their cognate transcription factors. Such changes in promoter structure have contributed to post-duplicational gene diversification by providing the divergent gene products with their own specific expression profiles. Steroid hormone receptor responsive elements are a classic example of this but a number of other studies of paralogous gene promoters have been reported which serve to illustrate the basic principles.

The murine immunoglobulin variable region heavy chain (V_H) genes may be grouped into 15 families on the basis of coding sequence homology, and the promoter sequences are conserved in a family-specific manner. However, these paralogous V_H gene promoters vary in terms of their transcriptional strength, at least *in vitro*, by as much as 60 to 70-fold. This variation appears to be due to several specific sequence differences between these promoters: (i) the presence/absence of a

TATA box, (ii) the presence/absence of initiator elements, (iii) the extent of divergence of the octamer sequence element from its consensus (ATGCAAAT), and (iv) the spacing between the octamer motif and the heptamer motif located between 2 bp and 20 bp upstream.

The human genome contains a cluster of 11 pregnancy-specific glycoprotein (PSG) genes on the long arm of chromosome 19q (PSG1, PSG2, PSG3, PSG4, PSG5, PSG6, PSG7, PSG8, PSG11, PSG12, PSG13). These genes are highly homologous and this homology extends to their promoter regions. The promoters of six PSG genes have been characterized and display differences based upon the minimal promoter required for optimal expression. The class 1 PSG genes require nucleotides −172 to −34 (relative to the transcriptional initiation site) whereas the comparable minimal region of class 2 PSG genes lies between −172 and −80. This difference has been attributed to the presence of an imperfect SP1 binding site between −148 and −141 in the class 1 promotes. The perfect SP1 element (CCCCGCCC) present in the class 2 promoters is altered either to CCCTGCCC or CCCCACC (N.B. both represent mutations in a CpG dinucleotide compatible with the mechanism of methylation-mediated deamination) in the class 1 promoters. The SP1 binding sites of the class 1 promoters bind SP1 with lower affinity and require additional activator element for optimal expression.

One of the best characterized differences between paralogous gene promoters is that manifested by one of the human small proline-rich protein genes. The SPRR1A, SPRR1B, SPRR2A, and SPRR2C genes possess an Ets binding site at position −55 (relative to the transcriptional initiation sites) which is critical for promoter activity. The corresponding site in the SPRR3 gene has however been lost through a C→T transition, although this loss appears to be compensated for, at least in part, by the presence of an Ets binding site at position −239. This single base-pair substitution has been shown to account for the lower rate of expression of the SPRR3 gene in cultured keratinocytes as compared with the SPRR1A and SPRR2A genes. This is despite the fact that the −239 Ets binding site in the SPRR3 gene has a higher affinity for its cognate transcription factor ESE-1 than the −55 sites in the SPRR1A and SPRR2A gene promoters. Fischer et al. (1999) found that removal of the −239 Ets binding site coupled with the restoration of the −55 binding site by a T→C mutation yielded a promoter activity comparable to that of the SPRR1A and SPRR2A promoters and three-fold higher than that of the wild-type SPRR3 promoter. The location of the Ets binding site in the SPRR3 gene therefore appears to be critical for determining the activation potential of ESE-1. In the SPRR3 promoter, three transcription factor binding sites (AP-1, ATF, and Ets) function cooperatively but these sites are not interdependent as in the SORR2A promoter. Thus, the loss of the proximal Ets binding site in the SPRR3 gene promoter during the evolution of the gene family may have been responsible for changing the promoter from one which required highly synergistic interactions between its cognate transcription factors to one which functions with less stringent, cooperative interactions.

Although the diversity of glycoproteins encoded by the HLA-B locus is greater than hat encoded by the HLA-A locus, the HLA-B locus promoter sequences are more homogeneous than those of the HLA-A locus. We may speculate that selection for diversity of B glycoproteins may be directly related to the conservation of the promoter sequences and *vice versa* for the A glycoproteins and their gene promoters. Although the human HLA-A (6p21.3) and HLA-B genes are coordinately expressed, they differ in their responsiveness to interferon. A functional interferon-responsive element (IRE) is present in the promoters of the HLA-B genes in the great apes but this IRE has been inactivated by an A→T transversion in the promoters of the HLA-A genes. Since this lesion is present in the HLA-A genes of orangutan, gorilla, chimpanzee and human, it is likely to have occurred before the divergence of the great apes.

The chorionic gonadotropin β gene (CGB; 19q13.32) has evolved by duplication of an ancestral luteinizing hormone β gene and is expressed exclusively in the placenta. It exhibits some 94% homology with the pituitary expressed luteinizing hormone-β (LHB; 19q13.32) gene between coding regions and 90% homology between promotes. Both LHB and CGB genes possess TATAAA boxes at identical locations. This motif directs the initiation of an LHB transcript with a 9 nucleotide 5' UTR. The CGB gene does not however use this box for transcriptional initiation but instead employs a TATA-less promoter (with putative initiator element) that serves to initiate transcription 357 bp upstream of the LHB transcriptional initiation site. This difference in transcription site usage is evolutionarily conserved, being found in rat, cow and human. The experimental exchange of motifs between the two promoters has allowed the identification of three distinct regions between −362 and +104 necessary for placental expression of the CGB gene. It would appear as if the placental expression pattern of this gene has resulted from the use of an alternative transcriptional initiation site in conjunction with the acquisition of multiple regulatory elements that exert their effects in combinatorial fashion.

The paralogous members of the human growth hormone gene family on chromosome 17q23 differ in terms of their promoter sequences : a T→C transition at position –112 of the CSH1 gene promoter serves to reduce binding of the pituitary-specific transcription factor Pit-1. By contrast, the paralogous GH1, GH2, and CSH2 genes possess a T at this location within the Pit-1 binding is thought to represent a negative control mechanism by virtue of its interference with the binding of SP1 to an adjacent site in the promoter. The T→C transition in the CSH1 gene may thus abrogate this negative control mechanism since it facilitates SP1 binding and promoter gene activation by SP1.

Several other examples of promoter differences in paralogous genes are known. Thus the human γB (CRYGB; 2q33-q35) and γC (CRYGC; 2q33-q35) crystalline genes possess similarly located TATA boxes but the CRYGC gene lacks the CCAAT box present in the CRYGB gene. The expression of the human main-type histone H1 (H1F1; 6p21.3) genes is coordinated with DNA replication whereas the regulation of the replacement H1 subtype H1° (H1F0; 22q13) gene is more complex. This difference appears to be reflected in the structures of the respective gene promotes with a CCAAT box present in the main-type histone H1 gene promotes being absent at the H1° gene promoter. Similarly, the promoters of the human mammaglobin 1 (MGB1) and 2 (MGB2) genes, which are closely linked to the related uteroglobin (UGB) gene on chromosome 11q13, are homologous for the first 132 bp but then exhibit major differences that are probably responsible for the different expression patters of these genes. Finally, the highly homologous human placental (ALPP; 2q37) and intestinal (ALPI; 2q37) alkaline phosphatase genes possess several nucleotide substitutions and deletions in their 5' flanking regions which could account for their differing tissue specificity.

Orthologous Genes Containing Dissimilar Regulatory Elements

Differences between orthologous gene promoter sequences have also been described which may explain differences in expression of the same gene in different species. In some cases, the orthologous promotes differ with respect to the presence or absence of specific *cis*-acting sequences or alternatively in terms of the number of such motifs. In other cases, single base-pair substitutions have been introduced that alter the affinity of the *cis*-acting sequences for their cognate transcription factors. Orthologous genes may therefore acquire different expression profiles in different species. Thus, the expression of the human G protein-coupled receptor 1 (GPR1) gene (15q21.6) which is hippocampus-specific contrasts with that of its rat counterpart which is not expressed in the hippocampus. The human regulatory myosin light chain (MYL5; 4p16.3) gene is expressed in human adult retina and fetal muscle but not in adult skeletal muscle suggesting that this gene is developmentally regulated. By contrast, the orthologue of this gene is abundantly expressed in the adult skeletal muscle of the African green monkey, providing a good example of a gene that is differentially expressed between humans and another primate. The tissue-specific expression of many genes is known to be determined by the presence of enhancer elements that bind tissue-specific transcription factors. The evolution of novel tissue specificities can therefore be investigated by comparison of the sequences of gene regulatory elements between different species.

The human apolipoprotein A1 (APOA1; 11q23) gene is expressed predominantly in the liver whilst its rabbit counterpart is expressed predcminantly in the intestine. The rabbit *Apoa1* gene promoter contains two *cis*-acting elements, E2 (GGAGAAGAGAGGTCA) and E3 (GAAAGTCTCTCTTCTGTT) both similar to enhancer sequences found in murine immunoglobulin genes, that are absent from the human APOA1 gene promoter. These sequences may help to explain the differing tissue specificity of this gene between the two species.

The rate tissue-type plasminogen activator gene differs form that of mouse and human (PLAT; 8p12) in that it is induced by gonadotropins via a cAMP-dependent pathway. The rat *Plat* gene promoter possesses a cAMP-responsive element (CRE) but, at the corresponding location in the murine and human gene promoters, a single base-pair change is present that serves to reduce drastically the binding affinity to CRE-binding protein. This difference is sufficient to account for the unresponsiveness of the human and murine gene promoters to forskolin and follicle-stimulating hormone.

The promoters of the murine and human lactoferrin (LTF; 3p21.2-p21.3) genes have numerous *cis*-acting regulatory motifs in common. However, in the bovine lactoferrin gene, several of these motifs (GATA-1, Oct-1, COUP, AP-2, and glucocorticoid and the acute phase response elements) are absent, providing a possible explanation of the relatively weak expression of bovine lactoferrin in comparison to human an mouse.

Both tumor necrosis factor α and lipopolysaccharide stimulate the expression of P-selection in murine endothelial cells but this does not occur in human endothelial cells. This difference is thought to be explicable in terms of structural differences in the promoters of the P-selectin genes between the two species. The promoter of the human P-selectin (SELP; 1q23-q25) gene possesses a unique NFkB binding site which is thought to be specific for p50 or p52 homodimers. Its murine counterpart contains two tandem NFkB binding sites and a variant activating transcription factor/cAMP response element which is similar to that found in the E-selectin gene required for tumor necrosis factor α and lipopolysaccharide-inducible expression.

The promoters of the human and murine XRCC5 DNA repair genes (2q35) differ with respect to the number of copies of a 21 bp near-perfect palindromic element (Kpb); the mouse *Xrc*5 gene possesses a single copy whereas the human XRCC5 gene possesses seven copies. Amplification of this *cis*-acting element occurred in the human lineage and this may help to account for the human XRCC5 promoter being at least three to five-fold stronger than its murine counterpart.

Calbindin-D_{9K} is a mammalian-specific cytosolic calcium-binding protein which is encoded by a gene (CALB3; Xp) that is ubiquitously expressed in the intestine. In the rat, this gene is also expressed in the uterus under the control of estrogen whose effects are mediated by an estrogen response element (ERE). By contrast, uterine expression is not found in human or baboon. In these primates, two nucleotide substitutions are present that have abolished the binding affinity of the ERE to the estrogen receptor and probably account for the loss of uterine expression.

The chorionic gonadotropin α-chain gene is expressed in the pituitary of all mammals but is expressed in the placenta only in primates and horses. Since horses and primates are only distantly related, and species such as the cow and rodents which are more closely related to horses than primates do not express the gene in placenta, it would appear as if the placental expression of the gene has been independently derived, perhaps by the *convergent evolution* of placenta-specific enhancer elements. Support for this postulate has come from a comparison of human and equine chorionic gonadotropin α-chain (CGA; 6q21.1-q23) gene promoter sequences. The human gene contains a trophoblast-specific element (TSE) and two copies of a cAMP response element (CRE) all of which are required for full placental expression. Both the CRE and TSE are bound by the leucine zipper-containing protein, CREB. By contrast, cAMP regulation of the equine

Cga gene is not mediated by CREB but instead by aACT, a GATA-related protein that binds to the promoter at a site distinct from the CREB-binding site. It would thus appear as if the conversion of the CGA gene from being pituitary-specific to being also placentally expressed was a consequence of independent evolutionary processes in primates and horses.

In the New World monkey, *Cebus apella*, the γ2-globin (HBG2) gene is expressed at a 20-fold higher level than the closely linked γ1-globin (HBG1) gene. The most obvious difference between the promoters of the two genes and the one most likely to account for the difference in expression, is a CCAAC motif instead of the canonical CCAAT found in the HBG1 promoter. Intriguingly, the HBG1 gene has been inactivated by deletions in the Ateliade whereas in the Pitheciini, the CCAAT box has been changed to CCGAT. In the New World monkey *Aotus azarae*, a single hybrid HBG gene has been created as a result of an unequal crossing over event; this gene possesses the promoter and 5'UTR of the HBG1 gene coupled to the coding region of the HBG2 gene. Why the HBG1 gene has been inactivated several times independently in platyrrhines in unclear; expression from both HBG1 and HBG2 genes has been preserved in catarrhines.

The human prolactin (PRL; 6p22) gene possesses two promoters, the proximal one responsible for directing expression in the anterior pituitary, the distal one for expression in the decidualized endometrium and the mammary gland. Sequences homologous to both promoters are present in the rat *Prl* gene promoter but the distal sequence is nonfunctional. Whether the human PRL gene has gained a functional promoter in the last 100 Myrs since the divergence of our common ancestors, or whether the distal promoter in the rat gene has ceased to function during this time, is unclear.

Other examples of inter-specific differences in orthologous promoter regions include the thymidine kinase 1 (TK1; 17q25.2-q25.3) gene promoter which contains functional CCAAT elements in humans, chickens and Chinese hamsters but not in mice and rats and the osteonectin (SPARC; 5q31-q32) gene promoter whose GGA-box sequences contribute to cell-type specific expression in human but not in the bovine. The mammalian SPARC gene also differs from its *Xenopus* counterpart in that the latter contains a TATA box but lacks a CGA-box.

Various species-specific sequence differences have also been reported in nuclear factor binding sites in the promoter regions of the

D7Rp2e genes in *Mus domesticus* and *M. pahari* that alter both the pattern of binding site occupancy and the ability of the bound factors to repress transcription.

Various examples of the recruitment of metabolic enzymes to a novel lens-specific role have been documented among the taxon-specific crystallins. Recruitment of these genes to a new lens function has been achieved by acquiring the potential for lens expression by the development of novel promoter elements. Two strategies appear to have been employed viz. the modification of (i) distinct regions of the same non-functional intronic sequence to perform a role in lens-specific expression and (ii) pre-existing promoters previously utilized for nonlens tissue expression.

Not surprisingly, some differences between orthologous promoters are apparently neutral and do not obviously affect promoter function. One example of this is the 54 bp insert in the promoter of the human liver arginase (ARG1; 6q22.3-q23.1) gene which is absent form the same gene in macaques.

Not all regulatory sequences occur upstream of the transcriptional initiation site. Some occur in introns but such regulatory elements are not necessarily conserved between orthologous genes. Thus, the 83 bp intron 1 of the human CD68 (17p13) gene contains a macrophage-specific enhancer but the equivalent intron of the orthologous murine (macrosialin; Ms) gene does not despite ~80% nucleotide sequence homology.

Bidirectional Promoters

There is now a growing list of divergently transcribed gene pairs arranged in head-to-head fashion separated by a directional promoter. Such an organization has often but not always arisen through a process of gene duplication followed by inversion. In true cases of a bidirectional promoter, the gene promoters overlap and contain common elements that can allow the coordinate regulation of expression of both genes. Bidirectional promoters are often associated with CpG islands which therefore serve as useful markers for their location.

Examples of bidirectional promoters are thought to include the human histone H2A and H2B (1q21-q23) genes, type IV collagen (COL4A1 and COL4A2, 13q34; COL4A3 and COL4A4, 2q36-q37; COL4A5 and COL4A6, Xq22) genes, and the TAP1 and LMP2 (6p21.3) genes. These gene pairs encode proteins that are involved in the same biological processes whether as components of multi-chain proteins (collagens) or protein complexes (histones) or as proteins with associated functions (both TAP1 and LMP2 have a role in antigen processing). This is no coincidence since evolutionarily related pairs of genes that have arisen by a process of gene duplication/ inversion would be predicted to have similar functions. As a consequence of their mode of creation, they will often possess the potential for coordinate regulation as long as the promoter region(s) are still intact and the genes remain closely linked. In principle, the newly created functional redundancy of the promoter elements can be reduced by element removal and subsequent element sharing while still retaining the potential for coordinate regulation. Alternatively, retention of the functional redundancy could allow diversification of promoter elements leading to independent regulation.

There are examples of divergently transcribed gene pairs that are not obviously related evolutionarily and which are unlikely to have arisen by gene duplication and inversion. Thus the human phosphoribosyl-amino-imidazole carboxylase (PAICS) and phosphoribosylpyrophosphate amidotransferase (PPAT) genes, which encode enzyme of the *de novo* purine biosynthesis pathway, are only 229 bp apart on chromosome 4q12 but encode products that are not structurally related. Similarly, the human

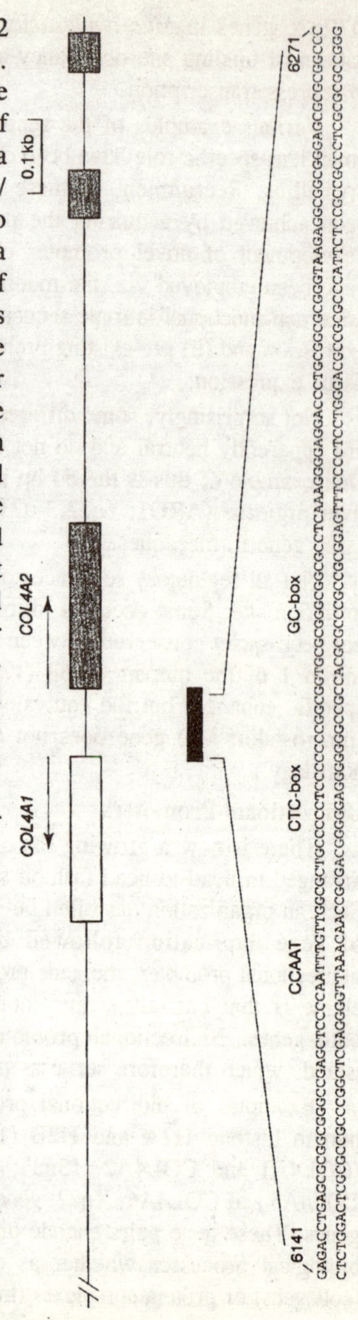

Fig. 6.2. Bidirectional promoter separating the human collagen a1(IV) and a2(IV) gene (COL4A1 and COL4A2).

isocitrate dehydrogenase 3 (IDH3G) and signal sequence receptor 8 (SSR4) genes on chromosome Xq28 are separated by a 133 bp sequence of CpG island character, share overlapping promoter elements but encode proteins which have no obvious common function or evolutionary history. Finally, some genes may be transcribed in opposite directions but do not necessarily share promoter elements nor encode proteins which share any obvious function, for example the human mini-chromosome maintenance 4 (MCM4) and DNA-activated protein kinase (PRKDC) genes which are separated by ~700 bp on chromosome 8q11.

5' and 3' Untranslated Regions of Genes

The 5' and 3' untranslated regions (UTRs) of vertebrate genes are often evolutionarily conserved but to a lesser extent than their corresponding coding sequences. Thus, whereas the average degree of nucleotide sequence identity shared between the coding regions of human and murine genes is ~85%, the 5' and 3' UTRs exhibit sequence identities of 67% and 69% respectively. In practical terms, this degree of sequence divergence is sufficient to allow the discrimination of different mammals simply by reference to their UTR sequences.

Within the 5' and 3' UTRs, *highly conserved regions* (HCRs) can be found. These were defined by Duret et al. (1993) as sequences of at least 100 bp that exhibit >70% homology between species, and which diverged more than 300 Myrs ago. Since such sequences would be expected, in the absence of selective pressure, to share only ~30% similarity, evolutionary conservation implies function. Comparison of mammalian and avian gens reveals that ~17% and ~30% of genes contain HCRs in their 5' UTRs and 3' UTRs respectively. When mammalian and fish genes are compared, the proportion of genes whose 3' UTRs possess an HCR falls to 5%. Since HCRs occur relatively infrequently within introns, the evolutionary constraints would appear to operate at the level of the mature mRNA. The HCRs in the 5' UTRs of the creatine kinase B, c-*jun* and actin genes span the CCAAT and TATA boxes and are therefore likely to play a role in transforming growth factor β3 and ferritin heavy chain genes span elements known to be involved in translation. Some HCRs in 3' UTRS are thought to play a role in mRNA degradation (e.g. c-*fos* and transferrin receptor genes) whereas others may be important for mRNA transport and translation. HCR in 3' UTRs appear to be preferentially associated with widely expressed genes especially those encoding DNA-binding proteins and cytoskeletal proteins. The human surfeit 2 and 4 (SURF2, SURF4; 9q34.1) genes differ from their murine counterparts in that whilst the mouse genes overlap by 133 bp in their 3' UTRs, the human genes are separated by 302 bp. The human SURF2 gene contains two alternative polyadenylation sites resulting in short 3' UTRS of 17 and 25 bp whereas the mouse *Surf2* gene contains a 359 bp 3' UTR. The much shorter human SURF2 3' UTR probably accounts for the absence of overlap with the human SURF4 gene.

One example of the emergence of a functional difference between the 3' UTRs of two paralogous human genes is provided by the α-globin (HBA1; 16p13.3) and ζ-globin (HBZ; 16p13.3) genes. The HBA1 and HBZ genes are coexpressed in the embryonic yolk sac. A switch to exclusive expression of the HBA1 gene in the fetus and adult involves the developmental silencing of the HBZ gene. Silencing is achieved both by transcriptional control but also through a post-transcriptional mechanism that serves to reduce the relative stability of HBZ mRNA. The HBA1 and HBZ genes both assemble an mRNP stability-determining complex on their 3' UTRs but these complexes form with different affinities on the two genes. The diminished efficiency of complex assembly on the HBZ 3' UTR results from a C→G transversion in a polypyrimidine tract that is common to both genes. This substitution is associated with a shortened poly(A) tail on the HBZ mRNA that may mediate accelerated HBZ mRNA decay.

Another example of afunctional difference between the 3' UTRs of evolutionarily related human genes is provided by the human alcohol dehydrogenase (ADH2; 4q22) gene which differs from the other paralogous ADH family members by virtue of a T→C transition within a canonical polyadenylation site 3' to the gene. This substitution appears to be at least in part responsible for the use of alternative polyadenylation sites leading to the formation of multiple ADH2 mRNAs.

Alterations in the 5' UTR may also influence the expression pathway. For example, the efficiency of mRNA translation of the renal ornithine decarboxylase (*Odc*) gene is significantly lower in *Mus pahari* as compared to *Mus domesticus*. This is thought to be due to the acquisition of several single nucleotide substitutions and a 12 bp deletion in the 5' UTR of the *Odc* gene in *M. pahari*. These sequence changes are predicted to alter the secondary structure of the mRNA molecule and this may influence translation efficiency.

Inter-specific Differences in Promoter Selection

Three distinct mRNA species are generated from the human aldolase A (ALDOA; 16q22-q24) gene via the differential incorporation of three exons encoding the 5' UTR. The production of the three

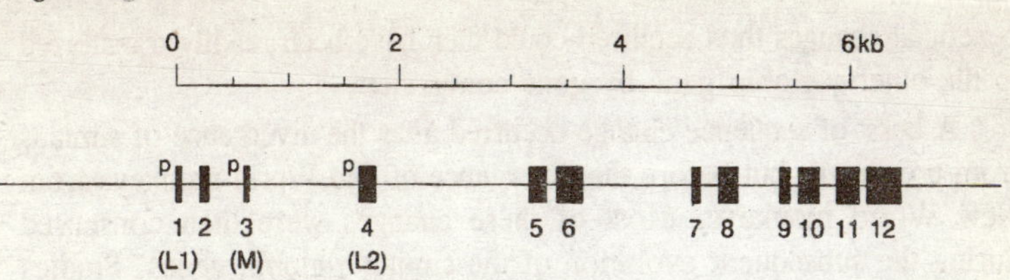

Fig. 6.3. Exon/intron distribution in the human aldolase A (ALDOA) gene.

mRNAs is controlled by three different promoters which are utilized singly, doubly or all together depending upon the expressing tissue. The DNA sequences corresponding to these promoters are present in the rat gene but the L1 promoter is not utilized. Thus the L1 promoter has either been acquired in the human lineage or, (perhaps more likely) lost in the rat lineage since the divergence of primates and rodents.

The vast majority of studies that have mapped transcriptional initiation sites have been performed on cultured cells and it is by no means clear that those sites identified by in vitro studies are actually always utilized *in vivo*. White et al. (1998) identified the *in vivo* transcriptional initiation sites of the cystic fibrosis transmembrane conductance regulator (CFTR; 7q31.3) gene in both human and mouse. Tissue-specific variation in the position of the transcriptional initiation sites was noted in both species but the sites were not conserved between equivalent tissues. This finding suggests that the precise mechanism of transcriptional initiation for a given gene may not be absolutely conserved between species.

Developmental Changes in Gene Expression

In higher primates (both platyrrhine and catarrhine), the γ-globin (HBG1, HBG2; 11p15.5) genes are expressed during fetal life whereas in nonprimates and prosimians, the genes are expressed in the embryo. The conversion of the HBG1 and HBG2 genes to a fetal pattern of expression must therefore have occurred after the divergence of simians from prosimians some 55 Myrs ago. Implicit in this conversion is the activation of the γ-globin gene in fetal life (a stage at which it was previously repressed) and repression of the γ-globin gene in embryonic life at which stage it was previously active. Prosimians are characterized by the possession of only one γ-globin gene whereas higher primates possess two copies. The promoter of one of the duplicated γ-globin genes may thus have been able to escape the influence of natural selection and in so doing accumulate mutations that served to alter the timing of its developmental expression. Any beneficial changes thus acquired could then have been readily transferred to the other γ-globin gene by gene conversion.

A burst of sequence change occurred after the divergence of simians from prosimians but before the divergence of Old World monkeys from New World monkeys. Most of these changes were then conserved during the subsequent evolution of the simian γ-globin genes. Studies of the γ-globin gene promoter in transgenic mice have shown that all the sequence changes necessary for changing to the fetal pattern of gene expression are located within a 4 kb fragment containing the γ-globin gene. However, within 260 bp of the transcriptional initiation site, there are 19 nucleotide changes specific to simians, 16 of which are located in or near highly conserved sequence motifs. Further, comparison of the human and galago (a prosimian) sequences reveals 57 nucleotide differences over the same region with the majority again being located near highly conserved sequence motifs. It may be some time before the precise sequence changes responsible for the change to fetal expression are unequivocally determined.

One way of approaching this question has been through the use of 'differential phylogenetic footprinting' and gel retardation analysis. By these means, Gumucio et al. (1994) identified several proteins (G1, G2, G3 and G4) that bound the galago sequence but did not bind the corresponding human sequence. Phylogenetic reconstruction and gel retardation analysis were used to demonstrate that the promoter sequence of the embryonically expressed γ-globin gene of the primate common ancestor would have bound proteins G1 and G2 some 4-6-fold more strongly than the promoter of the fetally expressed simian ancestor. The binding strength of these proteins correlated with repression of promoter activity *in vitro*, suggesting that the loss of the binding sites for these proteins in the ancestral simian γ-globin gene could have potentiated the conversion of this gene to a fetal onset of expression.

Promoter Polymorphisms

Promoter polymorphisms affecting the expression of the downstream gene are probably not infrequent. However, as yet, relatively few examples of promoter polymorphisms in human genes have been properly characterized by means of functional (e.g. reporter gene) studies. Examples of such polymorphisms include plasminogen activator inhibitor type 1 (PAI1, 7q21.3-q22.1), tumor necrosis factor α (TNF; 6p21.3), apolipoprotein AI (APOA1, 11q23.3), lipoprotein(a) (LPA, 6q27), lipoprotein lipase (LPL; 8p22), interleukin 6 (IL6, 7p21-p15), factor VII (F7; 13q34), hormone-sensitive lipase (LIPE; 19q13) and monoamine oxidase A (MAOA, Xp11.23). The presence of polymorphisms in gene promoter regions is not unusual *per se* since all gene regions harbor polymorphisms. Indeed, such variants are quite consistent with, and explicable in terms of, a neutralist model. However, it is possible that those polymorphisms in the promoter region which specifically affect gene expression confer, or have conferred, a selective advantage.

The PAI1 promoter polymorphism constitutes the insertion or deletion of a single G residue at position -675. The *ins* allele contains an interleukin 1- responsive element which is not present in the *del* allele suggesting that individuals homozygous for the *del* allele may exhibit an altered PAI-1 response during the acute phase reaction. Similarly, a common insertion polymorphism (G) at position -1607 in the human matrix metalloproteinase-1 (MMPI; 11q22-23) gene promoter creates a binding site for members of the Ets family of transcription factors which results in the increased transcription of the gene.

A highly unusual 76 bp length polymorphism (f = ~0.20/0.80) in the human antithrombin III (AT3; 1p31.3-qter) gene promoter results form the alternative presence of two apparently distinct sequences of 108 bp (L) and 32 bp (S) at the same position, ~345 bp upstream of the ATG translational initiation codon. How did this unusual polymorphism evolve and become established in the general population? Winter et al (1995) noted that the sequence flanking the L allele contained numerous homologous motifs suggestive of an ancient duplication event. Some residual homology was also noted between the L and S alleles but the L allele did not obviously contain duplicated sequence. The most parsimonious explanation was therefore the emergence of the S allele from the L allele by partial deletion followed by sequence divergence. The deletion event(s) could have occurred by homologous recombination mediated by the homologies between the L-specific sequence and the region immediately downstream. Interestingly, however, no difference in expression could be discerned between the two alternative alleles.

A repeat length polymorphism in the human solute carrier family member SLC6A4 (17q11.1-q12) gene also appears to be present in the gorilla although not in the chimpanzee. It is unclear whether this is an example of an ancient trans-species polymorphism or whether it has arisen independently in the two lineages.

Promoter Duplication

Promoters may occasionally be duplicated in the absence of the duplication of their associated genes. Thus the BRCA1 (17q21) and

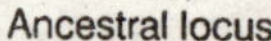

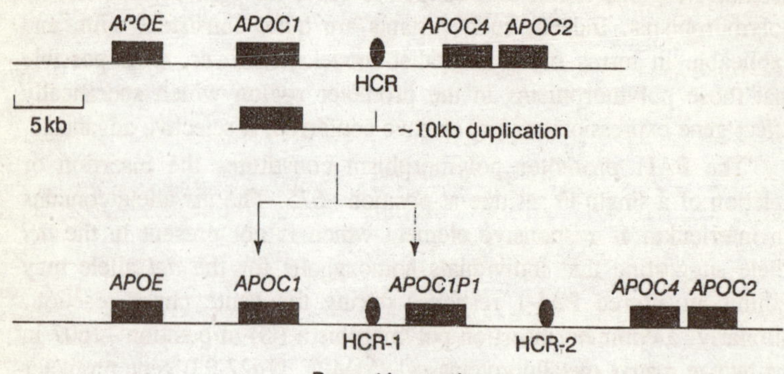

Fig. 6.4. Evolution of the human apolipoprotein E2/CI/CIV/CII loci.

1A1.3b (M17S2; 17q21.1) gene promoters are duplicated in tandem fashion but the functional significance of this observation is unknown.

Two copies of a motif known as the hepatic control region (HCR), located, respectively, 15 kb and 5 kb downstream of the human APOE and APOC1 genes on chromosome 19q13.2, arose by a regional duplication event which encompassed the APOC1 gene ~40 Myrs ago. Whilst the duplicated APOC1 gene has become a pseudogene (APOC1P1), the duplicated HCR (HCR2) has retained 85% homology with HCR1 and appears to be able in its new location to direct the transcription (albeit infrequent transcription) of a sequence resembling an exon which is spliced to APOC4 5' exons even in the absence of conventional promoter elements.

Functional Redundancy of Promoter Elements

The human 7SK RNA gene (RN7SK; chromosome 6) contains a proximal sequence element (PSE) between –49 and –65 and a distal sequence element (DSE) between –243 and –210 which display sequence similarity and functional homology. The PSE can retain function after extensive mutation but only if the DSE is intact. How this apparent functional redundancy has been maintained is unclear.

The promoter of the human ε-globin (HBE1; 11p15.5) gene displays functional redundancy of a different type; it contains eight YY1 binding sites, five binding sites for a putative stage selector protein and seven binding sites for a hitherto unidentified protein. Other probable examples of functional redundancy include the eight SP1 sites present in the human muscle phosphofructokinase (PFKM; 12q13.3) gene promoter.

Without detailed functional studies, however, it is unclear if the multiple cis-acting regulatory elements are truly redundant or if each additional copy leads to an incremental increase in transcription factor binding potential and hence promoter strength.

Recruitment of Repetitive Sequences as Promoter and Silencer Elements

Alu sequences

A number of examples of the recruitment of repetitive sequence elements, mostly *Alu* sequences, as gene promoter and silencer elements have been studied. Over evolutionary time, the insertion of *Alu* repetitive sequence elements in the vicinity of genes has served to introduce different motifs that have altered the expression level or tissue specificity of the associated gene either immediately or after subsequent fine tuning by natural selection. One such motif is a retinoic acid response element (RARE) in the human KRT18 (12q13) gene: three hexamer half sites, related to the consensus AGGTCA, arranged as direct repeats with a spacing of 2 bp. These sites are capable of binding the retinoic acid receptor and functioning as RAREs in transiently transfected cells. We may thus surmise that the random insertion of thousands of *Alu* sequences in the primate genome could have altered the expression of numerous genes over evolutionary time.

An *Alu* sequence in the last intron of the human CD8A (2p12) gene operates so as to modulate the activity of an adjacent T lymphocyte-specific enhancer. The *Alu* sequence appears to contain four functional transcription factor binding sites [Lyf-1 (2), bHLH (1), GATA-3 (1)]. Hambor et al. (1993) noted seven (non-CpG) nucleotide differences by comparing this *Alu* sequence with its probable source gene. Two of these differences were in the GATA-3 binding site and both were shown by site-directed mutagenesis to be necessary for its function. This was therefore proposed to be a possible example of positively selected change in an inserted *Alu* sequence. However, since the *Alu* sequence appears to be capable of modulating the activity of the enhancer through the formation of a cruciform (stem-loop) structure with another downstream *Alu* sequence, it is unclear what if any role the putative binding sites might have in modulating enhancer activity.

An *Alu* sequence in the promoter region of the gene (FCER1G; 1q23) encoding the gamma chain of the high affinity IgE receptor contains positive and negative cis-acting elements which contribute to the hematopoietic cell specificity of expression of this gene. An *Alu* sequence in the myeloperoxidase (MPO; 187q23.1) gene promoter contains four repeats related to the consensus recognition sequence for nuclear hormone receptors (AGGTCA). This sequence acts as a composite SP1-thyroid hormone-retinoic acid response element interacting with SP1 as well as the retinoic acid and thyroid hormone receptors. An *Alu* sequence may also have been recruited to perform a regulatory function in the human θ1-globin gene (HBQ1; 16p13.3). Finally, the estrogen responsiveness of the human breast cancer (BRCA1; 17q21) gene appears to have been conferred by an *Alu* repeat; located within the promoter region of the gene. *Alu* sequences have thus introduced a variety of different DNA sequence motifs capable of binding a range of *trans*-acting factors that have altered the expression level or tissue specificity of the associated genes.

Not all *Alu* sequences inserted into gene regions function as promoters or enhancers of transcription. Indeed, an *Alu* sequences in the third intron of the Wilms' tumor (WT1; 11p13) gene 12 kb downstream of the promoter, acts as a transcriptional silencer, repressing transcription of the WT1 gene in cells of nonrenal origin. Since this silencer can function in an orientation- and distance-independent fashion, Hewitt et al. (1995) suggested that it may have acquired silencer function rather than having simply possessed a silencer function intrinsic to the *Alu* sequence. Another example of a silencer is the RRE repetitive sequence element 1 kb upstream of the murine erythropoietin receptor gene. This sequence, one of the $\sim 10^5$ copies in the mouse genome, may exert its *cis*-mediated repressor effect on the *EpoR* gene by read-through transcription. Finally a 27 bp sequence that is important in the negative regulation of a murine immunoglobulin k light chain gene appears to have been derived from a B1 repetitive element.

A motif within an *Alu* sequence, 3.6 kb 5' to the transcriptional initiation site of the human parathyroid hormone (PTH; 11p15.1-p15.3) gene, contributes toward a negative calcium response element. This element possesses a 12 bp palindromic core (TGAGACAGGGTCTCA) and since it is common in the human genome courtesy of the wide distribution of *Alu* sequences, it may have provided the means for the wide distribution of Alu sequences, it may have provided the means for the expression of other genes to be down-regulated by extracellular calcium.

Wu et al. (1990) have proposed that an *Alu* repeat 2.2 kb upstream of the human ε-globin (HBE1; 11p15.5) gene abolishes down regulation of the gene mediated by a silencer element 4.5 kb upstream. These

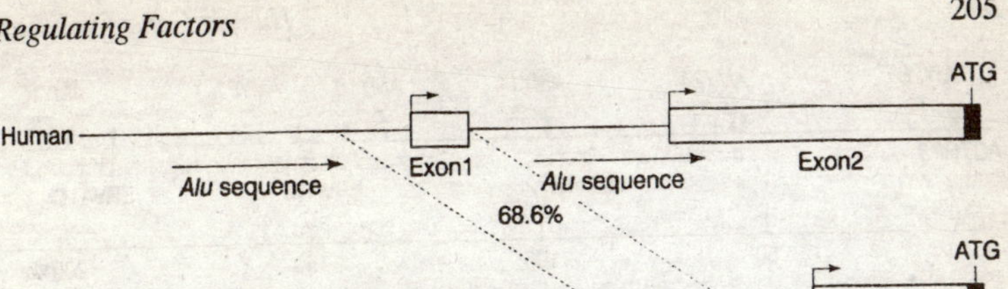

Fig. 6.5. Comparison of human and mouse APOBECl gene promoters).

authors suggested that the down-regulation was caused by transcriptional interference and that the *Alu* repeat somehow blocks transcription from the upstream element specifically in embryonic erythroid cells where it is transcriptionally active.

Upon insertion, *Alu* sequences may also provide alternative sites for transcriptional initiation, as found in the human apolipoprotein B mRNA-ending enzyme (APOBEC1) gene (12p13.1). The human APOBEC1 gene is expressed exclusively in the small intestine whereas in rodents, the gene is more widely expressed. Whilst the human gene contains two *Alu* sequences and to major transcriptional initiation sites, its murine counterpart lacks *Alu* repeats and contains a single transcriptinoal initiation site. Insertion of the *Alu* sequences must therefore have occurred during the last 100 Myrs since the divergence of the primate and rodent lineages. In the human gene, the first *Alu* repeat contains the first of the transcriptional initiation sites and lies upstream of a region exhibiting strong homology to the murine intestinal promoter. The second *Alu* repeat contains the second transcriptional initiation site and is located in the first intron. Comparison with the murine gene suggests that *Alu* sequence insertion may have split the human intestinal promoter leading to utilization of the downstream *Alu* sequence as an alternative site of transcriptional initiation. In this case, it would appear as if promoter function has simply adapted to the presence of the *Alu* repeat rather than being qualitatively altered by it.

Endogenous retroviral elements

Not all inserted sequences implicated in the influencing gene expression are *Alu* repeats. One example of the recruitment of non-*Alu* repetitive sequences within a mammalian promoter is provided by the human amylase genes. These genes are located in a 230 kb region of chromosome lp21; two genes (AMY2A and AMY2B) are expressed in the pancreas, three (AMY1A, AMY1B, AMY1C) are expressed in

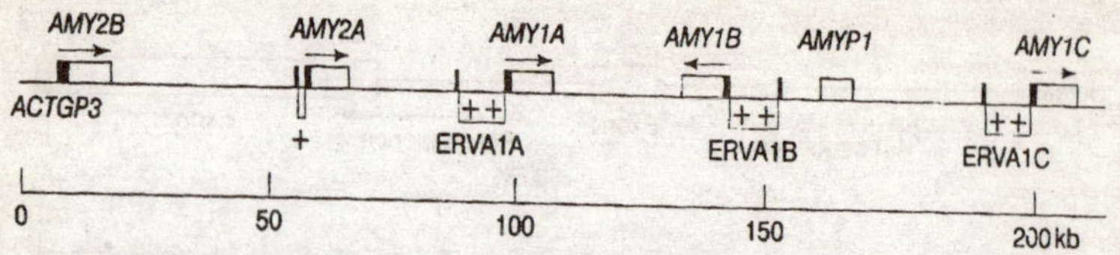

Fig. 6.6. Distribution of the amylase genes on human chromosome 1p21.

the salivary gland and one (AMYP1) represents a truncated pseudogene. A complete γ-actin processed pseudogene is located immediately upstream of AMY2B, the consequence of an insertion event ~40 Myrs ago, after the divergence of the New World monkeys but before the divergence of the New World monkeys from the human-ape lineage. Whether this pseudogene plays a role in regulating amylase gene expression is unclear but the observation that New World monkeys do not possess the pseudogene or express salivary amylase suggests that the pseudogene insertion could have been involved in the switch in tissue specificity of amylase gene expression. In the other four amylase genes, the actin pseudogene is interrupted by an endogenous retroviral sequence, the result of a second insertion which occurred in the human-ape lineage after the divergence of Old World monkeys and prior to the more recent gene duplications. The results of gene expression studies in transgenic mice were initially consistent with the view that the retroviral sequence was essential for salivary gland-specific expression the AMY1C gene. The AMY2A gene, which contains only a residual long terminal repeat (LTR) left by excision of the retroviral sequence, is expressed in the pancreas. Excision of the retroviral sequence from AMY2A thus appeared to be associated with reversion to a pancreas-specific pattern of expression. Transcription of the pancreatic amylase genes was found to be initiated at exon *a* whereas transcription of the salivary gland amylase genes appeared to be initiated from an untranslated exon within the actin pseudogene. It was therefore considered possible that insertion of the retroviral element activated a cryptic promoter within the actin pseudogene that specified the transcriptional initiation site for the expression of the salivary amylase genes. However, more recent studies in various Old World monkey species lacking the retroviral sequence have indicated that salivary amylase gene expression predated the retroviral insertion which cannot therefore be regarded as essential for amylase expression in the salivary gland.

There are now a number of other examples of endogenous retroviral elements which have been recruited to promoter function. These

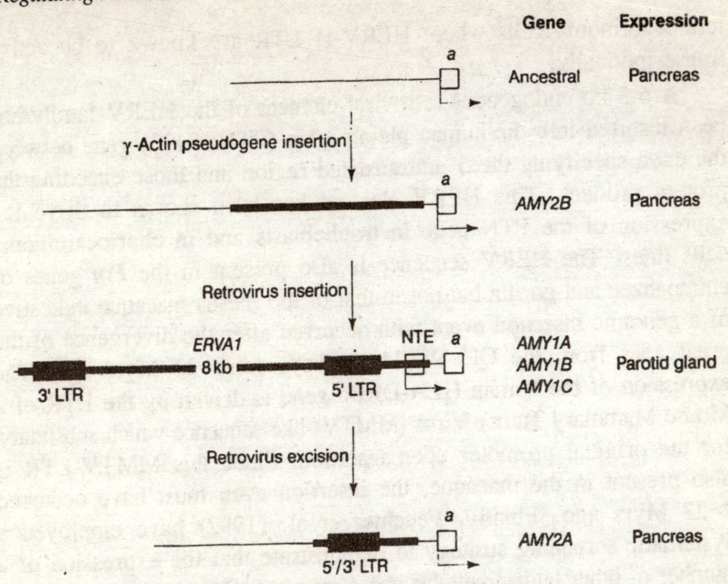

Fig. 6.7. Evolution of the human amylase genes.

elements have been aptly described as 'perpetually mobile footprints of ancient infections' and this mobility has sometimes been put to constructive use by evolution. For example, the hematopoietic cell-specific expression of the human zinc finger gene ZNF80 (3q13.3) is driven by the LTR of ERV9, a member of a low copy number family of endogenous retroviral elements. Since the ERV9 insertion was not found in the African green monkey, rhesus macaque or orangutan, the integration of this element must have occurred after the divergence of the orangutan from the other great apes but before the divergence of gorilla.

The LTR of a HERV-H-related sequence within an intron of a human phospholipase A2-like (*Pla2l*; 8q24) gene has been shown to be important for *Pla2l* gene expression. Since the retroviral LTR is also present in the orthologous genes of chimpanzee and gorilla but not in orangutan and lower primates, we may infer that it was integrated into the ancestral primate genome about 15-20 Myrs ago. Upon further analysis, it has become clear that the teratocarcinoma cell-specific *Pla2l* transcript is actually a fusion transcript between two once distinct genes, the HERV-4- associating 1 (HHLA1; 8q24) gene and the otoconin 90 (OC90; 8q24) gene. Presumably the LTR acts not only as a strong promoter but also as an inducer of transcriptional fusion, at least in teratocarcinoma cells where HERV-H LTR are known to be active transcriptionally.

A 6.3 kb endogenous retroviral element of the HERV family has been inserted into the human pleiotrophin (PTN; 7133) gene between the exon specifying the 5' untranslated region and those encoding the protein product. This HERV element has been shown to drive the expression of the PTN gene in trophoblasts and in choriocarcinoma cells lines. The HERV sequence is also present in the *Ptn* genes of chimpanzee and gorilla but not in that of the rhesus macaque indicative of a genomic insertion event with occurred after the divergence of the great apes from the Old World monkeys some 23 Myrs ago. The expression of the human HLA-DRB6 gene is driven by the LTR of a Mouse Mammary Tumor Virus (MMTV)-like sequence which substituted for the original promoter upon insertion. Since this MMTV-LTR is also present in the macaque, the insertion even must have occurred >23 Myrs ago. Finally, Feuchter et al. (1992) have employed a systematic screening strategy to demonstrate that the expression of a number of other human cellular genes may be influenced by endogenous retroviral elements; these include the cell division cycle 4-like (CDC4L) gene.

Retroviral insertion may affect gene expression even if the element is inserted outwith the promoter region. One example is the LTR of an RTV_L-H element which is present in the 3' UTR of a human placentally expressed gene (termed 'PLT'); the *Plt* mRNA undergoes alternative splicing at its 3' end with polyadenylation occurring within the LTR in one of the transcripts. This sequence is present in the *Plt* genes of the great apes and Old World monkeys and therefore must have been inserted prior to the divergence of these groups.

An LTR of an ERV9 element has been found at the 5' boundary of the human β-globin (HBB; 11p15) gene locus control region (LCR) just upstream of the DNAse I-hypersensitive site HS5. This LTR is composed of 14 tandem repeats containing recurring GATA, CACCC, and CCAAT motifs that are potentially capable of binding GATA-binding factor, BKLF/TEF2 and C/EBP transcription factors respectively. The orthologous sequence in gorilla has only five repeats whilst the repeat number is polymorphic in humans. Reporter gene studies have demonstrated that this LTR possesses both enhancer and promoter activity in erythroid cells. Moreover, in erythroid cells, the LTR activates transcription of the down stream retroviral R and H5 regions and of genomic regions still further downstream. The HS5

LTR may therefore play a role in regulating the transcription of the human β-globin LCR (which is preferentially transcribed in erythroid cells) which may in turn serve to open up the chromatin structure of the β-globin gene domain.

Retroviral insertions have also influence gene expression in other mammalian species. Thus the promoter of the rat oncomodulin gene (human counterpart OCM; 7p13-p11) contains a long terminal repeat of an intracisternal A particle (IAP, a family of endogenous retroviral elements) which has been recruited to perform a gene regulatory function. Since the mouse lacks the integrated IAP element, the IAP insertion must have occurred after the divergence of the two rodent species 40 Myrs ago.

LINE elements

LINE elements may also have served as mobile regulatory sequences altering the expression of target genes. They have a tendency to acquire promoter sequences from non LINE sources, with different sequence lineages acquiring different promoters. These regulatory sequences may then confer a novel specificity of expression upon genes with which they become associated. Thus, a LINE element at position -800 in the human factor IX (F9; Xq27.1-q27.2) gene promoter has been implicated in conferring high level liver-specific expression on the F9 gene. An enhancer some 20 kb 5' to the apolipoprotein(a) gene (LPA; 6127) which contains binding sites for Ets and Spl transcription factors, also resides within a LINE element.

The polyadenylation signal of the murine thymidylate synthase (*Tyms*) gene has been derived from an inserted LINE element. Similarly, the polyadenylation sites of several human genes including the β-tubulin (TUBB; 6p21.3) gene have been derived from MIR elements, mammalian wide interspersed repeats.

As we have seen only a proportion of LINE elements are transpositionally active, the remainder having been inactivated by truncation and rearrangement. Adey et al. (1994b) performed phylogenetic analysis to deduce the sequence of an ancestral murine transpositionally active LINE element. This element was then 'resurrected' by chemical synthesis and shown in vitro to possess promoter activity.

Minisatellites and microsatellites

Several minisatellites are known to have become recruited as gene regulatory elements. These include minisatellites 1 kb 3' to the polyadenylation signal of the human HRAS proto-oncogene (11p15.5;

28 bp repeat units), 600 bp 5' of the transcriptional initiation site of the human insulin (INS; 11p15.5; 14 bp repeat unit) gene and 4.1 kb upstream of the human insulin-like growth factor II (IGF2; 11p15) gene, and the minisatellite in the D_H/J_H intron of the human immunoglobulin heavy chain (IGHD/IGGJ; 14q32.33) gene cluster. The HRAS minisatellite binds members of the *rel*/NFkB family whilst the INS minisatellite binds the transcription factor Pur-1. The IGHD/IGHJ minisatellite binds a mycHLH protein closely related to USF/MLTF. This element may influence IGHD/IGHJ gene expression since sequestration of the transcription factor by the minisatellite inhibits transcriptional activation through a *bone fide* USF enhancer element. Since the HRAS, INS and IGHD/IGHJ minisatellites are absent from the analogous positions in orthologous non-primate genes, it would appear that evolution has recruited these elements during primate evolution. Such minisatellites may have provided the raw material for promoter and enhancer sequences which have then been optimized by selection. Alternatively if transcriptional effects emanating from these sequences are comparatively minor, selection would probably have been unable either to improve or remove them and they would have remained as transcriptional control elements with minor effect.

Microsatellites can also serve as regulatory elements and indeed some are conserved at orthologous positions in the genomes of different species. One example is the $(TCAT)_n$ repeat elements in the first intron of the human tyrosine hydroxylase (TH; 11p15.5) gene. This repeat is similar to the consensus thyroid response element (TRE) present in the human and rat TH genes and gel shift assay have provided evidence for the formation of specific complexes between the tetranucleotide repeat and proteins found in HeLa cell extracts. A $(CCTTT)_n$ pentanucleotide repeat polymorphism (51-72 copies) found within the promoter region of the human inducible nitric oxide synthase 2A (NOS2A; 17q11-q12) gene is also polymorphic in chimpanzees and gorillas but is monomorphic in orangutans and macaques; its influence, if any, on promoter function is not known.

mRNA Editing

The existence of mRNA editing represents something of an evolutionary puzzle since its selective advantage is not immediately obvious. The best understood example in humans is that involving the apolipoprotein B mRNA. ApoB-100, produced in the liver, is essential for the production of very low density lipoprotein whereas ApoB-48 is required for fat absorption in the small intestine. Both proteins are encoded by the same (APOB; 2p23-p24) gene. ApoB-48 mRNA is generated as a result of the introduction of an in-frame translational termination codon at the mRNA level by the deamination of cytosine to uracil (C6666T) in the first base of the CAA codon encoding Gln2153. A conserved 29 nucleotide element flanking the edited base (6662-6690) is found in mammals; this includes a regulator region, a spacer and a mooring sequence which is required for the mRNA editing process. By contrast to the situation found in mammals, chicken *Apob* mRNA is not edited although the various tissue-specific factors that serve to mediate the modification in mammals are present. The absence of mRNA editing in chicken is thought to be due to the presence of several single base-pair substitutions in the mooring region of the chicken *Apob* mRNA since the experimental introduction of A6671T, G6674T and C6680T substitutions into the chicken gene served to confer mRNA editing ability upon chicken cells.

The editing of the APOB mRNA is performed by a multi-protein complex ('editosome') whose catalytic component has been termed apobec-1. The APOBEC1 gene, localized to chromosome 12p13.1 in human, is not highly conserved when compared with its homologues in other mammals, consistent with its recent rapid evolution. Apobec-1 shows substantial sequence homology to cytidine/cytidylate deaminases and it would thus appear that a protein with a nucleoside as substrate has evolved from a protein with a nucleotide as a substrate. It is unlikely, however, that apobec-1 evolved simply by gene duplication and divergence since mRNA editing requires multiple factors which would have had to have coevolved in order to function as a cohesive complex. Even although the separation of apobec-1 and the cytidine/cytidylate deaminases is ancient, apobec-1 is only found in mammals suggesting that the ancestral apobec-1 protein might have had a function other than mRNA editing.

mRNA editing is not however unique to APOB mRNA. Different types of mRNA editing have also been found in several other human mRNAs viz. a T→C transition in the Wilms' tumor (WT1; 11p13) mRNA, a T→A transversion in the α-galactosidase (GLA; Xq21.3-q22) mRNA and C→T transition in the neurofibromatosis type 1 (NF1; 17q11.2) mRNA.

Coordinate Regulation

The clustering of certain genes may be important for their coordinate regulation by common control elements. Possible examples of this include the genes encoding the spermatid-specific nucleoprotamines (PRM1 and PRM2; 16p13.2), the platelet membrane glycoproteins (ITGA2B and ITGA3; 17q21-q22), the albumin family (ALB, AFP, AFM,CG; 4q11-q13), the pregnancy-specific glycoproteins (PSG1, PSG2, PSG3, PSG4, PSG5, PSG6, PSG7, PSG8, PSG11, PSG12, PSG13; 19q13.2) and the fibrinogens α, β, and γ (FGA, FGB and FGG; 4q31). In some cases, a head-to head arrangement may potentiate the use of bidirectional promoters.

Coordinate regulation does not however require close linkage as evidenced by the case of the α- (HBA1) and β-globin (HBB) genes on chromosomes 19p13.3 and 11p15.5 respectively. Another example is that of the human ribosomal protein genes which encode the 80-90 ribosomal proteins that together constitute the ribosome. The expression of the different ribosomal protein genes, which can account for 7-9% of total cellular RNA, is coordinately regulated in response to the cell's varying requirements for protein synthesis. However, the ribosomal protein genes are highly dispersed in the human genome indicating that their coordinate regulation must be brought about by the action of trans-acting factors rather than the influence of shared regulatory sequences.

Table 6.1. Chromosomal locations of chromosomally assigned human ribosomal protein genes

Ribosomal protein	*Gene symbol*	*Chromosomal location*
Ribosomal protein L3	RPL3	22
Ribosomal protein L4	RPL4	15
Ribosomal protein L5	RPL5	1
Ribosomal protein L6	RPL6	12q23-q24
Ribosomal protein L7	RPL7	8
Ribosomal protein L8	RPL8	8
Ribosomal protein L9	RPL9	4p13
Ribosomal protein L10	RPL10	X
Ribosomal protein L11	RPL11	1
Ribosomal protein L13	RPL13	16q24
Ribosomal protein L15	RPL15	3
Ribosomal protein L17	RPL17	18
Ribosomal protein L19	RPL19	17p11-q12
Ribosomal protein L22	RPL22	3q26
Ribosomal protein L23A	RPL23A	17q11
Ribosomal protein L24	RPL24	3
Ribosomal protein L27	RPL27	17

Ribosomal protein L27A	RPL27A	11
Ribosomal protein L28	RPL28	19q13
Ribosomal protein L29	RPL29	3q29-qter
Ribosomal protein L30	RPL30	8
Ribosomal protein L31	RPL31	2
Ribosomal protein L32	RPL32	3q13.3-q21
Ribosomal protein L35A	RPL35A	3q29-qter
Ribosomal protein L36A	RPL36A	14
Ribosomal protein L37	RPL37	5
Ribosomal protein L38	RPL38	17
Ribosomal protein L41	RPL41	12
Ribosomal protein S2	RPS2	16p13.3
Ribosomal protein S3	RPS3	11q13
Ribosomal protein S3A	RPS3A	4
Ribosomal protein S4	RPS4X/Y	Xq13.1/Yp11.3
Ribosomal protein S5	RPS5	19q13.4
Ribosomal protein S6	RPS6	9p21
Ribosomal protein S7	RPS7	2p25
Ribosomal protein S8	RPS8	1p32-p34
Ribosomal protein S9	RPS9	19p13.4
Ribosomal protein S10	RPS10	6
Ribosomal protein S11	RPS11	19q13
Ribosomal protein S12	RPS12	6
Ribosomal protein S13	RPS13	11
Ribosomal protein S14	RPS14	5q31-q33
Ribosomal protein S15A	RPS15A	16
Ribosomal protein S17	RPS17	11p-13-pter
Ribosomal protein S18	RPS18	6p21.3
Ribosomal protein S24	RPS24	10q22-23
Ribosomal protein S25	RPS25	11q23.3

Changes in Expression of Developmentally Significant Gene Products

The study of the molecular genetics of vertebrate development is still very much in its infancy. However, the identification of developmental control genes important in morphogenesis is proceeding apace as a result of studies of (i) model organisms including mouse, zebrafish, *Drosophila* and *Caenorhabditis elegans* and (ii) human dysmorphic syndromes and congenital malformations.

The spatial and temporal distribution patterns of expression of HOX genes have played an important role in the evolutionary emergence of novel body plants among the metazoa. The molecular basis of such a role has long been a puzzle but new studies are now providing a glimpse of how subtle genetic changes can have fairly dramatic effects on morphology. One of the best characterized examples is provided by a promoter mutation which appears to have morphological consequences for the evolution of the mammalian body plan: a 4 bp deletion in element C of the early enhancer region of the Hoxc8 gene of baleen whales. This lesion (TTAATTG→TT-G) is specific to baleen whales (five species tested) and is not found in the highly conserved Hoxc8 gene promoters of humans (HOXC8; 12q12-q13), rodents, artiodactyls or the toothed whales including sperm whales. In mice, the early enhancer region of the Hoxc8 gene promoter is required to initiate expression of the gene in the posterior region of the day 8.5 mouse embryo and to establish spatial domains of expression in the neural tube and mesoderm. After day 9.0, the late enhancer maintains anterior Hoxc8 early enhancer region containing the 4 bp deletion has been assayed in transgenic mouse embryos where it was found to direct the expression of the reporter gene to more posterior regions (4–5 somite levels posterior as compared with human or murine Hoxc8 enhancers) of the neutral tube but failed to direct expression to the posterior mesoderm. Similar results were obtained when site-directed mutagenesis was used to introduce the same lesion into the murine Hoxc8 early enhancer region. Thus the 4 bp deletion in the Hoxc8 gene promoter of baleen whales may have played a role in modifying the developmental program of these cetaceans. One wider implication of this work is that there may be additional mutations yet to discover in the cis-acting regulatory sequences of other Hox genes and these lesions could have contributed to the evolution of body plan diversity during mammalian evolution.

It is anticipated that mutational changes in a number of other genes encoding transcription factors that play a role in embryonic development will be characterized in the coming years thereby shedding new light on the molecular basis of morphogenesis. Possible candidate genes would include the Pax gene family, the Sox gene family, the engrailed (EN1, 2q13q21; EN2, 7q36) and *Wingless* (WNT1; 12q12-q13) genes, the brachyury (T; 6q27) gene and the snail family of transcription factors as well as the heat shock protein 90 (HSP90) family of signal transduction chaperonins. Sequence differences between orthologous developmental regulatory genes should provide in sights into the process of molecular evolution underlying morphological change.

Transcription Factors

Transcription Factor Families

Many mammalian transcription factors belong to families whose members bind to very similar or identical DNA sequence motifs. Thus, there are at least eight cAMP-responsive transcription factors of the CREB/ATF family that bind to the octanucleotide TGACGTCA. The evolutionary conservation of DNA sequence recognition can be fairly dramatic as for example in the case of the various members of the Brn-3 class of POU domain transcription factors found in both mammals and nematodes. On the other hand, some transcription factor families possess members that exhibit a considerable degree of divergence in terms of their DNA binding specificities. For example, at least four members of the nuclear factor 1 (NF10 family recognize sequences containing the trinucleotide TGG and at least eight members of the mammalian Ets family bind to an 11 bp purine-rich motif containing a conserved GGA core. The DNA binding specificity of different Ets family members is determined by the nucleotides at the 3' end of the Ets-binding site.

The evolutionary subdivisions of transcription factor families as revealed by phylogenetic analysis may be paralleled by functional subdivisions. If function is conserved within (although not between) subfamilies, then the functions of novel transcription factors may be to some extent predictable by comparison with other members of the same subfamily. One example of this is provided by the high mobility group (HMG) protein superfamily of DNA-binding proteins. These proteins possess one or more copies of an 80 amino acid domain termed the HMG box and have an evolutionary history that dates back 1000 Myrs. The HMG superfamily may be divided into two sub-families (i) the TCF/SOX family which comprises transcription factor proteins that contain a single sequence-specific HMG box and (ii) the UBF/HMG family of chromatin structure regulatory proteins which possess multiple HMG boxes that exhibit little if any sequence specificity. Representatives of the first family in the human genome include the chromosomally dispersed SRY-related HMG-box (SOX) genes (SOX1, 13q34; SOX2, 3q26-q27; SOX3, Xq26-q27; SOX4, 6p23; SOX5, 12p12; SOX9, 17q23; SOX10, 22q13; SOX11, 2p25; SOX20, 17p13; SOX22, 20p13), the lymphoid enhancer-binding factor 1 (LEF1; 4q23-q25) and the hepatocyte nuclear factor 1α (TCF1; 12q24) genes. Human representatives of the second family include the high mobility group (on-histone chromosomal) protein genes, HMG1 (13q12), HMG2 (4q31),

HMG4 (Xq28), HMG14 (21q22) and HMG17 (1p35-p36) and upstream binding transcription factor (UBTF; 17q21). The diversity of function exhibited by members of the HMG superfamily is a result of the action of a number of different evolutionary processes including gene duplication, intragenic duplication, exon shuffling and single base-pair substitution mediated divergence.

Duplications or amplifications of transcription factor genes are sometimes very ancient as is evidenced by the limited homology still evident between nuclear factor 1 (NF1) and the protein kinase family. On the other hand, some duplications have occurred during mammalian evolution [e.g. the transcription factors TCF1 (12q24) and LEF1 (4q23-q25) and the *hedgehog* gene family, *sonic* (SHH; 7q36) and Indian (IHH; 2q33-q35) and some as recently as 55 Myrs ago in the common ancestor of the simians e.g. the ZNF91 family (ZNF91; 19p12-p13.1).

Functional Conservation of Orthologous Transcription Factors

The general transcription factor, TBP (TFIID), binds to the upstream TATA box element and is essential for transcription in all eukaryotes. TBP is highly conserved throughout the eukaryotes and this conservation even extends to the archaebacteria. Functional conservation at the protein level is evidenced by the ability of yeast TBP to replace human TBP in *in vitro* transcription reactions and *vice versa*. Not surprisingly, therefore, yeast and human TBP have virtually identical sequence recognition characteristics, at least *in vitro*. However, *in vivo*, human TBP cannot properly substitute for yeast TBP and the yeast cells in question grow extremely poorly. This difference between the *in vitro* and *in vivo* situations is salutary and probably reflects subtle differences in the interactions of TBP with activator proteins or other components of the transcriptional initiation complex.

Interactions between cis-acting DNA sequence motifs and the *trans*-acting transcription factors binding them may thus be conserved over family long periods of evolutionary time. Another example is that of the rat pituitary-specific transcription factor Pit-1 which is able to bind to and activate the growth hormone gene promoter of the rainbow trout. Similarly, mammalian ets genes are functionally homologous to the *pointed* gene of *Drosophila*. Functional conservation is also evident with other factors such as *c-jun*, the serum response factor SRF, the CCAAT box-binding factor CP1 and the glucocorticoid and estrogen receptors; the mammalian proteins can substitute for their yeast counterparts in order to activate gene transcription in yeast. This functional conservation owes its existence to the fact that once specificity has been established between a given transcriptional activator and its cognate binding sites in the regulatory regions of different genes, both the structure of the DNA-binding domain of the activator and the sequence of the recognition site are evolutionarily constrained. Evolutionary change between both paralogous and orthologous transcription factors has nevertheless occurred in a number of different ways and some of these are illustrated in the following sections.

Functional Redundancy of Paralogous Transcription Factors

Gene duplication and amplification initially generate functional redundancy as in the case of the c-Ets-1 (ETS1; 11q23) and c-Ets-2 (ETS-2 (ETS2; 21q22) genes in mammals where the former appears to be dispensible and indeed replaceable by the later. Functional redundancy is also apparent in the MyoD family of transcription factors, for example between MYOD1 (11p15.1) and MYF5 (12q21) and between MYOD1 (11p15.1) and MYF6 (12q21). It has also been noted that the transgenic inactivation of several genes known to be important in mammalian development does not automatically lead to a deleterious phenotype, implying a degree of functional redundancy. We may surmise that developmental gene redundancy will have endured if its maintenance has served to confer a selective advantage. On the other hand, functional redundancy could also have originated merely as a consequence of certain genes being either especially prone to duplication or manifesting an increased probability of survival after duplication.

Functional redundancy does not however always endure and the relaxation of selective pressure consequent to gene duplication/amplification potentiates the diversification of paralogous transcription factors leading eventually to the emergence of families of related transcription factors with different DNA target sequence specificities and hence distinct functions.

Paralogous Transcription Factors

Perhaps the best characterized example of the divergence of paralogous transcription factors is provided by the nuclear receptor family. In this family, dimerized receptors typified by the estrogen receptor group, bind two 6 bp half sites of the sequence TGACCT whereas those of the glucocorticoid receptor group recognise the related sequence TGTTCT. The amino acids involved in discriminating between these motifs are located in the 'P-box' of the DNA recognition helix. The P-box of the glucocorticoid receptor group appears to have evolved from a progenitor resembling the present-day estrogen receptor. Thus, mutations in the P-box have altered the DNA-binding specificity of a

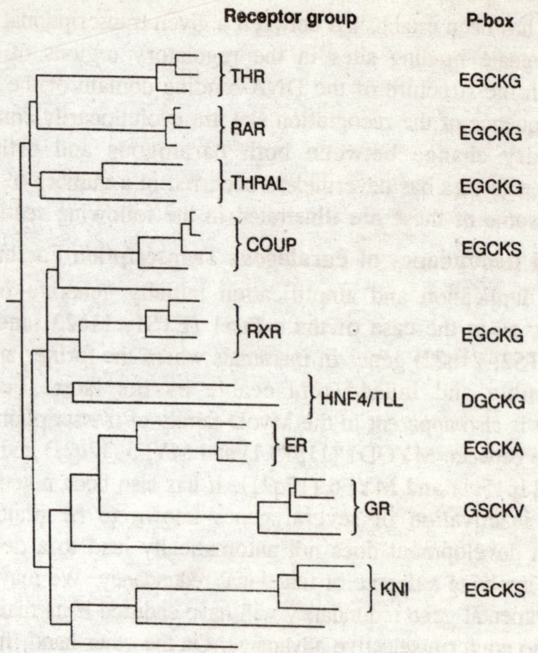

Fig. 6.8. P-box sequences of different receptor groups within the nuclear receptor family.

receptor with high affinity for TGACCT sites and low affinity for TGTTCT sequences, to proteins with the opposite specificity. Residues Glu439 and Ser440 are critical for conferring binding specificity on the receptor.

Another example of a functional change in the DNA-binding domain within a family of paralogous transcription factors is provided by Pax6 (PAX6; 11p13) which possesses an Asn at amino acid residue 47 in the third a-helix of the paired domain this amino acid recognizes the nucleotide T. All other known Pax genes encode proteins with a His at this position; this amino acid shows higher affinity toward the nucleotide G.

Orthologous Transcription Factors

Orthologous transcription factors have also diverged over evolutionary time. Divergence has occurred by, for example, incorporation of novel motifs or the amplification of existing motifs. Thus, the human transcription factor MOK2 (MOK2; 19q13.2-q13.3) contains 10 zinc-finger motifs in comparison to seven in its murine homologue. Similarly, the human erythroid-specific transcription factor

Eryth 1 contains different numbers of repeat motifs as compared with its chicken counterpart. Finally, gene sequences encoding TBP, the general transcription factor, exhibit a considerable degree of sequence simplicity as a direct result of simple repeat amplification, perhaps by replication slippage. The incorporation of new repeats and the consequent enlargement of TBP may have permitted novel interactions with domains of other proteins leading to the acquisition of new functions.

Alternative Splicing of Transcription Factor Genes

Alternative splicing provides the means to generate transcription factor diversity in the absence of gene duplication. Thus, alternative splicing of the PAX 8 (2q12-q14) gene results in the alternative presence or absence of a single Ser residue in the recognition helix of the paired domain which is critical for DNA binding. The two forms of Pax8 differ in their binding specificity.

Promoter Shuffling in Transcription Factor Genes

Promoter modularity arising from the shuffling of component motifs often occurs in the promoters of paralogous transcription factor genes. This provides the means for changes in the expression of single genes to lead to changes in the expression of many downstream target genes, a process which has contributed significantly to the evolution of complex gene expression networks.

Transcription Factor-binding Site Interactions

DNA sequence elements that play a role in gene regulation have evolved so as to provide appropriate binding sites for their cognate transcription factors. In many cases, element function (measured in terms of influence on transcription) is directly probability that a specific site is bound by the transcription factor protein. It follows that the same function can be achieved with a strong binding site and a small amount of protein as with a weak binding site and a large amount of protein. The extent of selective pressure on the DNA sequence is therefore likely to be determined by a combination of the cellular abundance of the protein, the functional activity of the protein and the initial binding strength of the DNA-protein interaction. Transcription factor binding may however also be influenced by the number of binding sites available on the promoter. Thus, the pituitary-specific growth hormone (GH1; 17q22-q24) gene of humans and rats contains two binding sites for the transcription factor Pit-1 whereas four such sequences occur in the promoter of the homologous Gh1 gene of *Oncorhynchus mykiss*, the rainbow trout Rat Pit-1 has been shown to be capable of binding to three regions of the trout gene promoter thereby driving expression of a downstream reporter gene.

Exon Shuffling in the Evolution of Transcription Factors

Transcription factors are sometimes encoded by genes that are evolutionarily unrelated yet share the same type of DNA-binding domain. Such genes may have arisen by exon shuffling, the process by which functional domains encoded by one or more exons have been dispersed to a variety of different proteins. That some DNA-binding domains are encoded by several exons implies that these exons must have been shuffled together as a single block. Matsuo et al. (1994) have suggested that the presence of short unconserved introns with different types of splice junction within the mammalian Oct-2 (POU2F2; chromosome 19) gene may have served to prevent recombination between the exons comprising the conserved POU domain without inhibiting the shuffling of this domain in its entirety between different transcription factor-encoding genes during evolution.

The divergence of the mammalian T-box family of transcription factors, which began before the separation of the vertebrate, arthropod and nematode lineages, has occurred both by the insertion or deletion of specific introns, or by *intron sliding* leading to variations in exon length.

MOLECULAR CLOCKS

The molecular clock hypothesis postulates that for any given macromolecule (a protein or DNA sequence) the rate of evolution is approximately constant over time in all evolutionary lineages. This hypothesis has stimulated much interest in the use of macromolecules in evolutionary studies, for two reasons. First, if macromolecules evolve at constant rates, they can be used to date species-divergence times and other types of evolutionary events, similar to the dating of geological times using radioactive elements. Moreover, phylogenetic reconstruction is much simpler under constant rates than under nonconstant rates. Second, the degree of rate variation among lineages may provide much insight into the mechanisms of molecular evolution. As noted under neutral mutation the rate of evolution is equal to the rate of mutation. Therefore, if the neutral mutation hypothesis is true and if the rate of neutral or nearly neutral mutation in a protein has not changed with time, the rate of evolution in that protein should be nearly constant. Thus, a large change in substitution rate may indicate a large change in evolutionary factors. For example, a large increase in the rate of evolution in a protein in a particular lineage may indicate adaptative evolution, relaxation of functional contraints (or less of function), or a large reduction in effective population size in that lineage. This issue will be discussed in more detail later.

Although the concept of a molecular clock has had a strong impact on the study of evolution, it has always been controversial. Indeed, there has been a wide range of views on this issue. In one, extreme, Ochman and Wilson (1987) suggested the existence of a universal clock of synonymous substitution that is applicable to all organisms. In the other extreme, Goodman (1976, 1981) and his associates denied even the existence of approximate constancy. Also, there has been a strong controversy on whether generation time can have a significant effect on the rate of molecular evolution. Classical genetic studies indicated that mutation rates are more comparable among organisms when measured in terms of generation than in terms of absolute time. For this reason, the molecular clock should run faster in organisms with a short generation time, for they will go through more generations per unit time than do organisms with a long generation. This has been known as the generation-time effect hypothesis. However, while DNA-DNA hybridization data supported this hypothesis protein sequence data seemed to indicate a nearly constant rate in terms of absolute time. To date the two opposing view still persist.

The rapid accumulation of DNA sequence data in recent years presents an unprecedented opportunity for studying various issues related to the molecular clock hypothesis. DNA sequence data allow a closer examination of the hypothesis than do protein sequence data (e.g., the existence of both coding and noncoding regions in DNA sequences) and can be interpreted more directly than DNA hybridization and immunological distance data. For this reason, many authors have used DNA sequence data to examine the molecular clock hypothesis. The discussion below will be based on DNA sequence data. For DNA hybridization studies, readers may refer to Catzeflis et al. (1987), Sibley and Ahlquist (1987), Caccone and Powell (1989) and Sibley et al. (1990).

Relative-Rate Test

The controversy over the molecular clock hypothesis often involves disagreements on dates of species divergence. To avoid this problem, Sarich and Wilson (1973) proposed a test that does not require knowledge of divergence times. This test called the *relative-rate test*. Suppose that we want to compare the rates in lineages A and B. Then, we use a third species, C, as a reference. The reference species should have branched off earlier than the divergence between species A and B. For example, to compare the rates in the human and orangutan lineages we can use a monkey species as a reference.

It is to see that the number of substitutions between species A and C, K_{AC}, is equal to the sum of substitutions that have occurred from point O to point A (K_{OA}) and from point O to point C (K_{OC}). This is,

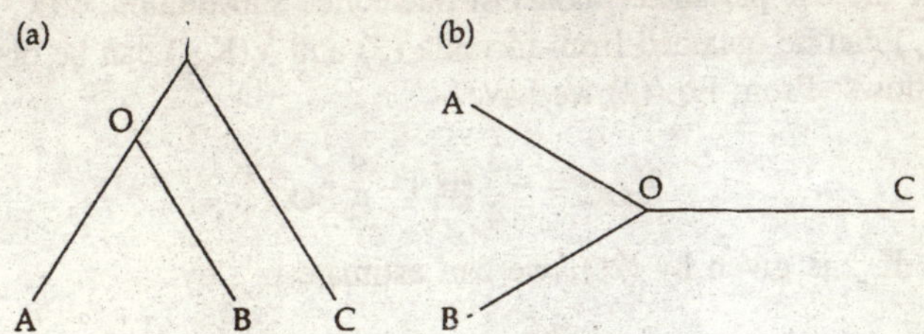

Fig. 7.1. (a) The rooted tree for species A, B, and C, assuming that C is the known outgroup. O denotes the common ancestor of species A and B. (b) An unrooted tree.

$$K_{AC} = K_{OA} + K_{OC} \quad \text{...(1a)}$$

Similarly,

$$K_{BC} = K_{OB} + K_{OC} \quad \text{...(1b)}$$

$$K_{AB} = K_{OA} + K_{OB} \quad \text{...(1c)}$$

Since K_{AC}, and K_{BC}, and K_{AB} can be directly estimated from the nucleotide sequences, we can easily solve the three equations to find the values of K_{OA}, K_{OB}, and K_{OC}:

$$K_{OA} = (K_{AC} + K_{AB} - K_{BC})/2 \quad \text{...(2a)}$$

$$K_{OB} = (K_{AB} + K_{BC} - K_{AC})/2 \quad \text{...(2b)}$$

$$K_{OC} = (K_{AC} + K_{BC} - K_{AB})/2 \quad \text{...(2c)}$$

We can now decide whether the rates of substitution are equal in lineages A and B by comparing the value of K_{OA} with that of K_{OB}. The time that has passed since species A and B last shared a common ancestor is, by definition, equal for both lineages. Thus, according to the molecular clock hypothesis, K_{OA} and K_{OB} should be equal, this is, $d=K_{OA}-K_{OB}$ should not be statistically different from 0. From Equations 2a and 2b, we obtain $K_{OA}-K_{OB}=K_{AC}-K_{BC}$. Therefore,

$$d = K_{AC} - K_{BC} \quad \text{...(3)}$$

A simple way to test whether an observed d value is significantly different from 0 is to compare it with the standard error; for example, if the absolute value of d is larger than two times the standard error, it may be considered significant at the 5% level. Since Var(x–y) =Var(x)+Var(y)–2Cov(x,y), Equation 3 implies

$$V(d) = V(K_{AC}) + V(K_{BC}) - 2\text{Cov}(K_{AC}, K_{BC}) \quad \text{...(4)}$$

Noting that $K_{AC}=K_{OA}+K_{OC}$ and $K_{BC}=K_{OB}+K_{OC}$ and the K_{OA} and K_{OB} are independent of each other because lineages A and B evolve independently, we have

$$V(d) = V(K_{AC}) + V(K_{BC}) - 2V(K_{OC}) \quad \text{...(5)}$$

Under the one-parameter model of nucleotide substitution, $V(K_{AC})$ and $V(K_{BC})$ can be obtained from formula (3) and $V(K_{OC})$ can be obtained as follows. From Eq. (2) we have

$$K_{OC} = -\frac{3}{4}\ell n\left(1-\frac{4}{3}p_{OC}\right)$$

Since K_{OC} is given by (2c), we can estimate p_{OC} by

$$p_{OC} = \frac{3}{4}\left[1-\exp\left(-\frac{4}{3}K_{OC}\right)\right] \quad ...(6)$$

Replacing p in Formula 3 by p_{OC}, we obtain an approximate formula for $V(K_{OC})$.

For the two-parameter model, V(d) can be derived in a similar manner, though the derivation is more complicated. For coding regions, one can treat synonymous and nonsynonymous substitutions separately and derive the mean and variance of d for each type of substitution using the procedure above and the formulas for the mean and variance of K_{ij}.

Sometimes we are not certain about the phylogenetic relationships among the three lineages under study. In this case, we may still test the molecular clock hypothesis. We use the unrooted tree because we do not know the root of the tree. However, Formula 2a-c can still be used to calculate the branch lengths from the common node O to each of the present-day sequences. To be favourable to the molecular clock hypothesis, we assume that the lineage with the largest branch length is the outgroup. For example, if K_{OC} is the largest, then species C is assumed to be the outgroup and the test can be done as above. If K_{OB} is the largest, then species B is taken as the outgroup, and by exchanging the subscript B and C in the above formulas, we can test the molecular clock hypothesis in a similar manner. Note, however, hat a failure to reject the hypothesis may not be taken as evidence for the hypothesis because we assume the most favourable condition for testing the hypothesis. Obviously, a rejection of the hypothesis can be taken as evidence against the hypothesis.

The above test is simple and will be used in this book. Tajima (1993) has recently proposed a simple alternative test. Muse and Weir (1992) have developed a maximum likelihood test, which is likely to be more powerful but computationally more demanding. The test given in Equation 5 has been extended to the case of more than two lineages by Li and Bousquet (1992).

A Local Clock in Mice, Rats, and Hamsters

Since the molecular clock hypothesis is controversial, the first question to ask is, "Does there exist a molecular clock in any group of organisms?" (Such a clock is known as a local clock.) The best organisms to look for the existence of a local clock are a group of organisms with similar physiology and life histories such as generation time. The muroid rodents (e.g., Mice and rats) and their relatives would be such a group for which there is abundance of DNA sequence data. Indeed, there have been two lines of evidence suggesting that the clock-behaviour is approximately maintained in this group of rodents. First, the DNA-DNA hybridization studies of Brownell (1983) and Catzeflis et al. (1987) revealed a constant rate among lineages in the murine (mouse and rat) family and the microtine (hamster) family. Second, in an analysis of nucleotide sequences using the relative-rates in the mouse and rat lineages. We review below the study of O'hUigin and Li (1992), who used extensive sequence data from mice, rats, and hamsters.

First, let us compare the substitution rates in the mouse and rat lineages, using the hamster lineage as a reference. The number of substitutions per synonymous site (K_S) is 30.3% between mouse and hamster and 31.1% between rat and hamster. The difference (0.8%) is not statistically significant because it is smaller than the standard error of K_S (1.0%). So, the synonymous rates in the mouse and rat lineages are nearly equal. The difference in the nonsynonymous rate (K_A), i.e., d=2.9% -2.7% = 0.2%, is equal to two times the approximate standard error of K_A and may be considered statistically significant. Thus, the nonsynonymous rate seems to be slightly faster in the mouse lineage than in the rat lineage.

Table 7.1. Numbers of nucleotide substitutions per 100 sites between species

Species pair	Synonymous sites		Nonsynonymous sites	
	K_S	L	K_A	L
Mouse-rat	18.0±0.7	4,229	1.8±0.1	15,217
Mouse-hamster	30.3±1.0	4,229	2.9±0.1	15,217
Rat-hamster	31.3±1.0	4,229	2.7±0.1	15,217
Mouse-human	53.4±1.5	4,229	5.2±0.2	15,217
Rat-human	51.6±1.5	4,229	5.0±0.2	15,217
Hamster-human	52.3±1.5	4,229	5.1±0.1	15,217

Second, we compare the substitution rates in the mouse (or rat) and hamster lineages, using the human lineage as a reference. The K_S value is 53.4% between mouse and human and 52.3% between hamster and human, so the difference 1.1% is smaller than the approximate standard error 1.5% and is not statistically significant. Similarly, the difference between the two lineages in the rate of nonsynonymous substitution is also not significant. The same conclusion can be drawn when the mouse lineage is replaced by the rat lineage. Thus, the mouse, rat and hamster lineages have evolved at nearly equal rates in terms of nucleotide substitution.

In conclusion, there appears to be an approximate molecular clock in these rodents, at least in terms of synonymous substitution. This clock may be used to date divergence times among these rodents. For example, since the K_S value is 18.0% between mouse and rat and is about 31.0% between mouse-rat and hamster, the hamster lineage is estimated to have branched off 0.31/0.18=1.7 times earlier than the mouse-rat divergence.

Lower Rates in Humans than in Monkeys

There has been a long-standing controversy over the hominoid rate-slowdown hypothesis, which postulates that the rate of molecular evolution has become slower in hominoids (humans and apes) after their separation from the Old World (OW) monkeys. This hypothesis, proposed by Goodman (1961) and Goodman et al. (1971), was based on rates estimated from immunological distance and protein sequence data. Sarich and Wilson (1967) and Wilson et al. (1977) contended that the slowdown was an artifact, owing to the use of an erroneous paleontological estimate of the ape-human divergence time. They conducted relative-rate tests using both immunological distance data and protein sequence data and concluded that there was no evidence for a hominoid slowdown. A similar conclusion was drawn from DNA hybridization studies; however, more recent DNA hybridization studies have produced conflicting conclusions. On the other hand, comparative analysis of DNA sequence data by Koop et al. (1986), Li and Tanimura (1987) provided strong support for the hominoid slowdown hypothesis and the hypothesis was accepted by many molecular evolutionists. However, Easteal (1991) has argued that the slowdown occurred only in the η-globin pseudogene, because when this pseudogene was removed from comparison, the rate of nucleotide substitution in the OW monkey lineage was no longer significantly higher than that in the human lineage.

To resolve this controversy, Seino et al. (1992) and Ellsworth et al. (1993) obtained more sequence data In Table 7.2, K_{13} and K_{23} are the distance between an OW monkey and a New World (NW) monkey and between the human and a NW monkey, respectively. For the introns compared K_{13}–K_{23} is positive, except that K_{13}–K_{23} is 0 for the ε-globin and interferon-α receptor introns and is slightly negative for the lipoprotein lipase intron. When all introns are considered together K_{13}–K_{23} is significantly faster than that in the human lineae. The same conclusion is obtained from the flanking sequence data. Thus, there is indeed evidence for the hominoid-slowdown hypothesis.

The intron sequence data suggest that the OW monkey lineage evolves 1.3 times faster than the human lineage, which is similar to that estimated (1.4) from the η-globin data. The flanking sequence data suggest that the rate ratio is larger than 2. However, since the latter data set is small, the ratio estimated from this set may not be reliable. Further data are needed to see whether the ratio varies among different RNA regions.

Table 7.2. Differences in the number of nucleotide substitutions per 100 sites and the relative rates of substitution between the Old World monkey (species 1) and human (species 2) lineages, with the New World monkey (species 3) as a reference

Sequence	Nucleotides compared	K_{12}	K_{13}	K_{23}	K_{13}–K_{23}	Rate ratio
η-Globin pseudogene	8,781	6.7	11.8	10.7	1.1±0.3	1.4
Intron						
IGF2	1,589	6.4	15.8	14.2	1.6±0.8	1.7
ε-globin	928	4.9	11.5	11.5	0.0±0.8	1.0
Insulin	862	9.7	17.0	15.9	1.1±1.3	1.3
Mast-cell carboxypeptide	1,275	5.5	13.3	12.5	0.8±0.8	1.3
Carbonic anhydrase 7	501	7.2	11.1	9.7	1.5±1.4	1.5
Interferon α receptor	885	7.6	14.0	14.0	0.0±1.1	1.0
Apolipoprotein C3	1,270	8.7	18.5	16.9	1.6±1.0	1.5
Lipoprotein lipase	1,168	7.9	13.6	13.8	–0.3±1.0	1.0
Total	8,478	7.1	14.7	13.9	0.8±0.3	1.3
Flanking and untranslated regions						
ε-globin	388	5.3	13.5	10.6	2.9±1.4	3.4
Insulin	548	9.8	15.8	12.6	3.2±1.5	2.0
Total	936	7.9	14.9	11.7	3.1±1.1	2.3

Higher Rates in Rodents than in Primates

From DNA hybridization data, Laird et al. (1969) and Kohne (1970) estimated the substitution rates between mouse and rat and between human and chimpanzee and concluded that the former rate is much higher than the latter. They attributed the higher rate in rodents to a shorter generation time, i.e., the generation-time effect. Sarich and Wilson (1973) argued that this difference in rate was based on questionable assumptions about the divergence times between species. In order to avoid controversies over the assumptions of divergence times, Wu and Li (1985) used artiodactyls and carnivores as outgroups to compare the rates of substitution in the rodent and primate lineages, but the validity of the outgroups was questioned.

To avoid the above criticism, Gu and Li (1992) used the chicken as an outgroup. They used amino acid sequences instead of DNA sequences because the chicken and mammalian lineages diverged about 300 million years ago, so that it is difficult to obtain reliable estimates of the divergences at synonymous sites. Let N_R be the number of residues at which the rodent and human sequences are different but the human and chicken sequences are identical and let N_H be the number of residues at which the human and rodent sequences are different but the rodent and chicken sequences are identical. For example, if the residues at a particular site are alanine in human and chicken but valine in rodent, then $N_R=1$ and $N_H=0$ for that site. However, if the residues at a site are different for the three sequences (e.g., alanine in human, valine in mouse, and leucine in chicken), then that site is noninformative and is not included in the comparison. For each protein N_R (N_H) is the sum of the N_R (N_H) values over the informative sites of the sequences. Under the assumption of equal substitution rates in the human and rodent lineages, N_R should be statistically equal to N_H for each protein. Among the 54 proteins compared, 35 proteins show a faster rate in the rodent lineage (i.e., $N_R>N_H$), show a faster rate in the human lineage ($N_H>N_R$), and the rest (7 sequences) show an equal rate. A sign test indicates a significantly faster rate in rodents than in humans ($P<0.005$). When all the proteins are pooled together, $N_R=600$ is clearly significantly larger than $N_H=416$ ($P<0.001$). Therefore, there is strong evidence for an overall faster substitution rate, from the common primate-rodent ancestor, in the lineage to mouse or rat (the rodent lineage) than in the lineage to humans (the human lineage).

Gu and Li also compared the frequencies of insertions and deletions (gaps) in the human and rodent sequences using the chicken sequences as references. They used two measures. One measure is simply the number of gaps in each sequence. This measure ignores the gap length. The other measure is the total number of amino acids involved in the gaps (i.e., the sum of gap lengths) in the sequences; the number in the rodent sequence is denoted by G_R and that in the human sequence by G_H. This measure is more reasonable because it gives higher weights to larger gaps. A binomial test was conducted for the gap data. Because the number of gap events in each sequence is small, no individual sequence shows a significant difference between the two lineages. When the data are pooled, in terms of the sum of gap length, $G_R=385$ is significantly larger than $G_H=108$ ($P<0.001$), though in terms of the number of gaps, the difference (44 vs. 31) is insignificant ($0.05<P<0.1$). As noted above, the sum of gap length is a more reasonable measure than the number of gaps. We may therefore conclude that the rate of insertion/deletion is higher in the rodent lineage than in the human lineage.

Table 7.3. Number of amino acid substitutions attributable to the rodent sequence (N_R) and to the human sequence (N_H), and the sum of lengths of gaps attributable to the rodent sequence (G_R) and to the human sequence (G_H)

No. of proteins			*Substitutions*		*Gaps*	
$N_R>N_H$	$N_R<N_H$	$N_R=N_H$	N_R	N_H	G_R	G_H
35	12	7	600	416	385(44)	108(31)
$P<0.005$			$P<0.001$		$P<0.001$	$(0.05<P<0.1)$

Comparing the rate of nucleotide substitution in the human and rodent lineages using marsupial genes as references, Easteal and Collet (1994) concluded that the rate of silent substitution did not differ between the rodent and human lineages, though the rate of nonsynonymous substitution was significantly higher in the rodent lineage than in the human lineage. They took the near equality of estimated silent rates in rodents and humans as evidence that the mutation rate is the same in rodents and primates, i.e., a molecular clock exists at silent sites. However, as silent distances between marsupial and rodent (or human) genes are large, estimates of these distances should be taken with caution. Table 7.4 shows a summary of their data for the coding regions. Easteal and Collet noted that the number of transition substitutions per 100 sites at twofold degenerate site (A_2) is similar to that at fourfold degenerates sites (A_4) for the human-rodent and human-marsupial comparisons (27.4 vs. 25.8 and 48.9 vs. 44.2) but is much higher than

Table 7.4. Numbers of transition substitutions (A_i) and transversion substitutions (B_i) per 100 nondegenerate, twofold degenerate, and fourfold degenerate sites in comparisons between human, rodent, and marsupial genes

Sites	*Taxa compared*		
	Human-rodent	*Human-marsupial*	*Rodent-marsupial*
Nondegenerate			
A_0	4.4±0.3	7.3±0.4	7.6±0.4
B_0	4.3±0.3	8.7±0.4	9.7±0.4
Twofold			
A_2	27.4±1.6	48.9±2.8	54.4±3.2
B_2	5.2±0.5	8.5±0.7	9.3±0.7
Fourfold			
A_4	25.8±1.8	44.2±3.6	40.6±3.6
B_4	20.6±1.4	45.7±2.8	52.4±3.2

the latter in the rodent-marsupial comparison (54.4 vs. 40.6). For this reason, they omitted the data for A_2, arguing that these data were biased and not as suitable as those for A_4. However, note that for fourfold degenerate sites the number of transitions per site (A_4) is about the same as the number of transversions per site (B_4) for both the human-rodent comparison and the human-marsupial comparison but is much lower than the latter in the rodent-marsupial comparison. A_4 tends to be higher than B_4 in mammalian genes. Thus, in the rodent-marsupial comparison, what is peculiar is A_4 rather than A_2, because A_2 is similar to B_4 whereas A_4 is much lower than B_4. Indeed, with the exception of A_4, all A_i and B_i for the rodent-marsupial comparison are larger than the corresponding values for the human-marsupial comparison, suggesting a higher rate in the rodent lineage than in the human lineage. Note that the number of transition substitutions per site (A) is estimated by the formula

$$A = -(1/2)\,\ell n\,(1-2P-Q) + (1/4)\,\ell n\,(1-2Q)$$

where P and Q are the proportions of transitional and transversional differences between the two sequences. Therefore, it is easier to obtain a reliable estimate of A when Q is small, which is the case for twofold degenerate sites, than when Q is large, which is the case for fourfold degenerate sites. For this reason, A_2 is preferable over A_2 and, as noted above, the A_2 values point to a higher synonymous rate in the rodent lineage than in the human lineage. Moreover, since the number of transversional substitutions (B) is given by the simple formula

$$B = -(1/2)\,\ell n\,(1-2Q)$$

it is easier to estimate B than A. Let us therefore consider B_4. For B_4, the difference between the rodent-marsupial and human-marsupial comparisons is 6.7%, which is significant because the standard error is 2.7%. The above results are consistent with the view of a higher rate in the rodent lineage than in the human lineage for synonymous substitutions as well as for nonsynonymous substitutions.

Since at the early stage of the primate-rodent divergence the rates in the two lineages should have been very similar, the rate differences between the two lineages should have occurred mainly in more recent times. This appears to be the case. In order to estimate the substitution rate between two species we must know their divergence time. However, divergence times are usually not well established, so we consider a range of estimates and an intermediate date for every pair of species compared. It is commonly thought that the divergence between the human and the Old World monkey lineages came after middle Oligocene times (some 30 Myr ago) and before early Miocene times (some 20 Myr ago), so we use a range of 20 to 30 Myr with an intermediate date of 25 Myr. The divergence time between the New World (NW) monkeys and the OW monkeys has been generally thought to be between 35 and 45 Myr ago and we use a range of 30 to 45 Myr with an intermediate of 35 Myr ago. The divergence time between mouse and rat has been estimated to be between 8 and 14 Myr ago, but Wilson et al. (1977) argued that it can be anywhere between 5 and 35 Myr ago. We use a range of 10 to 30 Myr with an intermediate date of 15 Myr. Table 7.5 shows the estimates of the substitution rate in introns. The intermediate estimates are 1.4×10^{-9}, 2.1×10^{-9} and 4.8×10^{-9} for the human-OW monkey, human-NW monkey, and mouse-rat comparisons, so the rate may be 2 to 4 times higher in rodents than in higher primates. However, these comparisons should be taken with caution because they involve assumptions of divergence times and are based on limited sequence data.

Table 7.5. Rates of substitution per site per year in introns

Species pair	*L* *(bp)*	*Percent divergence*	*Time* *(10^6 yrs)*	*Rate* *(10^{-9})*
Human vs. OW monkeys	8,478	7.1	25 (20-30)	1.5 (1.2-1.8)
Human vs. NW monkeys	8,478	14.7	35 (30-45)	2.1 (1.6-2.5)
Mouse vs. rat	4,038	14.4	20 (10-30)	4.8 (2.4-7.2)

Male-driven Sequence Evolution

Since Haldane (1935, 1947), it has been commonly believed that the mutation rate is much higher in the human male germ line than in the female germ line because the number of germ-cell divisions per generation is much larger in males than in females. However, direct estimation of mutation rates was difficult, and data supporting this view were limited, deriving mainly from sex-linked genetic diseases. Thus, it was unclear how high the ratio (α) of male to female mutation rates is. Another problem with this type of data is that no distinction between indels (insertions and deletions) and substitution mutations was made because the molecular changes causing the disease were not examined. Here we are more interested in substitution mutations.

An alternative approach proposed by Miyata et al. (1987) provides a simple means for estimating α. We consider a slightly different formulation. Let u_m and u_f be the mutation rates per nucleotide site per generation in the male and female germ lines, respectively, and $\alpha=u_m/u_f$. Note that an autosomal sequence is derived from the father and the mother with equal probability. So, the mutation rate per generation for an autosomal sequence is $A=(u_m+u_f)/2$. For a Y-linked sequence the mutation rate is $Y=u_m$ because it must come from the father. For an X-linked sequence the mutation rate is $X=(1/3)u_m$ + $(2/3)u_f$ because it is, on average, carried two-thirds of the time by the mother and one-third of the time by the father. Therefore, the expected mutation ratio of the Y chromosome to autosomes is

$$Y / A = 2\alpha / (1+\alpha)$$

the mutation ratio of the X chromosome to autosomes is

$$X / A = (2/3)(2+\alpha) / (1+\alpha)$$

and the mutation ratio of the Y chromosome to the X chromosome is

$$Y / X = 3\alpha / (2+\alpha)$$

Miyata et al. (1987) noted that as α increases from 1 to infinity Y/A increases from 1 to 2 while X/A decreases from 1 to $2/3 \approx 0.67$. From the sequences of a Y-linked and an autosomal pseudogene in humans they estimated Y/A=2.2, and from the synonymous rates in X-linked and autosomal genes from humans and rodents they estimated X/A=0.60. Since the two estimates exceed the upper and lower limits for the ratios, respectively, α is infinitely large. As this implies that the contribution of females to mutation is negligible, Miyata et al. concluded that the evolution of DNA sequences in mammals is male-driven.

This approach is ingenious, but the assumption that synonymous changes are selectively neutral may not hold. If synonymous changes are in fact subject to some, even very weak, selective contraints, then selection is more effective in reducing the rate in X-lined genes than in autosomal genes because in males an X-linked gene is present in a single copy and in females on of the two copies may be inactivated. A second problem with the data of Miyata et al. is that the genes they used were not homologous, but the mutation rates for nonhomologous sequences may be different, because the mutation rate depends on both the nucleotide composition (e.g., G and C are on average more mutable than A and T) and neighbouring nucleotides. The autosomal and Y-linked pseudogenes they used were indeed homologous and probably not subject to selection, but the data was limited. So, the reliability of their estimate was not certain. In fact, limited data from the coding regions of mammalian X-linked by Y-linked zinc finger protein (Zfx and Zfy) genes suggested that the rate at synonymous sites is only two times higher in Zfy than in Zfx, instead of the predicted ratio Y/X=3 for large α.

To obtain a better estimate Shimmin et al. (1993) sequenced the last intron (~1 kb) of the human, orangutan, baboon and squirrel monkey Zfx and Zfy genes, which are highly homologous. For each species pair, the Y sequences are more divergent than the X sequences. For example, for the human-orangutan pair the divergence is only 2.1% between the two X sequences but 4.0% between the two Y sequences.

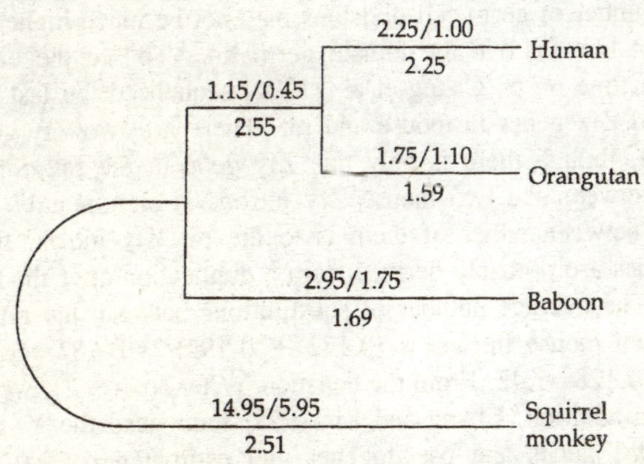

Fig. 7.2. The numbers above each branch are branch lengths estimated from the Y-linked and the X-linked intron sequences, respectively, and the number below each branch is the ratio of the preceding two nummbers. The least-squares method was used for the estimation.

As a more detailed analysis, they computed the length of each branch of the four-species tree. Clearly, in each branch the Y sequence has evolved faster than the X sequence, though the ratio Y/X varies considerably among branches, probably largely because the number of substitutions in each branch is relatively small. When all branches are considered together, they obtained Y/X=(14.95 + 2.95 + 1.15 + 1.75 + 2.25)/(5.95 + 1.75 + 0.45 + 1.10 + 1.00)=2.25.

The above Y/X ratio leads to an estimate of $\alpha \approx 6$ from the relation $Y/X=3\alpha/(\alpha+2)$. This indicates that in higher primates the mutation rate is considerably higher in the male germ line than in the female germ line. However, the α value is far lower than suggested by Miyata et al. (1987). The study by Ketterling et al. (1993) on the germ line origins of mutation in families with hemophilia B (which is caused by defects in the gene for clotting factor IX) gave an estimate of $\alpha \approx 35$. Although these two estimates have large standard errors, they strongly suggest that α is, at most, of the order of 10.

As in the case of synonymous changes, introns may be subject to selective constraints and this can distort the rate ratio. In the present case, however, selection is unlikely to be more effective against the Y intron than against the X intron, because Zfy and Zfx are expressed in males whereas only Zfx is expressed in females, so that Zfy may be dispensable whereas Zfx may not.

Since the above α value was estimated from primate sequences, it may not be applicable to other mammals. In particular, the generation times in rodents are much shorter than those in higher primates, so that the number of germ-cell divisions may not be much higher in the male germ line than in the female germ line. To see the effect of generation time on α, Chang et al. (1994) sequenced the last introns of Zfy and Zfx genes in mouse and rat. There are two Zfy genes in the mouse, though there is only one Zfy gene in the rat. Since the distance between the two mouse Zfy introns is significantly shorter than that between either of them is to the rat Zfy intron, the two mouse genes are probably derived from a duplication after the mouse-rat split. The average number of substitutions between the rat intron and the two mouse introns is (0.172 + 0.192)/2=0.182 and so Y/X=0.182/0.128=1.42. From the equation $Y/X=3\alpha/(\alpha+2)$, we obtain $\alpha=1.8$. In addition, Chang and Li (1995) sequenced the Y- and X-linked *Ube*1 genes and pseudogenes and estimated $\alpha=2.0$. These estimates suggest that the α value ($\sim$2) in mice and rats is considerably lower than that in higher primates. So, there appear to be generation-time effects on α.

Causes of Rate Variation among Lineages

Three factors have been proposed as the major causes of rate variation among evolutionary lineages. First, the efficiency of the DNA repair system may differ among lineages. This hypothesis was proposed by Britten (1986) to explain the difference in substitution rate between the primate and rodent lineages. There is evidence from cultured cells for supporting this hypothesis. Unfortunately, no *in vivo* data seem to exist and it is difficult to evaluate the importance in this factor.

Second, the generation-time effect hypothesis postulates a higher rate of evolution in organisms with a short generation time than in organisms with a long generation time, because in any arbitrary unit of time short-generation organisms will go through more generations and therefore more rounds of germ-cell divisions (DNA replications). As noted earlier, the rate of substitution is higher in monkeys than in humans and is even higher in rodents. These observations are consistent with the generation-time effect hypothesis because among these three groups of organisms, the generation time is longest in humans, intermediate in monkeys, and shortest in rodents. The hypothesis is further supported by the observation that the rate of substitution in the chloroplast genome is more than five times higher in grasses than in palms.

The generation-time effect hypothesis implicitly assumes that errors during DNA replication are the major source of mutation. This assumption can be tested by comparing the sex-ratio of mutation rate (α) and the sex-ratio of the number of germ-cell divisions (c). Using published data on gametogenesis, Li has estimated the numbers of germ-cell divisions per generation in females (n_f) and males (n_m) and the ratio $c=n_m/n_f$. In the development of a female mouse, the number of DNA replications in the germ line (i.e., from zygote to mature egg) is $n_f \approx 27$. In the development of male mouse, the number of germ-cell divisions from zygote up to the formation of stem spermatogonia is ≈ 29. Spermatogenesis requires 10 further DNA replications to the production of spermatids, initiating on day 6 after birth and occurring on average every 8.6 days in adult male mice. The span of high reproductivity in laboratory male mice is approximately from 2 to 8 months. If we assume that the average reproductive age of male mice in the wild is 5 months, then a stem spermatogonia would have gone through $(5 \times 30)/8.6 \approx 17$ divisions and the total number of DNA replications from zygote to age 5 months is $n_m \approx 29 + 10 + 17 = 56$. Therefore, $c \approx 56/27 = 2.1$. If the average reproductive age is 2 or 8 months, c becomes 1.6 and 2.4, respectively.

For the rat, $n_f \approx 28$. The number of cell divisions in the male germ line to form stem spermatogonia is ≈ 31, the spermatogenesis cycle occurs every ~12.9 days, and the period of maximal fertility of laboratory rats occurs between 100 and 300 days after birth. If we assume that the average reproductive age of males in the wild is 7 months, a stem spermatogonium would have gone through $(7 \times 30)/12.9 = 16$ divisions, and the total number of DNA replications in the male germ line is $n_m = 31 + 10 + 16 = 57$ and $c \approx 57/28 = 20$.

In humans, n_f=33. The data for n_m are more scanty than those for rodents and so the estimate of n_m is less reliable. The number of cell divisions from zygote to stem spermatogonia at puberty was estimated to be ~40. Spermatogenesis requires 5 further DNA replications, and the spermatogenesis cycle occurs every ~16 days or 23 cycles per year. If the average reproductive age of males is 20 years, then the number of DNA replications for stem spermatogonia from puberty (age 13) to 20 is $(20 - 13) \times 23 \approx 160$, $n_m \approx 40 + 5 + 160 = 205$, and $c \approx 6.2$. If the average reproductive age of male is 15, then $c \approx 28$.

Although these estimates are rough, they are similar to the estimates of $\alpha = 2$ in mice and rats and $\alpha = 3$ to 6 in higher primates. This rough agreement suggests that errors in DNA replication during germ-cell divisions are indeed the primary source of mutation and that the contribution of replication-independent factors such as oxygen radicals to mutation is considerably less important.

Third, the metabolic-rate hypothesis postulates a higher rate of evolution for organisms with a high metabolic rate than for organisms with a low metabolic rate. Martin and Palumbi (1993) found a negative correlation between adult body size and substitution rate in the η-globin pseudogone is higher primates and in the cytochrome b gene in mammals. They also reviewed published genetic distances obtained from restriction analysis of mtRNA and fossil dates for various vertebrates and found that the same general trend holds. As body size is unlikely to control the rate of DNA evolution, its relationship with substitution rate probably comes from its correlation with generation time and metabolic rate. Although generation time and metabolic rate can be used to explain some patterns of rate heterogeneity equally well, Martin and Palumbi (1993) noted that differences in metabolic rate could also explain some important exceptions to the generation-time model. For example, whales have a slow substitution rate relative to primates, despite their shorter generation times. They noted also that the separation of mtDNA rates into two broad groups comprising

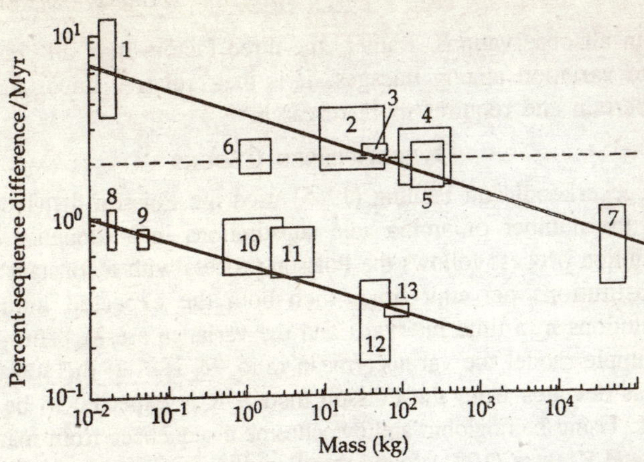

Fig. 7.3. Relationship between rate of mtDNA sequence divergence (% change per million years) for various vertebrates.

the homeotherms and the poikilotherms is incompatible with a simple generation time hypothesis. There are, however, exceptions to the metabolic rate hypothesis. For example, birds appear to have slower rates than do mammals, though birds have higher metabolic rates than do mammals, and the substitution rates are about the same in rats and mice, despite the fact that the body size of rats is more than twice that of mice.

The metabolic rate hypothesis is based on the argument that the rate of DNA damage is proportional to the metabolic rate because metabolism produces oxygen radicals, which are highly reactive molecules with free electrons that can damage DNA directly by attacking the sugar-phosphate backbone or the base, or indirectly via lipid peroxidates. This hypothesis may hold for mtDNA because about 95% of oxygen in cells is consumed by mitochondria, so that oxidative damage is greatest to mtDNA; this may partially explain why rates of nucleotide substitution are greater in mtDNA that in nuclear DNA. However, whether it holds for nuclear DNA remains to be seen because of the low consumption of oxygen in nucleus. The nuclear DNA sequence data used to support the metabolic rate hypothesis were limited to the η-globin pseudogene, and this analysis involved assumptions about generation times and sizes. Relative-rate tests on DNA-DNA hybridization data from birds did not support the metabolic rate hypothesis but showed generation-time effects.

It should be emphasized that the above three alternative hypotheses are not mutually exclusive and that each single hypothesis may not explain all observations. Rather, the three factors may all contribute to rate variation among lineages. It is their relative importance that is uncertain and requires more research.

Overdispersed Clocks

Zuckerkandl and Pauling (1965) used the Poisson distribution to study the number of amino acid substitutions in a sequence. If the substitution process follows the Poisson process with a constant rate of λ substitutions per unit time, then both the expected number of substitutions in a time interval t and the variance are λt. Thus, under this simple model the variance/mean ratio, R, is 1. If the substitution process deviates from the Poisson model, R is expected to be larger than 1. Using hemoglobin and cytochrome c sequences from mammals, Ohta and Kimura (1971) found that R was approximately 2. Thus, the substitution process did not seem to follow the Poisson process with a constant rate. As R is often called the dispersion index, the clock is said to be overdispersed if R is larger than 1. Because under strictly neutral mutations, the substitution process follows the Poisson process with the substitution rate λ equal to the mutation rate u, a significant deviation of R from 1 has usually been taken as evidence against the rate-constancy assumption and the neutral mutation hypothesis. However, as will be discussed below, there are factors that can inflate the R value.

Kimura (1983) noted that a star phylogeny in which all lineages diverged at the same time is an ideal situation for studying the variability of the molecular clock, i.e., for testing $R = 1$, because under the assumption of rate constancy the number of amino acid (or nucleotide) substitutions is expected to be the same in all lineages and the variance of the number of substitutions among lineages can be easily computed. Kimura (1983) analyzed the α-globin sequences from human, mouse, rabbit, dog, horse, and bovine, which were assumed to conform to a star phylogeny, and concluded that the molecular clock hypothesis could not be rejected from these data because $R=1.26$ was not significantly larger than 1.

In a series of papers Gillespie (1986a,b; 1987) developed a more sophisticated theory for studying R and applied it to analyze protein and DNA sequence data. He found that R is roughly in the range from 2.5 to 3.5, so that the clock appears to be overdispersed. However, the results may not be taken as evidence against the neutral mutation hypothesis for two reason. First, the assumption of a star phylogeny may not hold, i.e., some lineages may have in fact evolved for a longer time than others. Second, the mutation rate or the generation time may not have remained the same in all lineages, i.e., there may be generation-time effects. Gillespie (1989) called these two effects collectively the lineage effect and proposed to take this effect into account by using different weights for different lineages. In the case of three lineages, the situation is quite simple. Let N_i be the number of substitutions in the ith lineage. The weight w_i for the ith lineage is chosen so that

$$E(N_1 / w_1) = E(N_2 / w_2) = E(N_3 / w_3)$$

with the following constraint

$$w_1 + w_2 + w_3 = 3$$

In this model one can write

$$E(N_i) = w_i u; \; V(N_i) = w_i \sigma^2$$

and shown that

$$R = V(N_i) / E(N_i) = \sigma^2 / u$$

which is equivalent to the ideal case of a star phylogeny and is expected to be 1 if the substitution process in each lineage is a Poisson process with a constant rate.

Gillespie (1989) analyzed the coding sequences of 20 genes from human, mouse (or rat), and bovine. He treated synonymous and nonsynonymous substitutions separately and considered three different weights: (1) equal weight, (2) the weights obtained from nonsynonymous substitutions, and (3) the weights obtained from synonymous substitutions. The R values corresponding to the three weights are denoted by R_e, R_n, and R_s, respectively Let us first consider nonsynonymous substitutions. For the equal-weight case, $w_p = w_r = w_a = 1$, where the subscripts, p, r, and a refer to the primate, rodent, and artiodactyl lineages, respectively, and the R value varies from 0.01 to 32.22, with an average of 8.26. For the nonsynonymous weights, $w_p = 0.84$, $w_r = 1.28$, and $w_a = 0.89$ and the average R value is indeed reduced, but only from 8.26 to 6.95. For synonymous substitutions, R varies from 1.00 to 57.45 with an average of 14.41 for the equal-weight case but the average reduces to 8.67 for the synonymous-weight case ($w_p = 0.63$, $w_r = 1.61$, and $w_a = 0.76$).

All the above analyses neglected two effects: (1) the effect of corrections for multiple substitutions at the same site and (2) the correlation between comparisons, e.g., in the comparison of the two pairs (A, B) and (A, C), lineage A is used twice. To take the two effects into consideration, Bulmer (1989) derived a correction factor for the Jukes-Cantor model and found that it becomes serious when the degree of sequence divergence is 25% or higher. Gillespie (1989) conducted a simulation study and found that in the case of nonsynonymous substitution the correction factor was indeed not important and, as the above ratio of 6.95 remained about the same, he took it as evidence against the neutral mutation hypothesis. On the other hand, in the case of synonymous substitution, the correction factor was quite large, so that the above ratio of 8.67 could not be taken as evidence against the neutral mutation hypothesis.

Goldman (1994) used computer simulation to study Kimura's (1983) and Bulmer's (1989) estimators and concluded that R may be a poor statistic for analyzing the accuracy of Poisson models. In particular, he argued that even for the cases where R was found by Gillespie (1989) to deviate for beyond 1, the R value became nonsignificant if the assumption of a star phylogeny was removed but a new phylogeny was inferred by the maximum likelihood method with a constant rate. However, assuming a constant rate of evolution is equivalent to assuming that the root of the tree is on the longest branch. In other words, it presumes that the longest branch is not due to a faster rate of evolution. Under this assumption, one may not be able to reject the molecular hypothesis, even if it actually does not hold.

In the case of nonsynonymous substitution, the largest deviations from 1 occurred for prolactin, growth hormone, and GPHA. When these three genes are removed from comparison, the average R value reduces to 2.77 and the deviation of R from 1 becomes considerably smaller. Ohta (1995) extended the analysis to include 49 genes, including prolactin, growth hormone and GPHA, and found an average R value of 5.60, which is somewhat smaller than that (6.95) for the 20 genes.

The above results suggest that while the variation in synonymous clock among lineages is mainly due to lineage effects, the variation in the nonsynonymous clock is subject to both lineage effects and episodic effects, the former being systematic whereas the latter are erratic. It has been controversial as to how the overdispersed clock for nonsynonymous substitution arose thinks that the overdispersion is caused by the interplay between fluctuation of population size and weak selection. She argues that the population size of a species fluctuates in time and the effectiveness of natural selection varies accordingly. Once in a while, such as at the time of speciation, the population size becomes small and the effect of random drift predominates, so that slightly deleterious mutations may become fixed in the population.

Later, mutations that compensate the fixed, slightly deleterious ones may occur and become fixed in the population. Thus, fluctuations in population size and the prevalence of slightly deleterious mutation cause irregularities in the clock. On the other hand, Gillespie (1989) thinks that the population size of a species rarely becomes small enough for the random fixation of slightly deleterious mutations to occur and that the episodic effects are mainly caused by environmental changes, so that different mutations are favoured at different times—since environmental changes are episodic, so are events of nucleotide substitution. As both views involve many assumptions, the controversy is not resolved.

Ratios of Synonymous and Nonsynonymous Rates in Different Lineages

A quantity that is considerably easier to compute than the dispersion index is the ratio of the synonymous rate to the nonsynonymous rate. Ohta (1993, 1995) proposed to use this ratio to test her nearly neutral mutation hypothesis. She considered this ratio in the primate, artiodactyl, and rodent lineages. As noted above, rodents have a shorter generation time and are expected to have a faster molecular clock than do the other two lineages. However, since selection against slightly deleterious mutations should be more effective in rodents than in primates and artiodactyls because rodents have a larger population size and since nonsynonymous changes are more likely to be exposed to selection than synonymous changes, the generation-time effect is expected to be

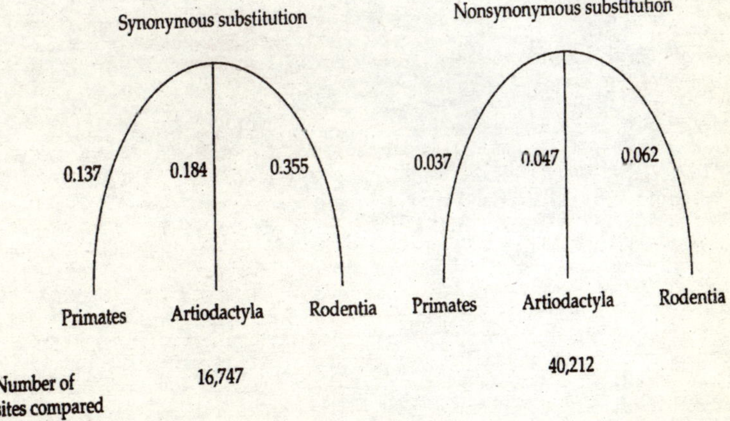

Fig. 7.4. Numbers of nucleotide substitutions per site on the primate, artiodactyl, and rodent lineages.

stronger for the synonymous rate than for the nonsynonymous rate. In other words, under the nearly neutral mutation hypothesis the ratio (r) of the synonymous rate to the nonsynonymous rate is expected to be larger in the rodent lineage than in the other two lineages.

To test the above prediction Ohta (1993, 1995) analyzed 49 genes. The ratio r is 0.137/0.037=3.7 in the primate lineage, 0.184/0.047=3.9 in the artiodactyl lineage, and 0.355/0.62=5.7 in the rodent lineage. Clearly, the ratio is highest in the rodent lineage and Ohta's prediction appears to hold.

8

SUBSTITUTING NUCLEOTIDES

The preceding chapter provides a conceptual basis for understanding the process of evolutionary change in nucleotide sequences. In reality, however, with the exception of viral genes, the substitution of one nucleotide for another may take thousands or even millions of years, so that it cannot be observed within a person's life time. Therefore, to detect nucleotide substitution in DNA sequences, we usually rely on comparison of sequences that have been derived from a common ancestral sequence. The number of nucleotide substitutions between two sequences is a quantity of foremost importance in molecular evolution because it is used to compute the rate of evolution, to estimate divergence time, and to reconstruct phylogenetic trees by many methods. This chapter is devoted to methods of estimating this number. To compare sequences, one must first align the sequences; however, this topic will be discussed last because it is more involved and not major aim of the chapter.

As noted in already, if multiple substitutions have occurred at any site, then the observed number of differences between the two sequences under study is smaller than the actual number of substitutions. Thus, to estimate the number of substitutions between two sequences we must correct for multiple hits. Our aim here is to study how to make such corrections. Since the pioneering work of Jukes and Cantor (1969), Halmquist (1972), and Kimura and Ohta (1972), many methods have been proposed for this purpose. A number of methods are present below:

The number of nucleotide substitutions between two sequences is usually expressed in terms of the number (K) of substitutions *per nucleotide site* rather than as the *total number* (N) of substitutions between the two sequences. If the number of nucleotide sites compared between the two sequences is L, then K=N/L. This quantity facilitates comparison of the degrees of divergence among sequence pairs that do not have the same sequence length (L). The K value is also commonly referred to as the evolutionary distance between two sequences.

Protein-coding and noncoding sequences should be treated separately, for they usually evolve at different rates. In the former case, it is advisable to distinguish between synonymous and nonsynonymous substitution, because they may evolve at markedly different rates and because a separation of the two types of substitution provides more information for studying the mechanisms of molecular evolution.

Noncoding Sequences

The following methods assume that all the sites in a sequence evolve at the same rate and follow the same substitution scheme; the case of nonuniform rates will be considered later. The number of sites compared between two sequences is denoted by L. Deletions and insertions are excluded from comparison.

Jukes and Cantor's Method

This simplest method is based on the one-parameter model of nucleotide substitution. Under this model the expected proportion of identical nucleotides between two sequences that diverged t time units ago is given by

$$I_{(t)} = \frac{1}{4} + \frac{3}{4} e^{-8\alpha t}$$

The probability that the two sequences are different at a site at time t is $p = 1 - I_{(t)}$ and so

$$p = \frac{3}{4}\left(1 - e^{-8\alpha t}\right)$$

or

$$8\alpha t = -\ln\left(1 - \frac{4p}{3}\right) \quad \text{...(1)}$$

The time of divergence between two sequences is usually unknown, and thus we cannot estimate α. Instead, we estimate K, the actual number of substitutions per site since the divergence between the two sequences. In the one-parameter model, K=2(3αt), where 3αt is the expected number of substitutions per site in one lineage. Using Equation (1) we estimate K by

$$K = -\left(\frac{3}{4}\right)\ln\left(1 - \frac{4p}{3}\right) \qquad \text{...(2)}$$

where p is assumed to be equal to the observed proportion of different nucleotides between the two sequences. When L is large, the sampling variance of K is approximately given by

$$V(K) = p(1-p) / [L(1-4p/3)^2] \qquad \text{...(3)}$$

Kimura's Two-Parameter Method

In this method, the differences between two sequences are classified into transitions and transversions. Let P and Q be the proportions of transitional and transversional differences between the two sequences, respectively. Then, the number of nucleotide substitutions per site between the two sequences is estimated by

$$K = \left(\frac{1}{2}\right)\ln(a) + \left(\frac{1}{4}\right)\ln(b) \qquad \text{...(4)}$$

where $a=1/(1-2P-Q)$ and $b=1/(1-2Q)$. The sampling variance is approximately given by

$$V(K) = [a^2P + c^2Q - (aP + cQ)^2] / L \qquad \text{...(5)}$$

where $c=(a+b)/2$.

Tajima and Nei's Method

Tajima and Nei (1984) proposed a method that does not require the assumption of equal frequencies of the four nucleotides. They noted that formula (2) can be written as

$$K = -b_1 \ln(1 - p / b_1) \qquad \text{...(6)}$$

where $b_1=1-\Sigma q_i^2$; q_i is the equilibrium frequency of the ith nucleotide (i=1, 2, 3, 4 corresponding to A, G, T, C). In the Jukes-Cantor model, $q_i=1/4$, so that $b_1=3/4$. Tajima and Nei showed that formula (6) also holds under the equal-input method, i.e., $\alpha_{ij}=\alpha_j$ for all $i \neq j$. When $\alpha_j=\alpha$ for all j, this model becomes identical with the Jukes-Cantor model. If the condition $\alpha_{ij}=\alpha_j$ does not hold for all $i \neq j$, formula (6) tends to give an underestimate. To remove this deficiency, Tajima and Nei proposed the following formula as a general estimator of K.

$$K = -b \ln(1 - p / b) \qquad \text{...(7)}$$

where $b=(b_1+p^2/h)/2$ and

$$h = \sum_{i=1}^{3} \sum_{j=i+1}^{4} \frac{(x_{ij})^2}{2q_iq_j}$$

in which x_{ij} is the proportion of pairs of nucleotides i and j between the two homologous DNA sequences. The sampling variance for this estimator is approximately given by

$$V(K) = b^2 p(1-p) / [b-p)^2 L] \quad ...(8)$$

Other Methods

Explicit analytic solutions have also been obtained for a number of other models. Gojobori et al. (1982a) solved Kimura's (1981a) 6-parameter model and Tamura and Nei (1993) solved an equal-input related model with 6 parameters, i.e., the fourth matrix with the condition $\beta_1=\beta_2=\gamma_1=\gamma_2=\beta$. The method of Lanave et al. (1984) has been interpreted as a 12-parameter method because their transition matrix is equivalent to the last matrix, but it is in effect a 9-parameter method because the estimation procedure involves some assumptions.

Instead of K (the number of nucleotide substitutions per site), one may consider other types of distance between sequences. For example, Lake (1994) has proposed the following distance. Let N_{ij} be the number of sites where the nucleotide is i in the first sequence and j in the second sequence and let J be the determinant defined by

$$J = \begin{vmatrix} N_{AA} & N_{AT} & N_{AC} & N_{AG} \\ N_{TA} & N_{TT} & N_{TC} & N_{TG} \\ N_{CA} & N_{CT} & N_{CC} & N_{CG} \\ N_{GA} & N_{GT} & N_{GC} & N_{GG} \end{vmatrix}$$

Further, let $F_i=A_iT_iC_iG_i$, where A_i is the number of occurrences of nucleotide A in sequence i, i=1, 2; T_i, C_i, and G_i are similarly defined. Then the distance between the two sequences is estimated by

$$d = -\frac{1}{4}\left[\ln(j) - \frac{1}{2}\ln(F_1F_2)\right] \quad ...(9)$$

This distance is equal to K only if the frequencies of the four nucleotides are equal to 1/4. However, the distance is additive and is therefore called the *paralinear* distance. Two other advantages of this distance is that it holds for the general (12-parameter) model of nucleotide substitution and is applicable even if the nucleotide frequencies change with time, i.e., they are nonstationary. Therefore, it may be suitable for phylogenetic reconstruction when the nucleotide composition of the sequences under study has changed with time. Lockhart et al. (1994) have proposed a distance, called the LogDet, that is similar to the paralinear distance.

An example

As an example of using some of the above methods to compute the number of nucleotide substitutions between two sequences (K), let us consider the first introns of human and owl monkey (a New World monkey) insulin genes. Since we are concerned with nucleotide substitutions, we exclude the gaps from comparison; in total, 163 nucleotides are compared, i.e., L=163. For the Jukes-Cantor method, we first count the number of differences between the two sequences, which is N=18. We then obtain p=N/L=18/163=0.110 as the proportion of nucleotide differences between the two sequences. Putting this p value into Equation (2), we obtain K=0.119 as an estimate of the number of nucleotide substitutions per site between the two sequences.

For Kimura's two-parameter method, we first count the number of transitional changes (14) and the number of transversional changes (4) and then obtain the proportions of transitional and transversional differences, i.e., P=14/163=0.086 and Q=4/163=0.025. Putting these values into Equation (4), we obtain K=0.122.

For Tajima and Nei's method, the computation is more complicated. First, we estimate the equilibrium frequencies of the four nucleotides A, G, T, and C, which will be denoted 1, 2, 3, and 4, respectively. Since there are 22 and 23 A nucleotides in the first and second sequences, respectively, the frequency of nucleotide A is estimated to be $q_1=(22+23)/(2\times163)=0.138$. Similarly, we estimate the frequencies of G, T, and C: $q_2=0.365$, $q_3=0.310$. From the alignment we can compute $x_{ij}=y_{ij}+y_{ji}$, in which y_{ij} is the proportion of nucleotide pairs where the first and second sequences have nucleotides

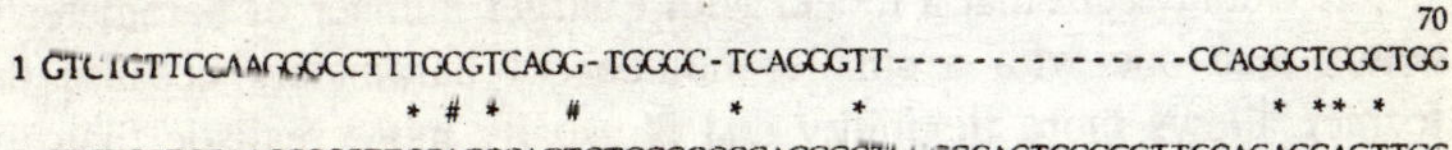


```
                                                                    140
1 ACCCCAGGCCCCAGCTCTGCAGCAGGGAGGACGTGGCTGGGCTCGTGAAGCATGTGGGGTGAGCCCAGG
        * *                 **                    #
2 ACCCCAGGTCTCAGC---------GGGAGGGTGTGGCTGGGCTC-TGAAGCATTT--GGGTGAGCCCAGG

                                                  196
1 GGCCCCAAGGCAGGGCACCTGGCCTTCAGCCTGCCTCAGCCCTGCCTGTCTCCCAG
    *   *                   #
2 GGCTC-AGGGCAGGGCACCTG-CCTTCAGC-GGCCTCAGC-CTGCCTGTCTCCCAG
```

Fig. 8.1. Alignment of the sequences of the first introns of human (sequence 1) and owl monkey (sequence 2) insulin genes.

i and j, respectively. For example, x_{12}=(2+2/163=0.025. Similarly, x_{13}=0.000, x_{14}=0.006, x_{23}=0.018, x_{24}=0.000, and x_{34}=0.061. Putting the q_i and x_{ij} values into Equation (7), we obtain K=0.124.

For the method of Lanave et al., we use a computer and obtain K=0.126.

For the method of Lake (1994), we have

$$J = \begin{vmatrix} 20 & 0 & 0 & 2 \\ 0 & 24 & 5 & 1 \\ 1 & 5 & 45 & 0 \\ 2 & 2 & 0 & 56 \end{vmatrix}$$

and A_1=22, T_1=30, C_1=51, G_1=60, and A_2=23, T_2=31, C_2=50, G_2=59. Thus, J=1,175,566, F_1=2,019,600, and F_2=2,103,350. Putting these values into Formula (9), we obtain d=0.147.

It is seen that the estimate of K increases somewhat with the sophistication of the method. This change occurs because the region is rich in G and C nucleotides (68% GC) and because there are more transitional changes than transversional changes (i.e., 16 vs. 4). As a consequence, the smallest (0.119) and largest (0.126) estimates differ by about 6%. Note, however, we do not know which estimate is most accurate because we do not know the true value; the question of accuracy is discussed below. Note also that the distance obtained by Lake's (1994) method (0.147) is considerably larger than those obtained the other methods, but, as mentioned above, Lake's method was not intended for estimating K.

Comparison of Methods

It would seem that a model with a larger number of parameter is better than one with a smaller number of parameters because the former allows more flexibility and is usually more realistic than the latter. In practice, this is not always the case, for two reasons. First, in addition to assumptions on the substitution scheme, it is usually necessary to make further assumptions in the derivation of a method. For example, Tajima and Nei's method assumes that the common ancestral sequence had the equilibrium nucleotide frequencies and that these frequencies are estimated from the two extant sequences, but such assumptions are only implicit in the one-parameter and Kimura's two-parameter methods, and, indeed, Formulas (2-5) do not involve nucleotide frequencies. Obviously, additional assumptions can cause estimation errors. Second, and more importantly, the effect of sampling errors, which arise because the number of nucleotides compared is finite, is usually stronger on a model with a larger number of parameters than one with a smaller number of parameters. The effect may be seen in the variance or the bias of an estimate. More seriously, sampling errors may render a method inapplicable, because all estimators of K involve logarithmic functions and the argument can become zero or negative. This chance is expected to increase with the number of parameters involved. Of course, as the sequence length increases, the effect of sampling errors should decrease and a model with more parameters is likely to become better than one with fewer parameters.

Gojobori et al. (1982a), Tajima and Nei (1984), Zharkikh (1994), and others have conducted extensive simulation studies to compare the performances of various methods for estimating K. The simulation results suggest that if K is 0.5 or smaller, then estimation of K is fairly simple, and even such simple methods as the Jukes-Cantor method and the Tajima-Nei method give fairly accurate estimates of K. For more divergent sequences, the methods of Lanave et al. (1984) and Gojobori et al. are preferable to the other methods, particularly for long sequences ($L \geq 1000$ bp). Tamura and Nei's (1993) 6-parameter method was not included in these studies, but it is likely to perform at least as well as that of Gojobori et al. because both methods have the same number of parameters and because the former allows more flexibility for unequal transitional and transversional rates. For example, the rates of C→A (transversion) and G→A (transition) are allowed to be unequal in Tamura and Nei's method but not in Gojobori et al. For divergent sequences ($K > 1$), however, the methods of Lanave et al., Gojobori et al., and Tamura and Nei may not be applicable and in this case one may have to use a simpler method such as that of Tajima and Nei (1984). Table 8.1 shows Zharkikh's (1994) simulation comparison of five estimation methods for $K = 1$; six nucleotide substitution models were used to simulate sequence divergences. If the sequence length is only 100 bp, the estimates obtained by the methods of Gojobori et al. and Lanave et al. (denoted as GIN and LA) actually are, on average, slightly worse than that by the Tajima and Nei (TN) method, though they are, on average, better than those obtained by the Jukes-Cantor method and Kimura's two-parameter methods. Moreover, the proportion of inapplicable cases is much higher for GIN and LA than for the other methods. These results are due to the fact that when the sequence is short, sampling effects tend to be stronger for GIN and LA than for the one- and two-parameter methods. As the sequence length increases to 1000 bp, GIN and LA start to show between estimates than the other methods and the proportion of inapplicable cases is considerably reduced for both GIN and LA. The simulation results show the difficulty of obtaining accurate estimates of K for divergent sequences. Thus, regardless of which method is used, an estimate should be taken with caution, if it is substantially larger than 1.

Table 8.1. Estimates of the number of nucleotide substitutions per site (K) by different methods from simulation data with sequence length L and the true value of K=1

Models	JC	K2	TN	GIN	LA
(a) L=100 bp					
JC	1.025	1.044	1.085	1.113	1.125
K2	0.935	1.050	1.091	1.094	1.107
EI	0.904	0.943	1.114	1.045	1.122
EIr	0.829	0.864	1.039	1.014	1.089
GIN	0.878	0.925	1.078	1.097	1.107
RAND	0.902	0.920	1.052	1.069	1.089
NA(%)0	0.0	1.1	2.2	24.4	33.9
(b) L=1000 bp					
JC	1.003	1.004	1.008	1.011	1.012
K2	0.901	1.006	1.011	1.016	1.017
EI	0.887	0.907	1.009	0.941	1.016
Elr	0.813	0.832	0.949	0.931	1.019
GIN	0.843	0.870	0.975	1.018	1.021
RAND	0.869	0.877	0.966	1.009	1.026
NA	0.0	0.0	0.3	3.7	4.8

Nonuniform Rates among Nucleotide Sites

All the above methods assume that the rate of substitution per site (λ) is uniform along the sequence. As noted already, this assumption is not realistic for functional sequences, and the decrease in similarity between two sequences may be considerably slower under nonuniform rates than under the assumption of a uniform rate. It is known that rate variation among sites can cause severe under-estimation of K if the variation is not taken into account.

As noted already, limited data indicate that the gamma distribution may be used to describe approximately the distribution of substitution rate in a sequence. Under this distribution and under the Jukes-Cantor model, the expected number of nucleotide substitutions between two sequences ($K = 2\bar{\lambda}t$) is given by

$$K = \frac{3a}{4}\left[\left(1 - \frac{4p}{3}\right)^{-1/a} - 1\right] \quad ...(10)$$

where p is the proportion of different nucleotides sites between the two sequences and $a = \bar{\lambda}^2 / V(\lambda)$.

For Kimura's (1980) two-parameter model, Jin and Nei (1990) showed

$$K = \frac{a}{4}\left[2(1 - 2P - Q)^{-1/a} + (1 - 2Q)^{-1/a} - 3\right] \quad ...(11)$$

where P and Q are, respectively, the proportions of transitional and transversional differences between the two sequences. Tamura and Nei (1993) have solved the nine-parameter case.

Although the gamma distribution seems to be suitable for describing the rate variation among sites in some sequences, it is unlikely to be applicable to all sequences. More empirical studies on the distribution of substitution rates among sites in sequence are needed for making better estimates of K.

Protein-Coding Sequences

Miyata and Yasunaga (1980), Li et al. (1985b), Nai and Gojobori (1986), and others have proposed methods for estimating the number of substitutions between two protein-coding sequences. In most of these methods, synonymous and nonsynonymous substitutions are treated separately. We shall first describe the method of Li et al. (1985b) and later the modification by Li (1993) and Pamilo and Bianchi (1993). In this method the initiation and termination codons are excluded from analysis because these two codons rarely change with time. Unless stated otherwise, the universal code is used in the following discussion.

First, classify nucleotide sites into nondegenerate, twofold degenerate, and fourfold degenerate sites. A site is *nondegenerate* if all possible changes at this site are nonsynonymous, *twofold degenerate* if one of the three-possible changes is synonymous, and *fourfold degenerate* if all possible changes at the site are synonymous. For example, the first two positions of the codon TTT (Phe) are nondegenerate, while the third position is twofold degenerate. In comparison, the third position of the codon GTT (Val) is fourfold degenerate. For simplicity, the third position in each of the three isoleucine codons (the universal code) is treated as a twofold degenerate site, although the degeneracy at this position is actually threefold. In mammalian mitochondrial genes, there are only two codons for isoleucine, and their third position is indeed a twofold degenerate site. Using the above rules, we first count the numbers of the three types of sites in each of the two sequences compared and then compute the averages, denoting them by L_0 (nondegenerate), L_2 (twofold), and L_4 (fourfold), respectively.

Next, compare the two sequences codon by codon and infer the nucleotide differences between each pair of codons. Classify each difference according to the type of site at which it has occurred. This classification can be done by considering all possible paths between the two codons. For two codons that differ by only one nucleotide, the difference is easily inferred. For example, the difference between the two codons GTC (Val) and GTT (Val) is synonymous, while the difference between the two codons GTC (Val) and GCC (Ala) is nonsynonymous.

For two codons that differ by more than one nucleotide, we need to consider all possible minimum evolutionary pathways that can lead to the observed changes. For example, for the two codons AAT (Asn) and ACG (Thr), there are two possible pathways:

Pathway I: AAT (Asn) ↔ ACT (Thr) ↔ ACG (Thr)

Pathway II: AAT (Asn) ↔ AAG (Lys) ↔ ACG (Thr)

Pathway I requires one synonymous and one nonsynonymous change, whereas pathway II requires two nonsynonymous changes. It is known that synonymous substitutions occur, on average, considerably more often than nonsynonymous substitutions, and we may assume that pathway I is more likely than pathway II. For example, if we assume a weight of 0.7 for pathway I and a weight of 0.3 for pathway II, then the number of synonymous differences between the two codons is estimated to be $0.7 \times 1 + 0.3 \times 0 = 0.7$, and the number of nonsynonymous differences is $0.7 \times 1 + 0.3 \times 2 = 1.3$. Here, the weights used are hypothetical. The weights for all possible codon pairs have been estimated by Miyata and Yasunaga (1980) from protein sequence data and by Le et al. (1985b) from DNA sequence data. If we assume that both pathways are equally likely, then for the above example the number of synonymous differences is $(1 + 2)/2 = 1.5$. Thus, the weighted and unweighted approaches may give somewhat different results. In practice, the differences in estimates between the two approaches are usually small, but they can be important for genes coding for highly conserved proteins such as histones and actins.

The nucleotide differences in each class are further classified into transitional (S_i) and transversional (V_i) differences (i=0, 2, 4). Note that, in the class of twofold degenerate sites, transitions are synonymous and transversions are nonsynonymous. There are no exceptions to this rule in the mammalian mitochondrial code. In the universal code, however, there are two exceptions: the first position of the arginine codons (CCA, CGG, AGA, and AGG) and the last position in the three isoleucine codons (AUU, AUC, and AUA). In all these cases, all synonymous changes are included in S_2 and all nonsynonymous changes are included in V_2.

Let $P_i=S_i/L_i$ and $Q_i=V_i/L_i$, which are, respectively, the proportions of transitional differences and transversional per i-fold degenerate site between the two sequences. Applying Kimura's two-parameter method to each type of sites, we can show that the mean numbers of transitional (A_i) and transversional (B_i) substitutions per ith type site are given by

$$A_i = \left(\frac{1}{2}\right)\ln(a_i) - \left(\frac{1}{4}\right)\ln(b_i) \qquad \text{...(12)}$$

$$B_i = \left(\frac{1}{2}\right)\ln(b_i) \qquad \text{...(13)}$$

and their variance are given by

$$V(A_i) = \left[a_i^2 P_i + c_i^2 Q_i - (a_i P_i + c_i Q_i)^2\right] / L_i \qquad \text{...(14)}$$

$$V(B_i) = b_i^2 Q_i (1 - Q_i) / L_i \qquad \text{...(15)}$$

where $a_i=1/(1-2P_i-Q_i)$, $b_i=1/1(1-2Q_i)$ and $c_i=(a_i-b_i)/2$. The total number of substitutions per ith type site, K_i, is given by

$$K_i = A_i + B_i \qquad \text{...(16)}$$

An approximate sampling variance of K_i can be readily obtained from Equation (5) by replacing P by P_i, Q by Q_i, and L by L_i. Note that A_2 and B_2 denote the numbers of synonymous and nonsynonymous substitutions per twofold degenerate site, $K_4=A_4+B_4$ denotes the number of synonymous substitutions per fourfold degenerate site, and $K_0=A_0+B_0$ denotes the number of nonsynonymous substitutions per nondegenerate site. Thus, the above formulas can be used to compare the rates of substitution at the three different types of sites.

The number of substitutions per synonymous site (K_S) is computed as follows. The estimated total number of synonymous substitutions at twofold degenerate sites of L_2A_2 and that at fourfold degenerate sites is L_4K_4. By following the convention to count each fourfold degenerate site as a synonymous site and to count one-third of a twofold degenerate site as synonymous and two-thirds as nonsynonymous (because two of the three possible changes are nonsynonymous), we obtain the total number of synonymous sites as $L_4+L_2/3$. From these numbers we can show that

$$K_S = (L_2A_2 + L_4K_4) / L_2 / 3 + L_4) \quad ...(17)$$

$$V(K_S) = 9[L_2^2V(A_2) + L_4^2V(K_4)] / (L_2 + 3L_4)^2 \quad ...(18)$$

In a similar manner, we obtain the mean and variance of K_A, the number of substitutions per nonsynonymous site.

$$K_A = (L_2B_2 + L_0K_0) / (2L_2 / 3 + L_0) \quad ...(19)$$

$$V(K_A) = 9[L_2^2V(B_2) + L_0^2V(K_0)] / 2L_2 + 3L_0)^2 \quad ...(20)$$

The above method is that of Li et al. (1985b). Since transitional substitutions tend to occur more often than transversional substitutions and since most transitional changes at twofold degenerate sites are synonymous changes, counting a twofold degenerate site as one-third synonymous and two-thirds nonsynonymous will tend to overestimate the rate of synonymous substitution at twofold degenerates sites. For this reason, Li (1993) and Pamilo and Bianchi (1993) proposed the following modification. Assume that the transitional rate at twofold degenerate sites is the same as that at fourfold degenerate sites. We then take the weighted average $(L_2A_2+L_4A_4)/(L_2+L_4)$ as an estimate of the average transitional rate at twofold and fourfold degenerate sites; L_2A_2 is the total number of transitional substitutions at twofold degenerate sites between the two sequences, L_4A_4 is the corresponding number at fourfold degenerate sites, and L_2+L_4 is the total number of twofold plus fourfold degenerate sites. The number of substitutions per synonymous site and its variance are then computed as

$$K_S = (L_2A_2 + L_4A_4) / (L_2 + L_4) + B_4 \quad ...(21)$$

$$V(K_S) = [L_2^2V(A_2) + L_4^2V(A_4)] / (L_2 + L_4)^2] + V(B_4)$$
$$-2b_4Q_4[a_4P_4 - c_4(1 - Q_4)] / (L_2 + L_4) \quad ...(22)$$

Similarly, we have

$$K_A = A_0 + (L_0B_0 + L_2B_2) / (L_0 + L_2) \quad ...(23)$$

$$V(K_A) = V(A_0) + [L_0^2V(B_0) + L_2^2V(B_2)] / (L_0 + L_2)^2$$
$$-2b_0Q_0[a_0P_0 - c_0(1 - Q_{10}] / (L_0 + L_2) \quad ...(24)$$

Formula (22) and (24) correct a typographic error in Li (1993).

Table 8.2 shows a comparison of the estimates obtained by the old (Le et al. 1985b) and new (modified) methods. For the 14 pairs of mouse-rat genes, the old method gives an average K_S value of 0.183, which is 27% higher than that (0.144) obtained by the new method. For the 45 pairs of human-rodent genes, the old method gives an average K_S value of 0.690, which is 22% higher than that (0.566) obtained from the new method. To see which estimates are more reasonable, we can compare them with the number of substitutions per fourfold degenerate site estimated from the same set of genes. Because all possible changes at a fourfold degenerate site are synonymous, that site is truly synonymous. As can be seen from Table 8.2 and 8.3, the K_S values (0.144 and 0.566) estimated by the new method are close to the numbers of substitutions per fourfold degenerate site (0.137 and 0.594), whereas those estimated by the old method (0.183 and 0.690) are considerably larger. The overestimation by the old method arises because transitions occur more frequently than transversions. For example, for the 14 pairs of mouse-rat genes, the numbers of transitions and transversions per fourfold degenerate site are 0.084 and 0.053, respectively, although at each site there is only one type of transition but two types of transversion.

Table 8.2. Numbers of substitutions per synonymous site (K_S) and per nonsynonymous site (K_A) between mouse and rat genes estimated by the old and new methods

	K_S		K_A	
Gene	*Old*	*New*	*Old*	*New*
Mouse vs. rat	0.183	0.144	0.0137	0.0148
Human vs. rodent	0.690	0.566	0.1154	0.1248

Table 8.2 shows that the old method gives a smaller K_A value than does the new method. However, the differences are small. The number of nondegenerate sites in a gene is usually considerably larger than that of twofold degenerate site, so the K_A value is largely determined by the nondegenerate sites.

The new method implicitly assumes that in a gene the transitional rate at a twofold degenerate site is similar to that at a fourfold degenerate site. This assumption holds approximately for the data in Table 8.3: for the mouse-rat comparison the number of transitional substitutions per twofold degenerate site (0.094) is somewhat higher than that at a fourfold degenerate site, while for the human-rodent

Table 8.3. Number of transitional and transversional substitutions per site at nondegenerate, twofold degenerate, and fourfold degenerate sites of codons between genes

Types of substitution	*Non-degenerate*	*Twofold degenerate*	*Fourfold degenerate*
Mouse vs. rat			
Transition	0.009	0.097	0.084
Transversion	0.004	0.011	0.053
Total	0.013	0.108	0.137
Human vs. rodent			
Transition	0.064	0.298	0.359
Transversion	0.061	0.061	0.235
Total	0.125	0.359	0.594

comparison the situation is reversed (i.e., 0.298 vs. 0.359). The new method also assumes that the transversional rate at a twofold degenerate site is similar to that at a nondegenerate site. This assumption holds well for the human-rodent comparison (i.e., 0.061 vs. 0.011), though it does not hold well for the mouse-rat comparison (0.004 vs. 0.011)

The methods of Miyata and Yasunaga (1980) and Nei and Gojobori (1986) follow the convention to count a twofold degenerate sites one-third synonymous and two-third nonsynonymous and give estimates of K_S and K_A similar to those of Li et al. (1985b). Thus, they also tend to overestimate K_S.

Sequence Alignment

Comparison of two homologous sequences involves the identification of the locations of deletions and insertions that might have occurred in either of the two lineages since their divergence from a common ancestor. This process is referred to as "sequence alignment". Note that here we assume that the two sequences under study are known to have been derived from a common ancestral sequence. This is different from the problem of testing whether two sequences are related or homologous. It is also different from the so-called homology-search problem in which one wants to know whether a sequence or part of a sequence is related to any of the sequences in a database such as GenBank.

Comparisons of two DNA or protein sequences usually cannot tell us whether an insertion has occurred in one sequence or a deletion has occurred in the other. Therefore, the outcomes of both types of events are collectively referred to as *indels* or *gaps*.

An alignment consists of a series of paired bases, one base from each sequence. There are three types of aligned pairs: (1) pairs of matched bases, (2) pairs of mismatched bases, and (3) pairs consisting of a base from one sequence and a gap (null base) from the other. Gaps are denoted by -. A matched pair implies that no substitution has occurred at the site since the divergence between the two sequences, a mismatched pair implies that at least one substitution has occurred, and a gap assumes that a deletion or an insertion has occurred at this position in one of the two sequences. Thus, an alignment represents a specific hypothesis about the evolution of the two sequences. For example, the alignment

TCAGA
TC - GT

represents the hypothesis that three of the five nucleotide sites have not undergone any change since the divergence of the two sequences, one site has undergone at least one substitution, and one site has a deletion or insertion. Note that this alignment also implies that the first, second, fourth, and fifth sites are homologous between the two sequences, while the third site might have been either inserted into the first sequence or deleted from the second sequence.

Although the process of alignment will be illustrated below by using DNA sequences, the same principle and procedure can be used to align amino acid sequences. As a matter of fact, one usually obtains more reliable alignments by using amino acid sequences than by using DNA sequences. One reason is that the former are usually better conserved than the latter. Another reason is that there are 20 amino acids but only four nucleotides, so that the probability for two non-homologous sites of the two sequences to be identical by chance is lower at the amino acid level than at the nucleotide level.

The aim here is to provide a description of the basic principles for sequence alignment rather than any actual algorithms, as many computer programs for alignment are now available; a commonly used simple alignment program is CLUSTAL V by Higgins et al. (1992).

Dot-Matrix Method

For two sequences that are highly similar and have experienced few gap events, a reasonable alignment can be obtained either by visual inspection or by an approach called the *dot-matrix* method. In this method, the two sequences to be aligned are written out as column and row headings of a matrix. Dots are put in the matrix when the nucleotides in the two sequences re identical. If the two sequences are

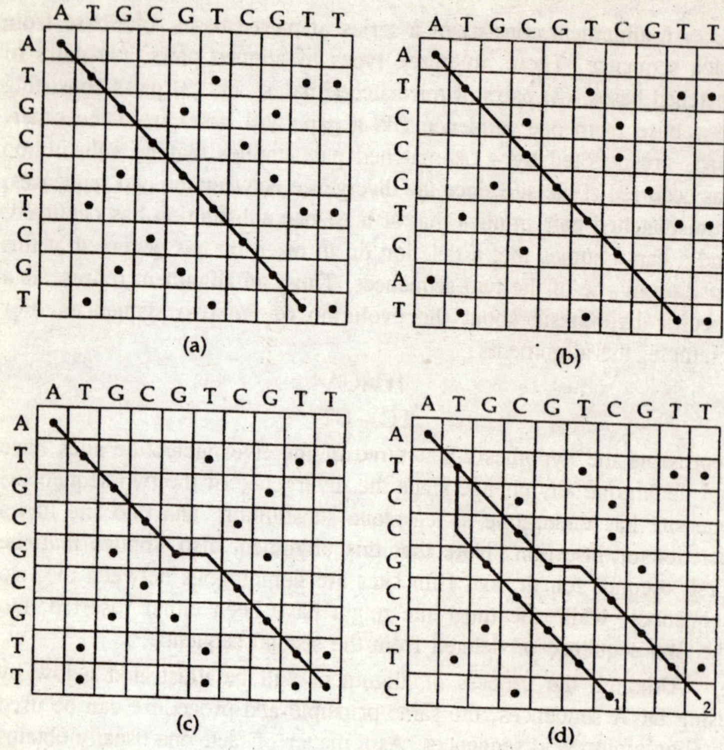

Fig. 8.2. Dot matrices for aligning nucleotide sequences. (a) the two sequences are identical; (b) the two sequences differ from each other but contain no gaps; (c) the two sequences contain a gap, but are otherwise identical to each other; (d) the two sequences contain both substitutions and gaps.

completely identical, there will be dots in all the diagonal elements of the matrix. If the two sequences are different but can be aligned without gaps, there will be dots in most of the diagonal elements. If a gap occurred in one of the two sequences, the alignment diagonal would be shifted vertically or horizontally. If the two sequences differ from each other by both gaps and substitutions, it may be difficult to identify the location of gaps and to choose between several alternative alignments. In such cases, visual inspection and the dot matrix method are not reliable, so a more rigorous method is required.

Similarity and Distance Methods

The basic principle for sequence alignment is either to maximize the number of matched pairs between the two sequences or to minimize the number of mismatched pairs, while at the same time keeping the number of gaps as small as possible. The former is called the similarity approach and was initiated in a landmark work by Needleman and Wunsch (1970); the latter is called the distance approach and was rigorously formulated by Sellers (1974) and Waterman et al. (1976).

We begin with the similarity approach. First we note that increasing the number of matches and reducing the number of gaps are two conflicting efforts. For example, consider the following two sequences:

A: TCAGACGAGTG

B: TCGGAGCTG

Since they differ in length, there must be at least one gap. One possible alignment is as follows:

```
        T C A G A C G A G T G
(I)     : :   : :       : :
        T C G G A - - G C T G
```

This alignment contains 6 matches and 1 gap of 2 nucleotides; among all the possible alignments with one gap, this alignment has the largest number of matches pairs. There are many other possible alignments, but among those with 2 gaps of one nucleotide each, the following two alignments have the largest number of matches, i.e., 7:

```
        T C A G A C G A G T G
(II)    : :   : :   :     : :
        T C G G A - G C - T G
        T C A G A C G A G T G
(III)   : :   : :   :     : :
        T C G G A - G - C T G
```

Which of the three alignments is the best? Obviously, we are facing a choice between having more point substitutions or having more gap events. Since comparing point substitutions with gaps is like comparing apples with oranges, we must find a common denominator with which to compare gaps and substitutions. This common denominator is called the *gap penalty*.

Let w_k be the penalty for a gap of k nucleotides. Then the Needleman-Wunsch similarity measure between two aligned sequences is defined as

$$S = x - \sum w_k z_k \qquad ...(25)$$

where x is the number of matched pairs and z_k is the number of gaps with length k. Many systems of gap penalty have been used. The linear system is most commonly used because it saves computer time. In this system $w_k = a + bk$, where a and b are nonnegative parameters. However, even with this system, how to choose a and b is a difficult problem. For example, if a=0.5 and b=2, then the similarity for alignment I is S=6–(0.5+2×2)=1.5 because there are 6 matches and 1 gap of 2 nucleotides. Since S=2 for both alignments II and III, they are favoured over alignment I. (This example shows that there may exist more than one optimal alignment under the same criteria.) However, if a=1.5 and b=1, then S=2.5 for alignment I, whereas S=2 for alignments II and III, so that alignment I is favoured. Therefore, which alignment is preferable depends on the penalty weights used. Gu and Li (1995) recently studied pseudogene sequences in humans and rodents and provided evidence that the linear penalty system is too heavy for long gaps. Their data suggested $w_k = a + b \ln(k)$ as a penalty function, which increases with k considerably more slowly than does the linear function $w_k = a + ak$.

In general, the similarity approach is to find an alignment with a similarity measured defined by

$$S = \mathrm{Max}\left(x - \sum w_k z_k\right) \quad \text{...(26)}$$

where the maximum is over all possible alignments. A commonly used algorithm for searching the S value and the optimal alignment is dynamic programming.

The distance approach is to find the alignment(s) with a distance measure defined by

$$D - \mathrm{Min}\left(y + \sum w'_k z_k\right) \quad \text{...(27)}$$

where y is the number of mismatched pairs and w'_k is a gap weight analogous to w_k.

Smith et al. (1981) showed that under certain conditions the above two approaches are equivalent. Consider two DNA sequences of lengths m and n, respectively. Since the total number of nucleotides in the two sequences are m+n, for any alignment we should have

$$n + m = 2(x + y) + \sum k z_k \quad \text{...(28)}$$

Putting this relation into Equation (26) we obtain

$$S = \mathrm{Max}\left[\frac{n+m}{2} - y - \sum (k/2 + w_k) z_k\right]$$

$$= \frac{n+m}{2} + \mathrm{Max}\left[-y - \sum (k/2 + w_k) z_k\right]$$

$$= \frac{n+m}{2} - \text{Min}\left[y + \sum (k/2 + w_k) z_k\right] \qquad ...(29)$$

Thus, to maximize the similarity measure defined by (25) is to minimize the following quantity:

$$y + \sum (k/2 + w_k) z_k$$

This is identical to minimizing the distance measure defined by Equation (27) if the following relation holds:

$$w'_k = k/s + w_k \qquad ...(30)$$

The above equation shows that in order to obtain the same alignment by the two approaches the penalty weights for the distance approach should be larger than those for the similarity approach. Since $w_k \geq 0, w'_k \geq k/2$.

Multiple Sequences

Alignment of multiple sequences by visual inspection is feasible only if the sequences under study are highly similar to one another and have not experienced many deletion or insertion events. When the similarity between sequences is not high, alignment of multiple sequences can be extremely difficult. One difficulty is to define a measure for the cost (or quality) of a multiple alignment and to choose gap costs consistent with the measure chosen proposed to use the sum of the branch lengths of the phylogenetic tree that relates the sequences; the length of a branch is the number of substitutions plus the gap costs for that branch. Another measure is the sum of the alignment costs imposed on each pair of the sequences in the multiple alignment this is called the SP (sum of pairs) measure. An advantage of the former measure is that it attempts to infer the minimum number of substitutions and gap events required for a given tree, but it is more difficult to compute than the SP measure. The latter, however, may be unduly influenced by closely related sequences (which include redundant information), because all pairwise alignments are treated equally. This problem can be circumvented by giving less weights to more closely related pairs.

The major difficulty in aligning multiple sequences is computational. This difficulty is particularly apparent for methods using dynamic programming, which is the most widely accepted algorithm, because an explicit, realistic measure for the cost of an alignment can be used.

It has long been recognized that alignment of multiple sequences should take into consideration the branching order of the sequences under study, for the order in which the sequences are aligned can influence the final outcome. If the phylogeny of the sequences is known, alignment can be made progressively from more related to less related sequences. If the phylogeny is not known, then sequence alignment and phylogenetic reconstruction should be done simultaneously—this is known as the integrated or unified approach. As this approach may require many cycles of tree reconstruction and sequence alignment, it is computationally intensive, and so most algorithms use heuristic approaches. For example, the algorithm of Hogeweg and Hesper (1984) and Konings et al. (1987) includes the following steps: (1) calculate all pairwise similarities by the Needleman-Wunsch algorithm modified by Fitch and Smith (1983), (2) construct an initial tree from these similarities, (3) realign the sequences progressively in order of their relatedness according to the inferred tree, (4) construct a new tree from the pairwise similarities calculated from the new alignment, and (5) reiterate the process, if the new tree is not identical to the previous one.

9

EVOLUTIONARY RESHUFFLING DNA SEQUENCES

A basic process in the evolution of DNA sequences is the change in nucleotides with time. This process deserves a detailed consideration because it is essential for understanding the mechanism of DNA evolution and because changes in nucleotide sequences are used both for estimating the rate of evolution and for reconstructing the evolutionary history of organisms. For this reason, many models have been proposed for studying this process. We shall study first the change in a single sequence and then the divergence between two sequences. The theory presented in this chapter provides a framework for estimating the number of nucleotide substitutions between two sequences.

NUCLEOTIDE SUBSTITUTION IN DNA SEQUENCE

To study the dynamics of nucleotide substitution, we must make assumptions regarding the probability of substitution of one nucleotide by another. Many such mathematical schemes have been proposed in the literature. We start with the simplest models: Jukes and Cantor's (1969) one-parameter model and Kimura's (1980) two-parameter model. More general models will be discussed later.

Jukes and Cantor's One-Parameter Model

This model assumes no bias in the direction of change so that substitutions occur randomly among the four types of nucleotide. For example, if the nucleotide under consideration is A, it will change to T, C, or G with equal probability. In this model, the rate of substitution for each nucleotide is 3α per unit time, and the rate of substitution in

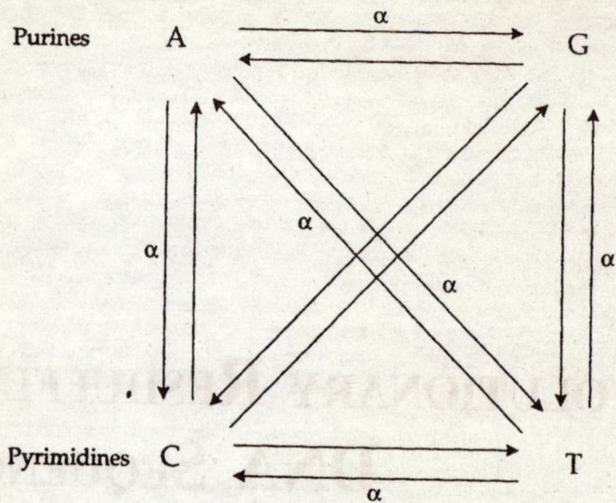

Fig. 9.1. One-parameter model of nucleotide substitution. In this model, the rate of substitution in each direction is α.

each of the three possible directions of change is α. Since the model involves only one parameter, α, it is also called the *one-parameter model*.

Let us assume that the nucleotide residing at a certain site in a DNA sequence is A at time 0. First, we consider the probability, $p_{A(t)}$, that this site will be occupied by A at time t. Since we start with A, the probability that this site is occupied by A at time 0 is $p_{A(0)}=1$. At time 1, the probability of still having A at this site is given by

$$p_{A(1)} = 1 - 3\alpha \qquad ...(1)$$

which reflects the probability, 1 – 3α, that the nucleotide has remained unchanged.

The probability of having A at time 2 is

$$p_{A(2)} = (1-3\alpha)p_{A(1)} + \alpha[1 - p_{A(1)}] \qquad ...(2)$$

To derive this equation, we consider two possible scenarios: (1) the nucleotide has remained unchanged, and (2) the nucleotide has changed to T, C, or G but has subsequently reverted to A. The probability of the nucleotide being A at time 1 is $p_{A(1)}$, and the probability that it has remained A at time 2 is 1 – 3α. The product of the probabilities for these two events gives us the probability for the first scenario, which constitutes the first term in Equation (2). The probability of the nucleotide not being A at time 1 is $1 - p_{A(1)}$, and its probability of changing to A at time 2 is α. The product of these two probabilities

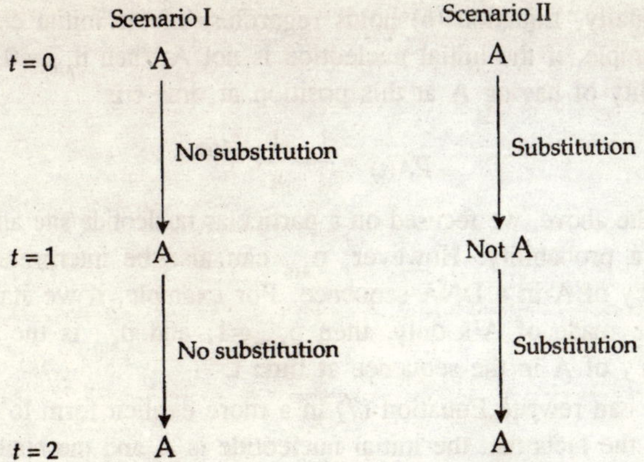

Fig. 9.2. Two possible scenarios for having A at a site at time t=2, given that the site had A at time 0.

gives us the probability for the second scenario, and constitutes the second term in Equation (2).

Using the above formulation, we can show that the following recurrence equation holds for any t:

$$p_{A(t+1)} = (1-3\alpha)p_{A(t)} + \alpha[1-p_{A(t)}] \qquad ...(3)$$

We can rewrite Equation (3) in terms of the amount of change in $p_{A(t)}$ per unit time as

$$p_{A(t+1)} - p_{A(t)} = -3\alpha\, p_{A(t)} + \alpha[1-p_{A(t)}]$$

or

$$\Delta p_{A(t)} = -3\alpha\, p_{A(t)} + \alpha[1-p_{A(t)}] = -4\alpha\, p_{A(t)} + \alpha \qquad ...(4)$$

So far we have considered a discrete-time process. It is more convenient to approximate this process by a continuous-time model, by regarding $\Delta p_{A(t)}$ as the rate of change at time t. With this approximation, Equation (4) is rewritten as

$$\frac{dp_{A(t)}}{dt} = 4\alpha\, p_{A(t)} + \alpha \qquad ...(5)$$

This is a first-order linear differential equation, and the solution is given by

$$p_{A(t)} = \frac{1}{4}\left(p_{A(0)} - \frac{1}{4}\right)e^{-4\alpha t} \qquad ...(6)$$

Since we started with A, $p_{A(0)} = 1$. Therefore,

$$p_{A(t)} = \frac{1}{4} + \frac{3}{4}e^{-4\alpha t} \qquad ...(7)$$

Actually, Equation (6) holds regardless of the initial conditions. For example, if the initial nucleotide is not A, then $p_{A(0)}=0$, and the probability of having A at this position at time t is

$$p_{A(t)} = \frac{1}{4} - \frac{1}{4}e^{-4\alpha t} \qquad ...(8)$$

In the above, we focused on a particular nucleotide site and treated $p_{A(t)}$ as a probability. However, $p_{A(t)}$ can also be interpreted as the frequency of A in a DNA sequence. For example, if we start with a sequence made of A's only, then $p_{A(0)}=1$, and $p_{A(t)}$ is the expected frequency of A in the sequence at time t.

We can rewrite Equation (7) in a more explicit form to take into account the facts that the initial nucleotide is A and the nucleotide at time t is also A.

$$p_{AA(t)} = \frac{1}{4} + \frac{3}{4}e^{-4\alpha t} \qquad ...(9)$$

If the initial nucleotide is G instead of A, then from Equation (8) we obtain:

$$p_{GA(t)} = \frac{1}{4} - \frac{1}{4}e^{-4\alpha t} \qquad ...(10)$$

Since all the nucleotides are equivalent under the JC model, $p_{GA(t)}=p_{CA(t)}=p_{TA(t)}$. In fact, we can consider a general probability, $p_{ij(t)}$, which is the probability that a nucleotide will become j at time t, given that the initial nucleotide is i. By using this generalized notation and Equation (9), we obtain

$$p_{ii(t)} = \frac{1}{4} + \frac{3}{4}e^{-4\alpha t} \qquad ...(11)$$

and from Equation (10)

$$p_{ij(t)} = \frac{1}{4} - \frac{1}{4}e^{-4\alpha t} \qquad ...(12)$$

where $i \neq j$.

Equations (11) and (12) are sufficient for describing the substitution process under the one-parameter model. For example, suppose that we wish to know the probability that the nucleotide at a given site will be A at time t. There are two possible situations. First, the nucleotide at the present time is A. In this case, i=A and Equation (11) shows that $p_{ii(t)}$ decreases exponentially from 1 to 1/4 as t increases from 0 to ∞. Second, the nucleotide at the present time is not A, but T (or C, or G). Then, i=T and j=A and Equation (12) shows that $p_{ij(t)}$

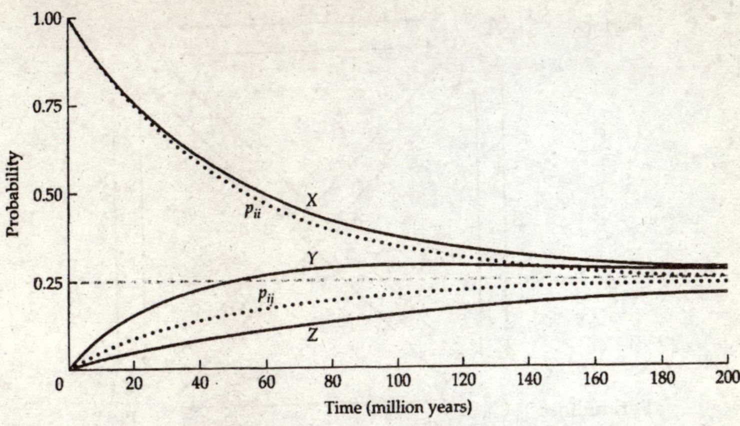

Fig. 9.3. Temporal changes in the probability of having a certain nucleotide, say A, at a given nucleotide site.

increases monotonically from 0 to 1/4 as t increases from 0 to ∞. Thus, regardless to the initial condition, the probability of having A at the site will eventually become 1/4. This also holds true for T, C and G. Therefore, under the JC model the equilibrium frequency of each of the four nucleotides is 1/4. (The equilibrium state is also referred to as the stationary state.) This result is of course in the average sense, because in finite sequences fluctuations in nucleotide frequencies are likely to occur.

Fig. 9.3 shows that the decrease in p_{ii} is linear when t is small but becomes nonlinear as t increases. The nonlinearity occurs because of back mutation, that is, the initial nucleotide may change to another nucleotide and then change back to the same nucleotide.

Kimura's Two-Parameter Model

The assumption that all nucleotide substitutions occur randomly, as in the JC model, is unrealistic in most cases. For example, transitions are generally more frequent than transversions. To take this fact into account, Kimura (1980) proposed a two-parameter model. In this scheme, the rate of transitional substitution at each nucleotide site is α per unit time, whereas the rate of each of the two types of transversional is β per unit time.

As in the one-parameter model, let $p_{A(t)}$ be the probability that the nucleotide at the site under consideration is A time t; $p_{T(t)}$, and $p_{G(t)}$ are similarly defined. The probability that the nucleotide at the site is A in the next generation is given by

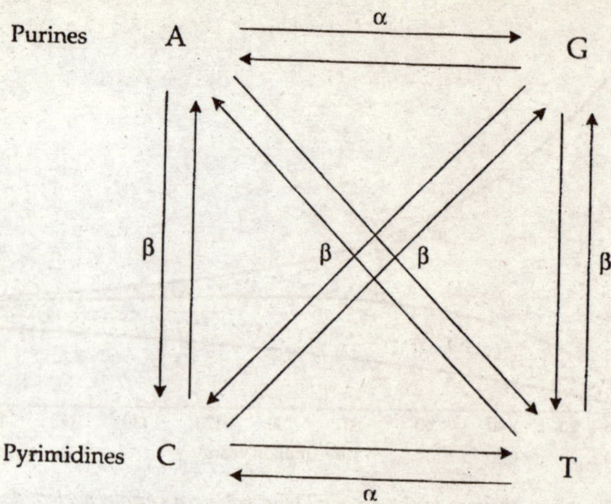

Fig. 9.4. Two-parameter model of nucleotide substitution. In this model, the rate of transition (α) may not be equal to the rate of each of the two types of transversion (β).

$$p_{A(t+1)} = (1-\alpha-2\beta)p_{A(t)} + \beta p_{T(t)} + \beta p_{C(t)} + \alpha p_{G(t)} \quad \text{...(13a)}$$

To derive this equation we need to consider four possible scenarios: First, the nucleotide at the site is A at time t and will remain unchanged at t+1. The probability that the nucleotide at the site is A at time t is $p_{A(t)}$ and the probability that it will remain unchanged at t+1 is (1–α–2β). The product of these two probabilities constitutes the first term in Equation (13a). Second, the nucleotide at the site is T at time t but will change to A at t+1, a transversional change. The probabilities for the two events are $p_{T(t)}$ and β, respectively, and their product constitutes the second term in Equation (13a). Third, the nucleotide at the site is C at time t but will change to A at t+1, a transversional change. The probabilities for the two events are $p_{C(t)}$ and β, respectively, and their product constitutes the third term in Equation (13a). Fourth, the nucleotide at the site is G at time t but will change to A at t+1, a transitional change. The probabilities for these two events are $p_{G(t)}$ and α, respectively, and their product constitutes the fourth term in Equation (13a). In the same manner, we can derive the following equations:

$$p_{T(t+1)} = \beta p_{A(t)} + (1-\alpha-2\beta)p_{T(t)} + \alpha p_{C(t)} + \beta p_{G(t)} \quad \text{...(13b)}$$

$$p_{C(t+1)} = \beta p_{A(t)} + \alpha p_{T(t)} + (1-\alpha-2\beta)p_{C(t)} + \beta p_{G(t)} \quad \text{...(13c)}$$

$$p_{G(t+1)} = \alpha p_{A(t)} + \beta p_{T(t)} + \beta p_{C(t)} + (1-\alpha-2\beta)p_{G(t)} \quad \text{...(13d)}$$

To solve the above equations we need to set the initial conditions. If the initial nucleotide is A, then $p_{A(0)}=1$ and $p_{T(0)}=p_{C(0)}=p_{G(0)}=0$, and a solution can be obtained for the above equations. Similarly, we can obtain a solution for the initial conditions $p_{T(0)}=1$ and $p_{A(0)}=p_{G(0)}=0$, and so on.

In the JC model, Equation (11) implies $p_{AA}(t)=p_{GG}(t)=p_{CC}(t)=p_{TT}(t)$. This equality also holds for Kimura's two-parameter model, because the four nucleotides are equivalent in this model. We shall denote this probability by $X_{(t)}$. It can be shown by using Equation (13a-c) that

$$X_{(t)}=\frac{1}{4}+\frac{1}{4}e^{-4\beta t}+\frac{1}{2}e^{-2(\alpha+\beta)t} \quad ...(14)$$

As in the one-parameter model, the solution is given in continuous rather than in discrete time.

Under the one-parameter (JC) model, Equation (12) holds regardless of whether the change from nucleotide i to nucleotide j is transition or a transversion. In contrast, in Kimura's two-parameter model, we need to distinguish between transitional and transversional changes because the two types of change occur with different probabilities. Denote by $Y_{(t)}$ the probability that the initial nucleotide and the nucleotide at time t differ from each other by a transition. For example, if the initial nucleotide is A and the nucleotide at time t is G, then they differ by a transition and $Y_{(t)}=p_{AG(t)}$. Because of the symmetry of the substitution scheme, $Y_{(t)}=p_{AG(t)}=p_{GA(t)}=p_{TC(t)}=p_{CT(t)}$. It can be shown that

$$Y_{(t)}=\frac{1}{4}+\frac{1}{4}e^{-4\beta t}-\frac{1}{2}e^{-2(\alpha+\beta)t} \quad ...(15)$$

The probability $Z_{(t)}$ that the nucleotide at time t and the initial nucleotide differ by a specific type of transversion is given by

$$Z_{(t)}=\frac{1}{4}-\frac{1}{4}e^{-4\beta t} \quad ...(16)$$

Note that each nucleotide is subject to two types of transversion. For example, if the initial nucleotide is A, then the two possible transversional changes are A→C and A→T. Therefore, the probability that the initial nucleotide and the nucleotide at time t differ by either of the two types of transversion is twice the probability given by Equation (16). Note that $X_{(t)}+Y_{(t)}+2Z_{(t)}=1$.

Fig. 9.3 shows a comparison between the one- and two-parameter models. The line denoted by X is computed from Equation (14). The rates used are $\alpha=10\times10^{-9}$ and $\beta=2.5\times10^{-9}$ substitutions per site per year; since the total rate of substitution per site is $\alpha+2\beta=15\times10^{-9}$, the rate of transition is two-thirds of the total rate. The dotted line denoted by p_{ii} is computed under the one-parameter model; the rates used are $\alpha=\beta=5\times10^{-9}$, so the rate of transition is one-third of the total rate. Although the total rate of substitution per site is the same for the two curves (the two models), X decreases more slowly than p_{ii}. To understand why this difference occurs, let us assume that the initial nucleotide is A, so that X and p_{ii} represent the probabilities of having A at the site at time t under the two-parameter and one-parameter models, respectively. The probability of changing from A to non-A is the same in the two models, but the probability of changing from A to G and backward to A is higher in the two-parameter model because it has a higher rate of transition ($A\leftrightarrow G$) than does the one-parameter model. Because of this difference in the probability of backward substitution, X decreases more slowly than p_{ii}. However, both X and p_{ii} will approach 1/4 as $t\rightarrow\infty$. Thus, as in the one-parameter model, in Kimura's two-parameter model the equilibrium value for the frequency of each nucleotide is 1/4.

Another interesting difference between the two models is as follows. In Fig. 9.3, lines Y and Z represent the probabilities that the nucleotide at time t differs from the initial nucleotide by a transition and by a specific transversion, respectively, under Kimura's two-parameter model. As in the one-parameter model, Z increases monotonically to the equilibrium value 1/4. But, surprisingly, Y is not monotonic and, in fact, can become greater than 1/4, the equilibrium value. To see why this can happen, let us study Equation (15). Note that $Y_{(t)}$ is $>1/4$ if

$$\frac{1}{4}e^{-4\beta t} > \frac{1}{2}e^{-2(\alpha+\beta)t}$$

or

$$\frac{1}{2} > e^{-2(\alpha-\beta)t}$$

This condition holds if $\alpha>\beta$ and

$$t > \frac{1}{2}\ln(2)/(\alpha-\beta) \approx 0.347/(\alpha-\beta) \qquad ...(17)$$

Thus, if the rate of transition (α) is higher than the rate of each of the two types of transversion (β), the probability (Y) of having a transitional difference between the initial nucleotide and the nucleotide at t can become larger than 1/4 as t becomes larger than $0.693/(\alpha-\beta)$. Of course, as $t\rightarrow\infty$, Y will become 1/4.

Other Models of Nucleotide Substitution

The two-parameter model is more realistic than the one-parameter model but also involves many restrictions. For example, it requires that the four types of transition occur with the same rate, but, this may not be the case. So we need to consider more general models. To do so, it is more convenient to use matrix theory. Let us number the four nucleotides A, T, C, and G as 1, 2, 3, and 4, respectively. Let M be a 4×4 matrix in which the i,j-th element ($i \neq j$), m_{ij}, represents the probability that the nucleotide will change to j in the next generation, given that the nucleotide is i at the present generation, and the i,i-th element, m_{ii}, represents the probability that the nucleotide will remain unchanged in the next generation. For example, for Kimura's two-parameter model, M is defined as

$$M = \begin{bmatrix} 1-\alpha-2\beta & \beta & \beta & \alpha \\ \beta & 1-\alpha-2\beta & \alpha & \beta \\ \beta & \alpha & 1-\alpha-2\beta & \beta \\ \alpha & \beta & \beta & 1-\alpha-2\beta \end{bmatrix} \quad ...(18)$$

If we define the vector $p_{(t)}$ as $p_{(t)}=(p_{A(t)}, p_{T(t)}, p_{C(t)}, p_{G(t)}$, then Equation (13a-d) can be written as

$$p_{(t+1)} = p_{(t)}M \quad ...(19)$$

This equation can be studied by using linear algebra.

The matrix M in Equation (18) satisfies the following two conditions: (1) all elements of the matrix are nonnegative and (2) the sum of the elements in each row is 1. Such matrices are called stochastic matrices. The above formulation has in effect assumed that the substitution process is a Markov chain, the essence of which is as follows. Consider three time points in evolution, $t_1<t_2<t_3$. The Markovian model assumes that the status of the nucleotide site at t_3 depends only on the status at t_2 but not the status at t_1, if the status of the site at t_2 is known. In other words, if the nucleotide at the site is specified at t_2, then no additional information about the status of the site in any past generation is required (or is of help) for predicting the nucleotide at t_3. For example, if the nucleotide at time t is A, then $p_{A(t)}=1$ and $p_{T(t)}=p_{C(t)}=p_{G(t)}=0$, and by using Equation (19) one can compute the probability $p_{i(t+1)}$ that the nucleotide will be i at time t+1, i=A, T, C, or G; and by iteration of the same procedure, one can compute $p_{i(\tau)}$ for any future generation τ. For the theory of Markov chains, readers may consult Feller (1968).

The above description shows that Kimura's two-parameter model can be represented by the first matrix in Table 9.1; the one-parameter model is a special case of this matrix with the condition $\alpha=\beta$. By increasing the number of parameters the substitution model can be made more complex and realistic. Several such models are shown in Table 9.1. In each of these matrices the four elements on the diagonal from the upper-left corner to the lower-right corner are the probabilities that the nucleotide at the site remains unchanged in one generation, the four elements on the diagonal from the upper-right corner to the lower-left corner are the rates of transitional substitution per generation, and the other eight elements are the rates of transversional substitution per generation.

Table 9.1. Models of nucleotide substitution

O/S	*A*	*T*	*C*	*G*
(a)	Two-parameter model (Kimura 1980)			
A	$1-\alpha-2\beta$	β	β	α
T	β	$1-\alpha-2\beta$	α	β
C	β	α	$1-\alpha-2\beta$	β
G	α	β	β	$1-\alpha-2\beta$
(b)	Four-parameter model (Blaisdell 1985)			
A	$1-\alpha-2\gamma$	γ	γ	α
T	δ	$1-\alpha-2\delta$	α	δ
C	δ	β	$1-\beta-2\delta$	δ
G	β	γ	γ	$1-\beta-2\gamma$
(c)	Six-parameter model (Kimura 1981a)			
A	$1-2\alpha-\gamma$	γ	α	α
T	δ	$1-2\alpha-\delta$	α	α
C	β	β	$1-2\beta-\varepsilon$	ε
G	β	β	ζ	$1-2\beta-\zeta$
(d)	Nine-parameter model			
A	$1-g_T\beta_1-g_C\gamma_1-g_G\alpha_1$	$g_T\beta_1$	$g_C\gamma_1$	$g_G\alpha_1$
T	$g_A\beta_1$	$1-g_A\beta_1-g_C\alpha_2-g_G\gamma_2$	$g_C\alpha_2$	$g_G\gamma_2$
C	$g_A\gamma_1$	$g_T\alpha_2$	$1-g_A\gamma_1-g_T\alpha_2-g_G\beta_2$	$g_G\beta_2$
G	$g_A\alpha_1$	$g_T\gamma_2$	$g_C\beta_2$	$1-g_A\alpha_1-g_T\gamma_2-g_C\beta_2$
(e)	General model			
A	$1-\alpha_{12}-\alpha_{13}-\alpha_{14}$	α_{12}	α_{13}	α_{14}
T	α_{21}	$1-\alpha_{21}-\alpha_{23}-\alpha_{24}$	α_{23}	α_{24}
C	α_{31}	α_{32}	$1-\alpha_{31}-\alpha_{32}-\alpha_{34}$	α_{34}
G	α_{41}	α_{42}	α_{43}	$1-\alpha_{41}-\alpha_{42}-\alpha_{43}$

The second matrix in Table 9.1 is Blaisdell's (1985) four-parameter model. This model differs from the two-parameter model in two important respects: First, the rates of the two transitions A→G and T→G (denoted by α) may not be equal to the rates of the two transitions G→A and C→T (denoted by β). In contrast, in the two-parameter model the rates of four types of transition are equal to α. Second, the rate of change from nucleotide i to nucleotide j may not be equal to the rate of change from j to i; for example, the rate of the transversion A→T is γ whereas the rate of the transversion T→A is δ. In contrast, in the two-parameter model the rates of forward and backward changes are equal, i.e., $m_{ij}=m_{ji}$. Obviously, having more parameters allows more flexibilities. For example, in the two-parameter model the equilibrium frequencies of the four nucleotides are the same and equal to 1/4, whereas in Blaisdell's four-parameter model the equilibrium frequencies of the four nucleotides can be different from one another and are given by

$$\hat{p}_A = 2(\beta+\gamma)\delta / (\lambda_1\lambda_2) \quad \text{...(20a)}$$

$$\hat{p}_T = 2(\beta+\delta)\gamma / (\lambda_1\lambda_3) \quad \text{...(20b)}$$

$$\hat{p}_C = 2(\alpha+\delta)\gamma / (\lambda_1\lambda_3) \quad \text{...(20c)}$$

$$\hat{p}_C = 2(\alpha+\gamma)\delta / (\lambda_1\lambda_2) \quad \text{...(20d)}$$

where $\lambda_1=-2(\gamma+\delta)$, $\lambda_2=-(\alpha+\beta+2\gamma)$, and $\lambda_3=-(\alpha+\beta+2\delta)$.

The third matrix in Table 9.1 represents Kimura's (1981a) six-parameter model, which has been solved by Gojobori et al. (1982a). The fourth matrix represents a nine-parameter model, in which g_A, g_T, g_C and g_G denote the equilibrium frequencies of A, T, C, and G, respectively, so the rate of substitution to a nucleotide depends on the equilibrium frequency of the nucleotide. Because the four equilibrium frequencies must sum up to 1, they actually represent only three independent parameters. In addition, α_1, α_2, β_1, β_2, γ_1, and γ_2 provide six independent parameters. So, the total number of independent parameters is nine. This model contains several submodels. For example, if $\alpha_1=\alpha_2=\beta_1=\beta_2=\gamma_1=\gamma_2=\alpha$, the model becomes Felsenstein's (1981) and Tajima and Nei's (1982) "equal-input mode," meaning that the mutation rates to a particular nucleotide from the other three nucleotides are the same. This model, which has four-independent parameters, was extended by Hasegawa et al. (1985) to a five-parameter model, which is the fourth matrix in Table 9.1 with the conditions $\alpha_1=\alpha_2$ and $\beta_1=\beta_2=\gamma_1=\gamma_2=\beta$. In this model, α and β represent the rates of transitional substitution and transversional substitution, respectively. This model was in turn extended by Tamura and Nei (1993) to a six-parameter model by assuming that the rate of transition (α_1) between purines (A and G) is different from that (α_2) between pyrimidine (C and T), i.e., the fourth matrix in Table 9.1 with the condition $\beta_1=\beta_2=\gamma_1=\gamma_2=\beta$.

Although a 4×4 matrix can have 16 parameters, a substitution model can have at most 12 independent parameters because the sum of the elements in each row of the substitution matrix must be 1. Thus, the last matrix in Table 9.1 represents the most general model of nucleotide substitution, which has been studied by Lanave et al. (1984), Barry and Hartigan (1987), Lake (1994), Lockhart et al. (1994), and Zharkikh (1994). The general model contains any model with fewer independent parameters as a special case. For a mathematical treatment of various model of nucleotide frequencies of the four nucleotides can be computed by the formula of Wright (1969) and Tajima and Nei (1982).

A substitution process is said to be time-reversible if the probability of starting from nucleotide i and changing to nucleotide j in a time interval is the same as the probability of starting from j and going backward to i in the same time duration. In mathematical terms, time reversibility requires $p_{ij(t)}\hat{p}_i = p_{ji(t)}\hat{p}_j$ for all i and j and t; $\hat{p}_i$ is the equilibrium frequency of nucleotide i. In particular, for t=1, the condition becomes $\alpha_{ij}\hat{p} = \alpha_{ji}\hat{p}_j$ for all i and j in the last matrix of Table 9.1. For example, time-reversibility holds in Kimura's two-parameter model because $\alpha_{ij} = a_{ji}$ and $\hat{p}_i = 14$ for all i and j. It can be shown that time-reversibility holds under the nine-parameter model in Table 9.1 but not under Blaisdell's four-parameter model or Kimura's six-parameter model. As will be seen below, time-reversibility simplifies the study of nucleotide sequence evolution.

Divergence between DNA Sequences

After the separation of two nucleotide sequences, each sequence will start accumulating nucleotide substitutions and the two sequences will diverge from each other. Initially, every substitution will increase the dissimilarity between the two sequences. However, this may no longer be the case when the degree of divergence becomes appreciable, so that multiple substitutions (also commonly dubbed as *multiple hits*) may occur at the same site. For example, if at a site the nucleotide had changed from T to A in sequence 1, then a change from T to A in sequence 2 will decrease rather than increase the degree of dissimilarity between the two sequences. This type of substitution is called *parallel substitution*. Some other types of substitution, such as back substitutions and convergent substitutions may also decrease the

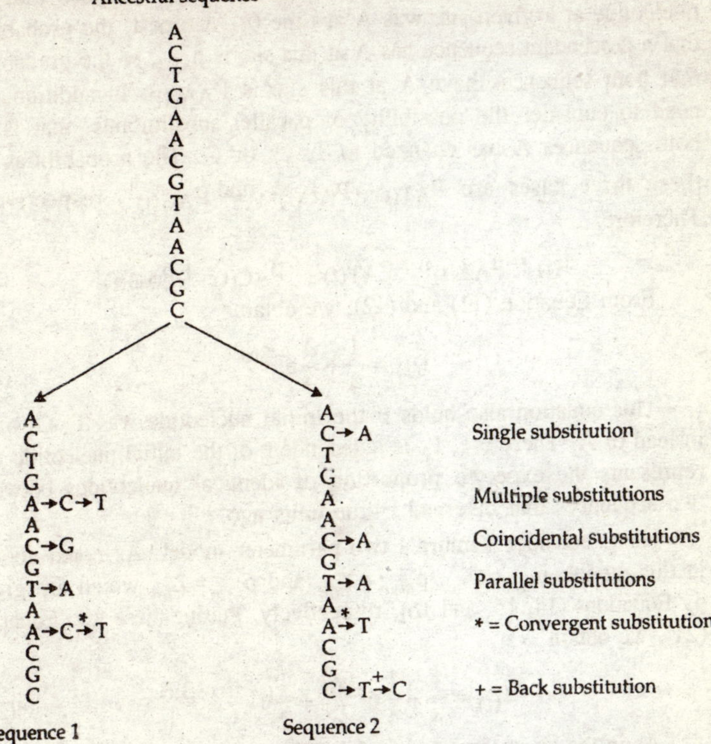

Fig. 9.5. Two homologous DNA sequences that descended from an ancestral sequence and accumulated mutations since their divergence from each other.

divergence between the two sequences. These possibilities need to be taken into account when considering sequence divergence.

In studying the divergence between two sequences one can consider either sequence similarity or dissimilarity, using the results obtained above for a single sequence. One important point to be emphasized is that the rate of change in both quantities depends not only on the rate of substitution but also on the pattern (model) of nucleotide substitution. We shall use the simple tree in which the two descendant sequences were derived from an ancestral sequence t time units ago.

Sequences Similarity

A common measure for sequence similarity is the proportion of identical nucleotides between the two sequences under study. The expected value of this proportion is equal to the probability, $I_{(t)}$, that the nucleotide at a given site at time t is the same in both sequences.

Let us start with the one-parameter model. Suppose that the nucleotide at a given site was A at time 0. At time t, the probability that a descendant sequence has A at this site is $p_{AA(t)}$, so the probability that both sequences have A at this site is $p_{AA(t)}^2$. In addition, we need to consider the possibility of parallel substitutions, that is, in both sequences A has changed to T, C, or G. The probabilities for these three cases are $p_{AT(t)}^2$, $p_{AC(t)}^2$, and $p_{AG(t)}^2$, respectively. Therefore,

$$I_{(t)} = p_{AA(t)}^2 + p_{AT(t)}^2 + p_{AC(t)}^2 + p_{AG(t)}^2 \qquad \text{...(21)}$$

From Equation (11) and (12), we obtain

$$I_{(t)} = \frac{1}{4} + \frac{3}{4} e^{-8\alpha t} \qquad \text{...(22)}$$

This equation also holds if the initial nucleotide was T, C, or G instead of A. Therefore, $I_{(i)}$ is independent of the initial nucleotide and represents the expected proportion of identical nucleotides between two sequences that diverged t time units ago.

Next, consider Kimura's two-parameter model. As noted above, in this model, $p_{AA(t)} = X_{(t)}$, $p_{AG(t)} = Y_{(t)}$, and $p_{AT(t)} = Z_{(t)}$, which are given by Equations (14, 15, and 16), respectively. Putting these into Equation (21), we obtain

$$I_{(t)} = \frac{1}{4} + \frac{1}{4} e^{-8\beta t} + \frac{1}{2} e^{-4(\alpha+\beta)t} \qquad \text{...(23)}$$

As in the one-parameter model, this equation also holds for T, C, or G and so $I_{(t)}$ represents the expected proportion or identical nucleotides between two sequences that diverged t time units ago.

Finally, let us consider Blaisdell's four-parameter model. In this model the four nucleotides are no longer equivalent, and $I_{(t)}$ depends on the initial nucleotide. To take this fact into account, we shall use $I_{i(t)}$, where the subscript i signifies that the initial nucleotide was i. For example, if i=T, then

$$I_{T(t)} = p_{TA(t)}^2 + p_{TT(t)}^2 + p_{TC(t)}^2 + p_{TG(t)}^2$$

The explicit formula for $I_{T(t)}$ will not be presented here because it is considerably more complicated than that for the two-parameter model. Since $I_{i(t)}$ now depends on i, the expected proportion, $I_{(t)}$, of identical nucleotides between the two descendant sequences depends on the nucleotide composition of the ancestral sequence. Let p_i be the proportion of nucleotide i in the ancestral sequence. Then,

$$I_{(t)} = p_A I_{A(t)} + p_T I_{T(t)} + p_C I_{C(t)} + p_G I_{G(t)} \qquad \text{...(24)}$$

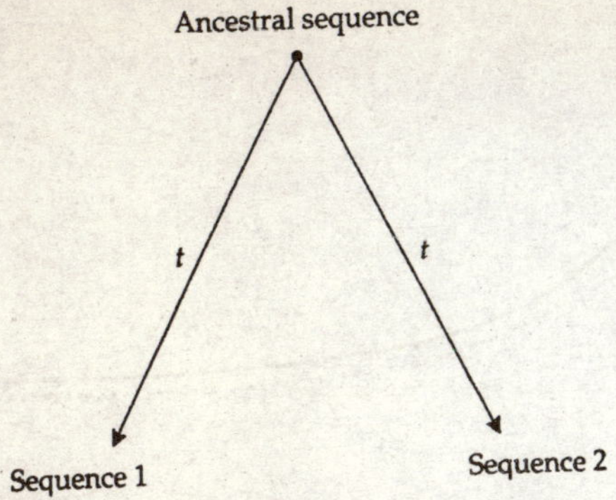

Fig. 9.6. Divergence of two sequences from a common ancestral sequence t time units ago.

If the frequencies of the four nucleotides in the ancestral sequence were at equilibrium, then $p_i = \hat{p}_i$, where $\hat{p}_i$, i=A, T, C, or G, are given by Equations (20a-d).

We now discuss some properties of $I_{(t)}$. A comparison of Equation (22) with Equation (11) shows that in the one-parameter model $I_{(t)}=p_{ii(2t)}$. Therefore, in this model the probability for two sequences that diverged t time units ago to have the same nucleotide at a given site is the same as the probability that in a single sequence the nucleotide at a given site at time 2t is the same as the nucleotide at time 0. In other words, in this model we can pretend that we start from sequence 1 (or 2) and go backward for t time units to the ancestral sequence and then go forward for t time units to sequence 2 (or 1). The one-parameter model has this property because in this model the substitution process is time reversible. Similarly, in Kimura's two-parameter model time is also reversible, no such simple relationship holds for Blaisdell's four-parameter model because in that mode the substitution process is not time reversible.

Fig. 3.7 shows the expected proportion, $I_{(t)}$, of identical nucleotides between two sequences diverged t years ago. The dotted line is computed under the one-parameter model with $\alpha=5\times10^{-9}$ substitutions/site/year, while the solid line is computed under Blaisdell's four-parameter model with $\alpha=10\times10^{-9}$, $\beta=6\times10^{-9}$, $\gamma=4\times10^{-9}$, and $\delta=2.5\times10^{-9}$ substitutions/site/year. $I_{(t)}$ decreases faster in the one-parameter model than in Blaisdell's model. This is because in the one-parameter model substitution occurs randomly among the four nucleotides so that the chance of back substitutions in a sequence and the chance of parallel substitutions in the two sequences are lower

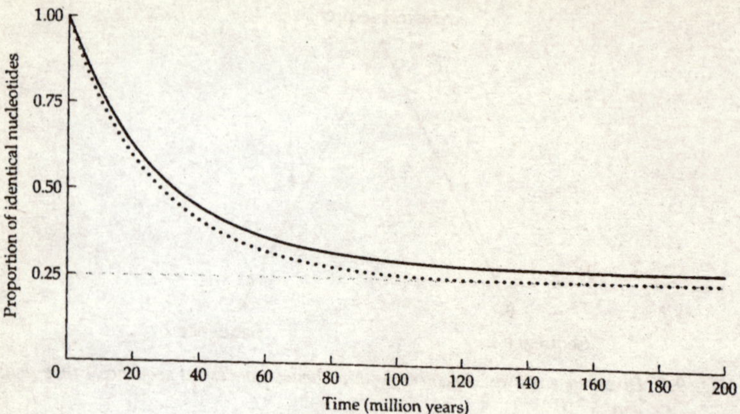

Fig. 9.7. Temporal changes in the expected proportion of identical nucleotides between two sequences that diverged t years ago.

than those in Blaisdell's model. As t increase, $I_{(t)}$ does not decrease to 0 but to an equilibrium value. Indeed, as can be seen from Equation (21), as $t \rightarrow \infty$, $I_{(t)}$ will become

$$\hat{I} = \hat{p}_A{}^2 + \hat{p}_T{}^2 + \hat{p}_C{}^2 + \hat{p}_G{}^2 \qquad ...(25)$$

where $\hat{p}_i$ is the equilibrium frequency of nucleotide i. For the one-parameter model, $\hat{p}_i = 1/4$ for i=A, T, C and G, and so $\hat{I} = 0.25$ at equilibrium. For Blaisdell's model with the parameter values used $\hat{p}_A = 0.16$, $\hat{p}_T = 0.25$, $\hat{p}_C = 0.37$, $\hat{p}_G = 0.22$ and so $\hat{I} = 0.27$, which is only slightly larger than 0.25 (the value for the one-parameter model), although the parameter values are chosen so that the equilibrium frequencies of the four nucleotides differ considerably from equality. Since the two sequences become random sequences as $t \rightarrow \infty$, the above results show that the expected proportion of identical nucleotides between two random sequences is at least 25%.

Sequence Dissimilarity

A simple measure of sequence dissimilarity is the proportion, D, of different nucleotides between two sequences. However, since D is simply 1—I, it does not deserve further analysis. We shall instead consider sequence dissimilarity in terms of transitional and transversional differences, which are more informative than D.

First, consider the expected proportion, $P_{(t)}$ of transitional differences between two sequences that diverged t time units ago. Obviously, $P_{(t)}=0$ at time 0. To see how $P_{(t)}$ increases with time, let us first use the one-parameter model. In this model $P_{(t)}$ is equal to the probability that at a given nucleotide site the two sequences differ by a transition at time t. Assume that the site was initially occupied by, say, A. To compute $P_{(t)}$ we consider all possible transitional differences at time t: (1) sequence 1 has A and sequence 2 has G or vice versa, and (2) sequence 1 has T and sequence 2 has C or vice versa. The probability for the first possibility is $2p_{AA(t)}p_{AG(t)}$ and that for the second possibility is $2p_{AT(t)}p_{AC(t)}$, where $p_{ij(t)}$ is given by Equation (11) if i=j or Equation (12) if i≠j. Therefore,

$$P_{(t)} = 2p_{AA(t)}p_{AG(t)} + 2p_{AT(t)}p_{AC(t)}$$

$$= \frac{1}{4} - \left(\frac{1}{4}\right)e^{-8\alpha t} \qquad ...(26)$$

A much simpler way to derive Equation (26) is to use the property of time-reversibility of the one-parameter model, that is, $P_{(t)}=p_{ij(2t)}$, which can readily be obtained from Equation (12) by replacing t by 2t. Similarly, in Kimura's two parameter model we have $P_{(t)}=Y_{(2t)}$, which can be obtained from Equation (15) by replacing t by 2t, that is,

$$P_{(t)} = \frac{1}{4} + \left(\frac{1}{4}\right)e^{-8\beta t} - \left(\frac{1}{2}\right)e^{-4(\alpha+\beta)t} \qquad ...(27)$$

Kimura (1980) obtained this formula by a different approach. Other substitution models are more complicated and are not considered here.

Next, let us consider the proportion, $Q_{(t)}$, of transversional differences between two sequences that diverged t time units ago. For Kimura's two parameter model, we can use the property of time-reversibility and obtain $Q_{(t)}=2Z_{(2t)}$, where the factor 2 comes from the fact that there are two types of transversion. Since $Z_{(2t)}$ is given by Equation (16),

$$Q_{(t)} = \frac{1}{2} - \left(\frac{1}{2}\right)e^{-8\beta t} \qquad ...(28)$$

This formula was also first obtained by Kimura (1980) by a different method. For the one-parameter model, $\beta=\alpha$ in the above formula.

Fig. 9.8 shows the expected proportion of transitional difference (P) and the excepted proportion of transversional differences (Q) between two sequences diverged t years ago. In the one-parameter model both P and Q increase monotonically to 1/4 and 1/2, respectively. In the two-parameter model, Q increases monotonically to the equilibrium value 1/2 whereas P can become greater than the equilibrium value 1/4 for a long period of time. $P_{(t)}$ can exceed 1/4 for a reason similar to that for Y(t).

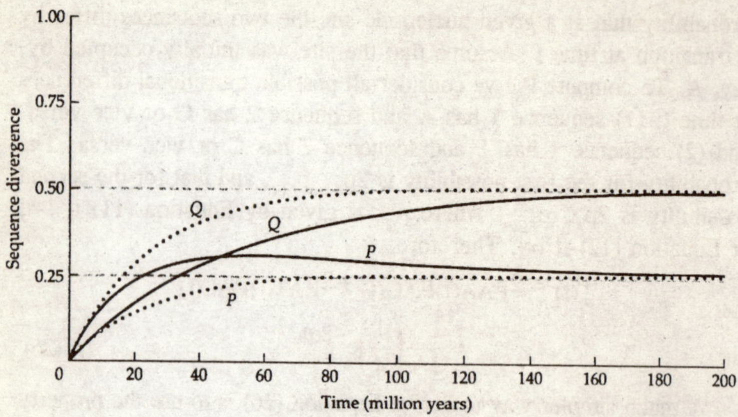

Fig. 9.8. Temporal changes in the expected proportion of transitional differences (P) and the expected proportion of transversional differences (Q) between two sequences that diverged t years ago.

NONUNIFORM RATES AMONG NUCLEOTIDE SITES

All the above treatments assumed that the rate of substitution per site (λ) is the same for all sites in the sequence, i.e., the sequence is homogeneous. This assumption is realistic only for nonfunctional sequences. For functional sequences, the substitution rate usually varies among sites. For example, some amino acid residues in cytochrome c appear to be invariant among species and amino acids on the surface of α globin evolve 10 times faster than those residues surrounding the heme group. Heterogeneity in substitution rate may have a strong effect on the observed degree of sequence divergence.

To see how rate heterogeneity may affect the degree of sequence divergence, let us consider a simple hypothetical case in which the sequences consist of only two types of sites: the substitution rate per site per year is $\lambda_1=3\alpha_1$ for the first class and $\lambda_2=3\alpha_2$ for the second class, and the proportions of the two classes of sites are f_1 and f_2, respectively. We treat the two classes separately. For simplicity, we use the one-parameter model. For two sequences that diverged t years ago, the similarity for the ith class sites (i=1 or 2) can be readily obtained from Equation (22) and is given by

$$I_i = \frac{1}{4} + \frac{3}{4}e^{-8\alpha_i t} \qquad \text{...(29)}$$

The average similarity over the entire sequence is

$$\bar{I} = f_1 I_1 + f_2 I_2 \qquad \text{...(30)}$$

Note that the average substitution rate per site for the above case is $3\alpha=3(f_1\alpha_1+f_2\alpha_2)$. For a homogeneous sequence with the substitution rate 3α, the sequence similarity (I) at time t is given by Equation (22), i.e.,

$$I=\frac{1}{4}+\frac{3}{4}e^{-8\alpha t} \qquad ...(31)$$

Fig. 9.9a shows a comparison of the two cases. The solid line represents the case of a sequence with two classes of sites with the proportions $f_1=0.05$ and $f_2=0.95$ and with the rates $\alpha_1=0$ and $\alpha_2=5\times10^{-9}$ substitutions per site years. The dotted line represents the case of a homogeneous sequence with $\alpha=f_1\alpha_1+f_2\alpha_2=4.75\times10^{-9}$. Although for any given t the two pairs of sequences have undergone the same number of substitutions, their I (sequence similarity) values differ appreciably when t is large. For the dotted line, I will decrease to the equilibrium value 0.25. In the solid line the first class sites (5%) are invariable (i.e., $\alpha_1=0$), so that $I_1=1$ for all t's while I_2 (the I value the second class sites) will decrease to the equilibrium value 0.25; therefore, the average I value will reach the equilibrium value $0.05 + 0.95 \times 0.25 = 0.29$. Thus, as $t\rightarrow\infty$, the difference between the two curves will increase to $0.29 - 0.25 = 0.04$.

In practice, the rate variation among sites is likely to be more complicated than the above hypothetical case. However, how the rate varies among sites in a sequence has been well studied; this question is currently a subject of much interest. Available data on amino acid substitution in protein sequences and on nucleotide substitutions in DNA sequences suggest that, to a first approximation, the substitution rate

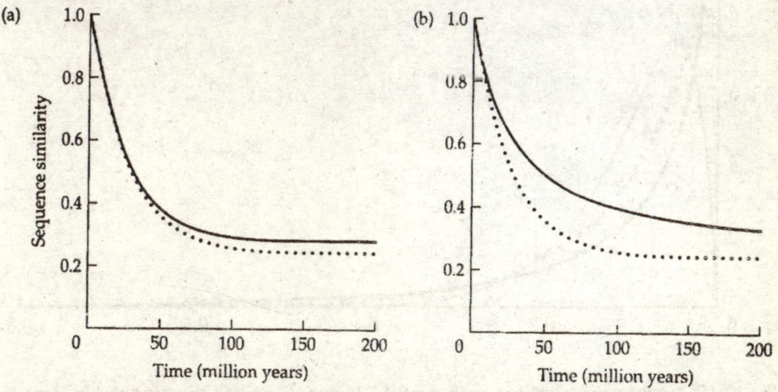

Fig. 9.9. Sequence similarity between two sequences diverged t years ago.

λ varies among sites according to the gamma distribution among sites, i.e.,

$$g(\lambda) = [b^a / \Gamma(a)]e^{-b\lambda}\lambda^{a-1}$$

where $a = \bar{\lambda}^2 / V_\lambda$ and $b = \bar{\lambda} / V_\lambda$, $\bar{\lambda}$ and V_λ being the mean and variance of λ, respectively. The shape of the distribution is largely determined by the parameter a. When a=1, g(x) decreases exponentially from g(0) to 0 as x increases from 0 to infinity. When a<1, g(x) approaches ∞ as x decrease to 0, meaning that a large proportion of sites in the sequence have a substitution rate close to 0. Note that g(x) decreases more slowly for the case of a<1 than for the case of a=1 or, in other words, when a<1, the distribution is more dispersed and the sequence is more heterogeneous. Finally, when a>1, G(0) is 0 and the distribution is more concentrated, implying that the sequence is less heterogeneous. More precisely, the heterogeneity is determined by the coefficient of variation, defined as $\sigma / \bar{\lambda}$, where σ is the standard deviation, i.e., the square root of $V(\lambda)$. Since $a = \bar{\lambda}^2 / V_\lambda$, the heterogeneity is inversely related to a. Indeed, roughly speaking, a=0.5,

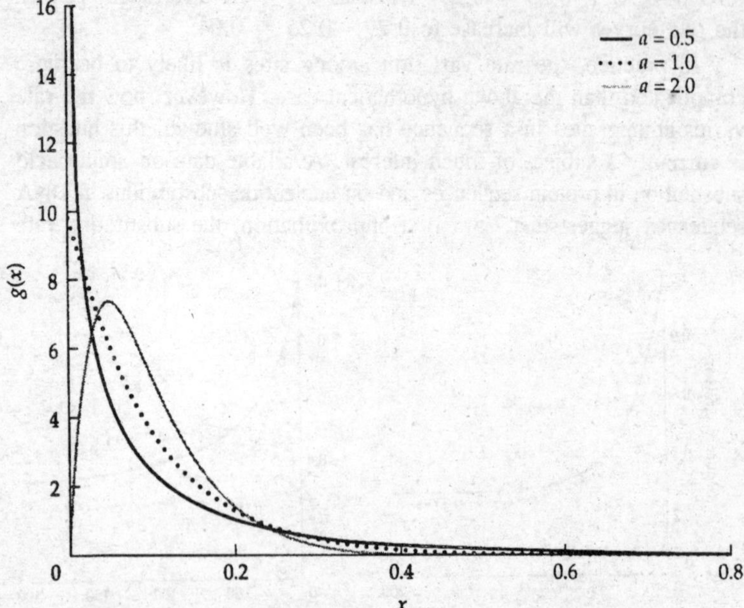

Fig. 9.10. Gamma distributions with a=0.5, 1, and 2. g(x) is the probability density function and a is the inverse of the square of the coefficient of variation.

1.0 and 2.0 means, respectively, strong, intermediate, and weak rate heterogeneity over sites. In the following we shall assume that λ follows the gamma distribution because there is some empirical support for this distribution and because the distribution can assume various shapes, as discussed above.

If we use the one-parameter model, the I(λ) value for a site substitution rate $\lambda=3\alpha$ is given by Equation (22). Therefore, the mean I value over the sites between the two sequences is

$$\bar{I} = \int_0^\infty I g(\lambda) d\lambda = \frac{1}{4} + \frac{3}{4}\left(\frac{a}{a+8\bar{\alpha}t}\right)^a \qquad ...(32)$$

where $\bar{\alpha} = \bar{\lambda}/3$.

In Fig. 9.10b the solid line represents the $\bar{I}$ value for the case with $\bar{\alpha} = 5\times10^{-9}$ and a=1, and the dotted line represents the I value for the case of homogeneous sequence with $\alpha=5\times10^{-9}$. For any given t the two pairs of sequences (i.e., the two curves) should have undergone the same number of substitutions, but the solid curve decreases with t considerably more slowly than the dotted curve and the difference between the two curves becomes large as t increases. Thus, rate heterogeneity can have a strong effect on the degree of sequence divergence between two diverging sequences. However, as t approaches infinity, both curves decrease to 0.25, as can be seen from Equation (22) and (32).

10

Evolution by Gene Duplication

The evolutionary significance of gene duplication was first recognized by Haldane (1932) and Muller (1935), who suggested that a redundant duplicate of a gene may acquire divergent mutations and eventually emerge as a new gene. However, few examples of duplicate genes were discovered before the advent of biochemical and molecular biology techniques. The development of protein sequencing methods in the 1950s provided a tool for the study of long-term evolution, and by the late 1950s the α and β chain of hemoglobin were recognized to have been derived from gene duplication. Later, isozyme, DNA-RNA hybridization, and cytogenetic studies added much further evidence for the frequent occurrence of gene duplication in evolution. Using data from these various types of studies, Ohno (1970) put forward the view that gene duplication is the only means by which a new gene can arise. Although other means of creating new genes or new functions are now known and are discussed later in this chapter, Ohno's view remains largely valid.

There is now ample evidence that gene duplication is the most important mechanism for generating new genes and new biochemical processes that have facilitated the evolution of complex organisms from primitive ones. For example, the genome of vertebrates contains many gene families that are not found in invertebrates, and many gene duplications apparently occurred in the early evolution of chordates. Gene duplication is also important for generating many copies of genes of the same function, thereby enabling the production of a large quantity of RNAs or proteins. Furthermore, internal (partial) gene duplication plays a major role in increasing the functional complexity of genes in evolution. Many proteins of present-day organisms show internal repeats of amino acid sequences, and the repeats often correspond to the functional or structural domains of the proteins. This observation suggests that the genes coding for these proteins were formed by internal gene duplication. This type of duplication allows the improvement of the function of a protein by increasing the number of active sites or the aquisition of an additional function by modifying a redundant segment. Many complex genes in present-day organisms might have evolved from small, simple primordial genes via internal duplication and subsequent modification.

The discovery of split genes prompted Gilbert (1978) to suggest that recombination in introns provides a mechanism for the exchange of exon sequences between genes, a phenomenon that has been commonly known as *exon shuffling*. Many examples of exon shuffling has been found, indicating that this mechanism has played a significant role in the evolution of eukaryotic genes with new functions. However, an exon-shuffling event is likely to be biologically significant only if it includes a structural or functional domain. Moreover, a domain-shuffling event may occur without the involvement of any exon. Thus, it is perhaps more meaningful to talk about domain shuffling rather than exon shuffling.

Types of DNA Duplication

A duplication may involve (1) part of a gene, (2) a single gene, (3) part of a chromosome, (4) an entire chromosome, or (5) the whole genome. These types of duplication are called, respectively, (1) partial or internal gene duplication, (2) complete gene duplication, (3) partial chromosomal duplication or partial polysomy, (4) chromosome duplication, aneuploidy, or polysomy, and (5) genome duplication or polyploidy. The first four categories are also known as regional duplications, because they do not affect the entire haploid set of chromosomes.

Ohno (1970) argued that genome duplication has generally been more important in evolution than regional duplication, because in the latter case only parts of the regulatory system of structural genes may be duplicated—such an imbalance may disrupt the normal function of the duplicated genes. However, as will be discussed later, partial and complete gene duplications apparently also have played a very important role in evolution; for example, this importance is suggested by the fact that genes controlling the same function are often closely linked. Partial polysomy, if it involves many genes, and polysomy are likely to cause a severe imbalance in gene product, and their chance of being incorporated into the population is small. Indeed, experiments in *Drosophila* have shown that the viability of terminal partial polysomy declines with the length of the triplicated region, becoming generally lethal when more than one-half of an autosomal arm is present in three doses. Also, it is well known that in man trisomies of larger chromosomes are lethal conditions, and even those of the smaller ones result in sterility; e.g., trisomy for chromosome 21 causes Down's syndrome. Thus, polysomy and partial polysomy are probably not important in evolution. Genome duplication was probably common in animal evolution before sex chromosomes became differentiated. In organisms with well-differentiated sex chromosomes such as mammals, birds, and reptiles, genome duplication disrupts the mechanism of sex determination and is likely to be quickly eliminated from the population; thus, it probably has contributed little in the evolution of bisexual organisms. In contrast, in plants genome duplication is a widespread phenomenon. Chromosome duplication and genome duplication as mechanisms of increasing the size of a genome will be discussed else where.

Domain Duplication and Gene Elongation

A protein *domain* is a well-defined region within a protein that either performs a specific function, such as substrate binding, or constitutes a stable, compact structural unit within the protein that can be distinguished from all the other parts. The former is referred to as a *functional domain*, and the latter a *structural domain* or *module*. Defining the boundaries of a functional domain is often difficult because functionality is in many cases conferred by amino acid residues that are located in different regions of he polypeptide. A structural module, on the other hand, consists of a continuous stretch of amino acids.

The above distinction is important when considering possible evolutionary mechanisms by which multidomain proteins have come into existence. If a functional domain coincides with a module, its duplication will increase the number of functional segments. In contrast, if functionality is conferred by amino acid residues scattered among different modules, the effects of a duplication of a single module may not be functionally desirable. The internal repeats found in many proteins often correspond to either structural modules or functional domains residing within a single module.

Several possible relationships may exist between the structural domains and the arrangement of exons in the gene. In some genes, a more or less exact correspondence exists between the exons of the gene and the structural domains; this relationship exists for many globular proteins. When such a relation exists, duplication of an exon will result in duplication of the corresponding structural domain. In the majority of cases, however, a more complex relationship exists between protein domains and exons, such as shown in the situations. In the case of duplication of an exon results in duplication of two domains, whereas in the case of duplication of an exon only duplicates part of a domain, thus being more likely to disrupt the protein function and less likely to survive.

A survey of gene in eukaryotes shows that internal duplications have occurred frequently in evolution. This increase in gene size, or *gene elongation*, is one of the most important steps in the evolution of complex genes from simple ones. Theoretically, elongation of genes can also occur by other means. For example, a mutational change converting a stop codon into a sense codon can also elongate the gene. Similarly, insertion of a DNA segment into one of the exons or the occurrence of a mutation obliterating a splicing site will achieve a similar result. These types of molecular changes, however, would most probably disrupt the function of the elongated gene, because the added regions would consists of an almost random array of amino acids. Indeed, in the vast majority of cases, such molecular changes have been found to be associated with pathological manifestations. For instance, the hemoglobin abnormalities Constant Spring and Icaria resulted from mutations turning the stop codon into codons for glutamine and lysine, respectively, thus adding 30 additional residues to the α chains of these variants. By contrast, duplication of a structural domain is less likely to be problematic. Indeed, such a duplication can sometimes even enhance the function of the protein produced, for example, by increasing the number of active sites. A well-known example is the haptoglobin α2 allele in humans, which was formed by a nonhomologous crossing-over within different introns of two α1 genes, probably a slow and a fast electrophoretic variant (α1S and α1F). The internal duplication nearly doubled the length of the polypeptide (changing from 84 to 143 amino acids) and it seems to increase the stability of the haptoglobin-hemoglobin complex and the efficiency in rendering the heme group of hemoglobin susceptable to degradation. The α2 gene is probably of recent origin, at least more recent than the human-chimpanzee split, but it has a fairly high frequency (30-

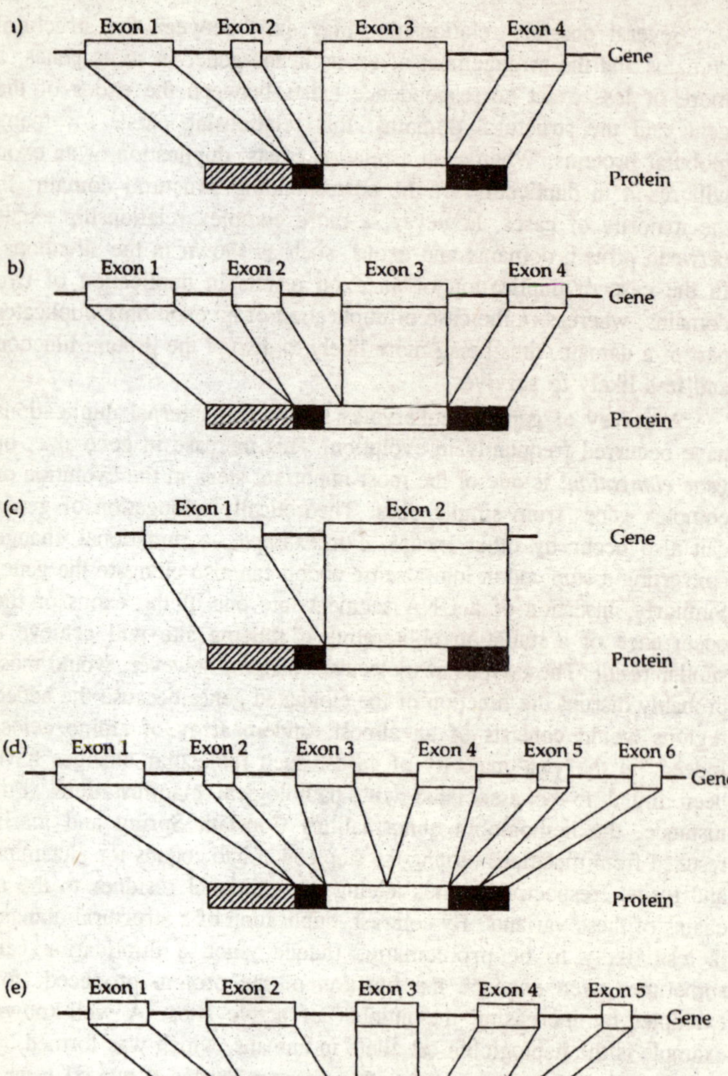

Fig. 10.1. Possible relationships between the arrangement of exons in a gene and the structural domains of the protein it encodes: (a) each exon corresponds exactly to a structural domain; (b) the correspondence is only approximate; (c) an exon encodes two or more domains; (d) a single structural domain is encoded by two or more exons; and (e) lack of correspondence between exons and domains.

70%) in Europe and in parts of Asia. It is likely that the α2 allele will replace the α1 alleles.

The examples below are to illustrate the point that gene elongation during evolution has depended largely on the duplication of domains.

α2 Type I Collagen Gene

A remarkable example of gene elongation by duplication is given by the α2 type I [α2(I)] collagen gene. Collagen is the main supportive protein of skin, bone, cartilage, and connective tissue in vertebrates. The major part of the mature collagen consists of three polypeptide chains, called α chains, arranged in a long rodlike helical configuration. This region contains about 338 repeats of the amino acid triplet Gly-X-Y, where X and Y are often proline and hydroxyproline. The chicken α2(I) collagen gene has been sequenced it has a length of approximately 38 kilobases and contains 52 exons, among the largest numbers of exons ever observed in a gene. The triplet-helical coding region of the pro-α2(I) collagen gene is composed of 42 triple-helical exons and 2 "junction" exons. Of the 42 triple-helical exons, 5 contain 45 bp; 23, 54 bp 5, 99 bp; 8, 108 bp; and 1, 162 bp—remarkably, all in multiples of the 9 bp that code for the triplet Gly-X-Y! For example, exon 8 is 54 bp long and encodes 18 repeats of Gly-X-Y as shown below:

GGT CCT CCT GGG TTT CAA GGT GTT CCT GGT GAG
Gly Pro Pro Gly Phe Gln Gly Val Pro Gly Glu
CCT GGT GAA CTT GGT CAA ACA
Pro Gly Glu Pro Gly Gln Thr

It is quite possible that all of these exons were derived from an ancestral exon of 54 bp by multiple duplications and recombinations. Many of the 54-bp exons are contiguous and could be the result of a series of exon duplications. On the other hand, a 99-bp exon could have arisen from an unequal crossing-over between two exons. One might argue that exons of 108 bp arose by precise delction of the intervening sequence and fusion of two 54-bp exons. However, the precise deletion of an intron appears to have occurred rarely in evolution, though a number of such cases have been found, e.g., in a rat insulin gene. It seems more likely that an unequal crossing-over between a 99-bp exon and a 54-bp exon gave rise to a 108-bp exon and a 45-bp exon. At any rate, because of its regular repetitive structure, partial gene duplication can occur easily in this gene. Such duplication may not disrupt the gene function if the duplicated part is a multiple of the 9-bp sequence coding for Gly-X-Y, which seems to be necessary for forming a helical structure.

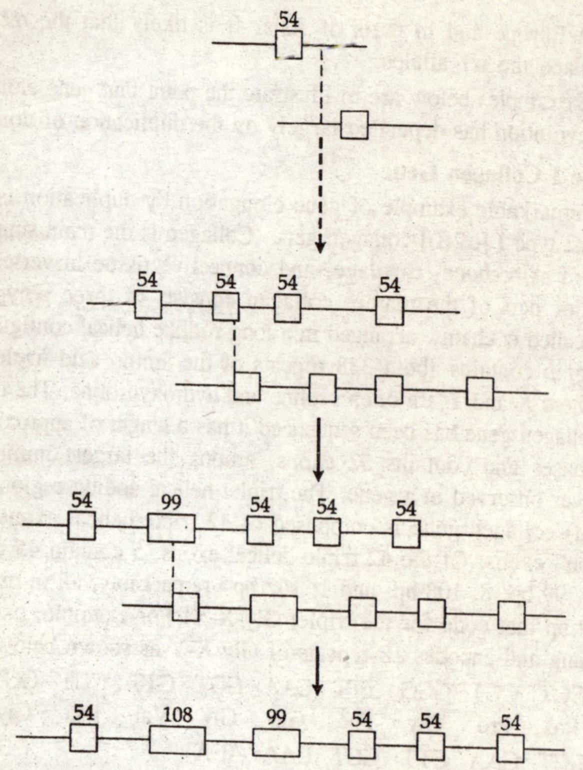

Fig. 10.2. A hypothetical scheme for the evolution of the exons in the DNA region coding for the helical region in the α2(I) collagen polypeptide.

Apolipoprotein Genes

The apolipoproteins are the protein components of plasma lipoproteins, which are the major carriers of cholesterol, phospholipids, and triglycerides in vertebrate blood. In mammals, nine different types of apolipoproteins (A-I, A-II, A-IV, B C-I, C-II, C-III, D and E) have been identified and characterized. The major function of the apolipoproteins is lipid binding so as to make lipoproteins water soluble. Most apolipoproteins have, in addition, acquired specialized functions in lipid metabolism. For example, apoC-II is the physiological activator of lipoprotein lipase, an enzyme that hydrolyzes triglycerides to release fatty acids, which are either used for energy or are re-esterified and stored in adipocytes. The apoA-I, apoA-II, apoC-I, apoC-II, apoC-III, and apoE genes have the same genomic structure, each containing a total of three introns at the same locations. The apoA-IV gene has lost the first intron, but has retained the other two. Thus, these seven

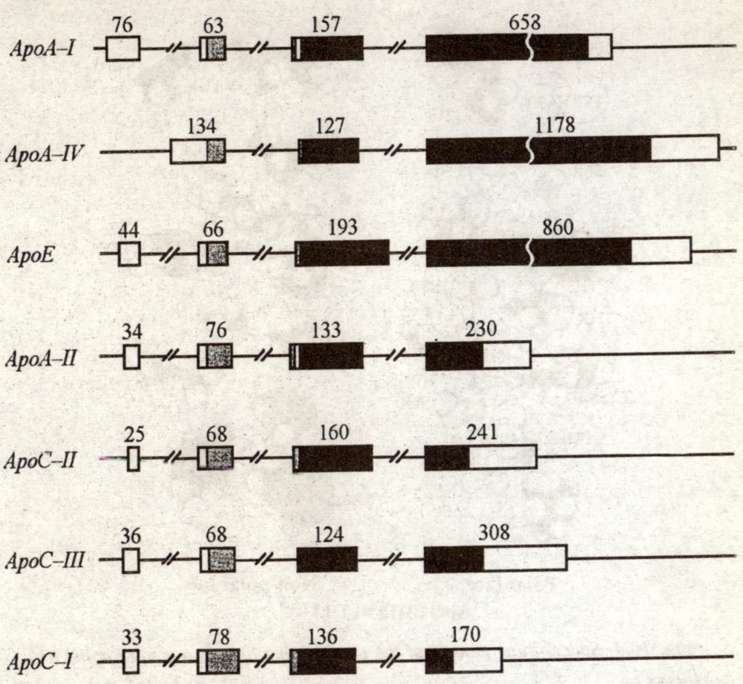

Fig. 10.3. Structural organization of the human apolipoprotein A-I, A-II, A-IV, C-I, C-II, C-III, and E genes.

genes clearly belong to one multigene family. Note, however, that the apoB and apoD gene have very different genomic structures and probably do not belong to the same family.

From analysis of amino acid sequence data, Barker and Dayhoff (1977), Fitch (1977a), and McLachlan (1977) found that apoA-I contains multiple repeats of a 22 amino acid unit (22-mer) each of which is a tandem array of two 11-mers. The 22-mer repeat unit has been suggested to be a structural element that builds an amphipathic α helix with a proline at the amino end. An important feature of the amphipathic helix model is the presence of two clearly defined faces, one hydrophobic, that is inserted between the fatty-acyl chains of the phospholipid molecules, and the other hydrophilic, that interacts with the phospholipid head groups and the aqueous phase. Another feature of the model is that the negatively charged residues, Glu and Asp, tend to occupy positions along the center of the polar face, while the positively charged residues, Lys and Arg, are located on the lateral edges of the polar face.

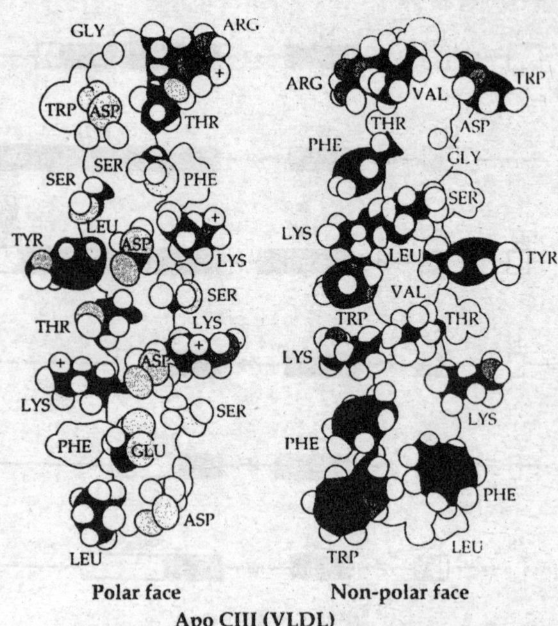

Fig. 10.4. Space-filling models of the amphipathic helical region of apoC-III.

DNA-sequencing work has provided better data for defining the repeats in apoA-I and has further revealed similar repeat patterns in apoA-II, apoA-IV, apoC-I, apoC-II, apoC-III and apoE. In each gene the repeats start from the first codon in exon 4 and end close to the end of this exon. Most of the repeats are full 22-mers, each of which is composed of two half-repeats (11-mers) and is divided into groups A and B. The main reason for considering the 22-mer as the fundamental unit of repeat is that most 11-mers are more similar to the 11-mer one unit removed than to the adjacent 11-mer.

Some remarkable features emerge from the alignment shown in Fig. 10.5. Most prolines (P) appear in the first column of group A. Columns 6 and 10 of both groups are mostly occupied by hydrophobic amino acids (M, methionine; V, valine; L, leucine; I, isoleucine; F, phenylalanine; Y, tyrosine; W, tryptophan), the rest being indifferent amino acids (G, glycine; A, alanine; S, serine; T, threonine; N, asparagine; Q, glutamine; H, histidine; C, cysteine). In both groups, column 3 is predominantly occupied by hydrophobic amino acids, columns 4 and 5 by acidic amino acids (D, aspartic acid; E, glutamic acid), and column 9 by basic amino acids (R, arginine; K, lysine); in

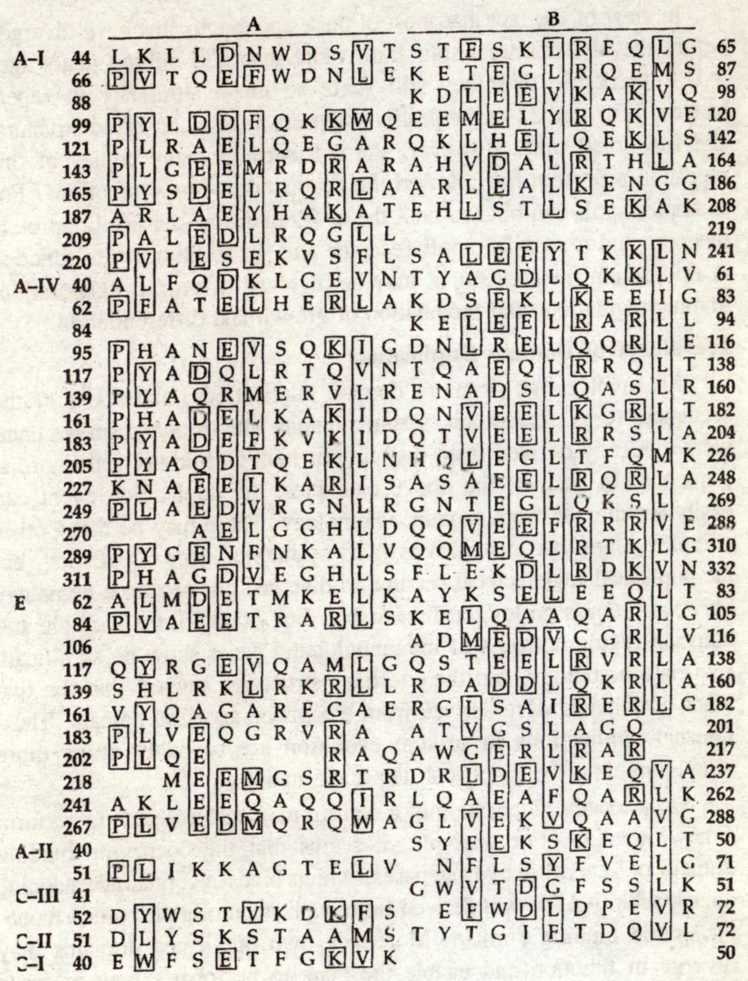

Fig. 10.5. Internal repeats in exons 4 of human apolipoproteins A-I, A-IV, E, A-II, C-III, C-II, and C-I.

column 11 there is no clear pattern. In group A, column 2 is predominantly occupied by hydrophobic amino acids, but in group B, this column has no predominant pattern. Thus, there are major similarities and only minor differences between groups A and B.

The large variation in size among the soluble apolipoproteins used to be a puzzle because they were apparently derived from a common ancestral gene. It is now clear that the size variation is largely due to differences in the number of repeats in the last exon i.e., exon 4.

In view of the fact that most of these apolipoproteins have diverged before the bird-mammal split, i.e., more than 270 million years ago, it is remarkable that they still share so much similarity in repeat pattern. Thus, the 22-mer repeat unit appears to be a conserved structural and functional unit. However, the existence of many copies of this repeat in a protein has allowed the acquisition of new functions. For example, apoE can bind to both the apoB/apoE low-density lipoprotein receptor and a specific apoE receptor and, in addition to its function in lipoprotein metabolism, it may be involved in neural regeneration, immunoregulation, and modulation of growth and differentiation.

Prevalence of Domain Duplication

All involve one or more domain duplications, and some of the sequences (e.g., ferredoxin, serum albumin, and tropomyosin α chain) were evidently derived from doubling or multiplication of a primordial sequence. In each of the above examples, the duplication event can easily be inferred from sequence homology. There may be many other complex genes that have also evolved by internal gene duplication, but the duplicated regions have become so diverged that sequence homology between them is no longer discernable. For example, the variable and constant region domains of immunoglobulin genes show no significant sequence homology, but their tertiary structures suggest that the two kinds of domains were derived from a common ancestral domain. Thus, internal duplications in protein evolution are probably much more ubiquitous than the empirical data have indicated.

Occasionally, duplication of a domain may enable a gene to perform a new function. For example, it seems that the coenzyme binding domain of several dehydrogenases, which binds nicotinamide adenine dinucleotide (NAD), was derived from duplication of a primordial mononucleotide binding domain. Moreover, two duplicated domains may diverge in function and enable the gene to perform a new or more complex function. For instance, the variable and constant region domains of immunoglobulin genes were probably derived from a common ancestral domain, but they now have distinct functions—antigen recognition by the variable regions and effector functions by the constant regions. Many complex genes might have arisen in this manner.

Formation of Gene Families and Acquisition of New Functions

A complete gene duplication produces two identical copies. How they will evolve varies from case to case. The copies may, for instance, retain their original function, enabling the organism to produce a larger quantity of certain RNA species or proteins. Or, one copy may be incapacitated by the occurrence of deleterious mutations and become a functionless pseudogene. More importantly, however, gene duplication may result in the emergence of genetic novelties or new genes. This will happen if one of the duplicates retains its original function, while the other accumulates molecular changes such that, in time, it can perform a rather different task.

Repeated genes are divided into two types: variant and invariant repeats. Invariant repeats are identical or nearly identical in sequence to one another. In several cases, the repetition of identical sequences can be shown to be correlated with the synthesis of increased quantities of a gene product that is required for the normal function of the organism. Such repetitions are referred to as *dose repetitions*. Dose repetitions are quite common whenever a metabolic need for producing large quantities of specific RNAs or proteins arises. Representative examples include the genes for rRNAs and tRNAs, which are required for translation, and the genes for histones, which constitute the chief protein component of chromosomes and are synthesized in large quantities during cell division. Another good example of dosage repetition is the amplification of an esterase gene in a California *Culex* mosquito, leading to insecticide resistance.

Variant repeats consists of copies of a gene that, although similar to each other, differ in their sequence to a lesser or greater extent. Representatives of variant repeats are various types of isozymes. Variant repeats may differ to some extent in function or regulation, so that they contribute to the fine-tuning of physiological functions of an organisms.

All the genes that belong to a certain group of repeated sequences in a genome are referred to as a *gene family* or *multigene family*. Members of a gene family usually reside in close proximity to each other on the same chromosome. In some cases, some functional or nonfunctional family members may be located on other chromosomes.

When duplicate genes become very different from each other in either function or sequence, it may no longer to convenient to assign them to the same gene family. The term *superfamily* was coined by Dayhoff (1978) in order to delineate between closely related and distantly related proteins. Accordingly, proteins that exhibit at least 50% similarity to each other at the amino acid level are considered to be members of a family, while homologous proteins exhibiting less than 50% similarity are considered to be members of a superfamily. For example, the α globins and the β globins are classified into two separate families, and together with myoglobin they form the globin superfamily. However, the two terms cannot always be used strictly according to Dayhoff's criteria. For example, human and carp α-globin chains exhibit only a 46% sequence similarity, which is below the limit for assignment to the same gene family. For this reason, the classification of proteins into families and superfamilies is determined not only according to sequence resemblance, but also by considering auxiliary evidence pertaining to functional similarity or tissue specificity. The genomes of higher eukaryotes contain numerous superfamilies of genes, e.g., the collagens, the actins, the immunoglobulins, and the serine proteases. In contrast to the case of a gene family, the members in a superfamily usually have diverged a long time ago and may have quite distinct functions. A well-known example is that of trypsin and chymotrypsin, which diverged about 1500 Myr ago. These two digestive enzymes have acquired distinct functions; trypsin cleaves polypeptide chains at arginine and lysine residues, whereas chymotrypsin cleaves polypeptide chains at phenylalanine, tryptophan, and tyrosine residues. Examples of gene superfamilies will be present in later sections.

The number of genes in a gene family varies widely. Some genes are repeated within the genome a few times, and these are referred to as "lowly repetitive." Others may be repeated 50 times or more within the genome and are called "highly repetitive." Between these two extreme are moderately repetitive genes. In the following section, rRNA and tRNA genes will be used to illustrate highly repetitive invariant genes. Lowly repetitive genes will be represented by isozymes and the colour-sensitive pigment proteins.

RNA-Specifying Genes

The mitochondrial genome of mammals contains only one 12S and one 16S rRNA gene. This is apparently sufficient for the mitochondrial translation system, because the genome has only 13 protein-coding genes. The mycoplasmas, which are the smallest self-replicating prokaryotes, contain two sets of rRNA genes. The genome of *Escherichia coli* is 4-5 times larger and it contains 7 sets of rRNA genes. The number of rRNA genes in yeast is approximately 140, and the numbers in fruit flies and humans are even larger. *Xenopus laevis* has a larger genome size and a larger number of rRNA genes than humans. Thus, there exists a good correlation between the number of rRNA gens and genome size. This is also true for the number of tRNA genes.

A conspicuous exception to the above rule is that *Tetrahymena* has a genome size considerably larger than that of *Saccharomyces cerevisiae* but has only one set of rRNA genes. This is, however, the copy number in the germinal nucleus, i.e., the micronucleus. In the derivation of macronuclei from the micronucleus the copy number is amplified to about 200. Thus, a large number of rRNAs can be produced in a cell, because macronuclei serve as somatic nuclei during vegetative growth and are responsible for virtually all of the transcriptional activity during growth and division.

Highly repetitive genes such as the rRNA genes are generally very similar to each other in sequence. One factor responsible for the homogeneity may be purifying selection, because such genes may have very specific functional or structural requirements. However, homogeneity often extends to regions having no known function (e.g., untranscribed regions), and so the maintenance of homogeneity necessitates invoking other mechanisms such as unequal crossing-over and gene conversion.

Isozymes

In addition to invariant repeats, the genomes of higher organisms contain numerous multigene families whose members have diverged to various extents in regulation and function. Good examples are families of genes coding for isozymes such as those for lactate dehydrogenase, aldolase, creatine kinase, pyruvate kinase, and carbonic anhydrase. *Isozymes* are enzymes that catalyze the same biochemical reaction but may differ from each other in tissue specificity, developmental time, electrophoretic mobility, or biochemical properties. They are encoded by different loci, usually duplicated genes, as opposed to *allozymes*, which are distinct forms of an enzyme encoded by different alleles at a single locus.

The study of multilocus isozyme systems has greatly enhanced our understanding of how cells with identical genetic materials can differentiate into hundreds of different types that constitute the complex body organization of vertebrates. Although isozymes still serve the same function, different members of the family may become better adapted for utilization of substrates in different tissues and this increases the fine tuning of cell physiology.

A well-known example of isozymes is the two genes encoding the A and B subunits of lactate dehydrogenase (LDH) in vertebrates. These

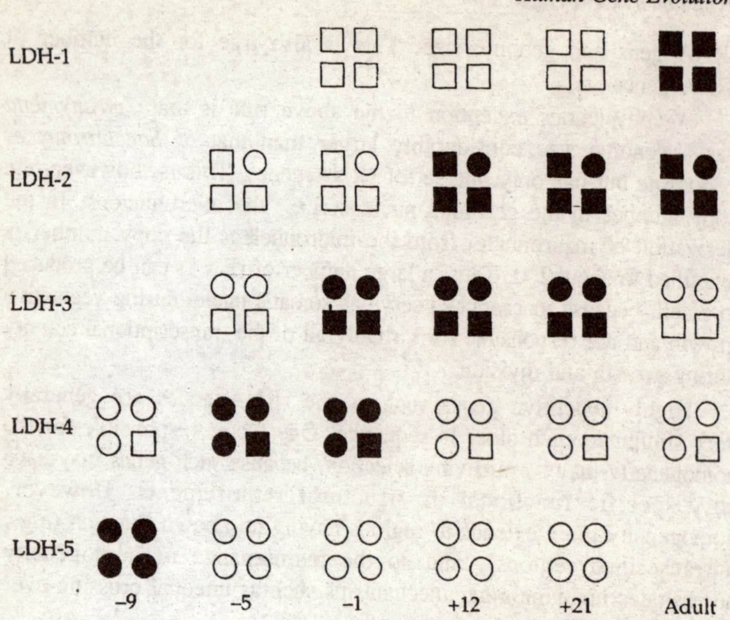

Fig. 10.6. Developmental sequences of five lactate dehydrogenase (LDH) isozymes in the rat heart.

two subunits form five tetrameric isozymes, A_4, A_3B, A_2B_2, AB_3, and B_4, all of which catalyze either the conversion of lactate into pyruvate in the presence of the coenzyme nicotinamide adenine dinucleotide (NAD^+), or the reverse reaction in the presence of the reduced coenzyme (NADH). It has been suggested that LDH-B_4 and the other isozymes rich in B subunits, which have a high affinity to NAD^+, function as a true lactate dehydrogenase in aerobically metabolizing tissues such as heart, whereas LDH-A_4 and the isozymes rich in A subunits, which have a high affinity metabolizing tissues such as skeletal muscle. The more anaerobic the heart is, specifically in the early stages of gestation, the higher the proportion of LDH isozymes rich in A subunits will be. Thus, the two genes have become specialized to different tissues and to different developmental stages. As the two subunits are present in almost all vertebrates studied to date, the duplication that produced the two genes probably occurred before or during the early stage of vertebrate evolution.

An interesting feature of the above example is that the two subunits can form heteromultimers, which increase the versatility of the enzyme. Many other examples of multimeric enzymes that are composed of polypeptides encoded by duplicated genes can be found among multilocus isozymes.

Colour-Sensitive Pigment Proteins

Humans, apes, and Old World monkeys possess three colour-sensitive pigment proteins. The blue pigment is encoded by an autosomal gene, while the red and green pigments are each encoded by an X-linked gene amino acid sequences of the red and green pigments are 96% identical, but their similarity with the blue pigments is only 43%. The blue pigment gene and the ancestor of the green and red pigment genes diverged about 600 million years ago. In contrast, the close linkage and high homology between the red and green pigments point to a very recent gene duplication. Because New World monkeys have only one X-linked pigment gene, whereas Old World monkeys and humans have two or more, the duplication probably occurred in the ancestor of apes and Old World monkeys after their divergence from the New World monkeys about 35-40 million years ago. As a consequence of this duplication, humans, apes, and Old World monkeys can distinguish three colours.

Interestingly, in some New World monkey species, e.g., squirrel monkeys, there are three highly polymorphic alleles at the X-linked

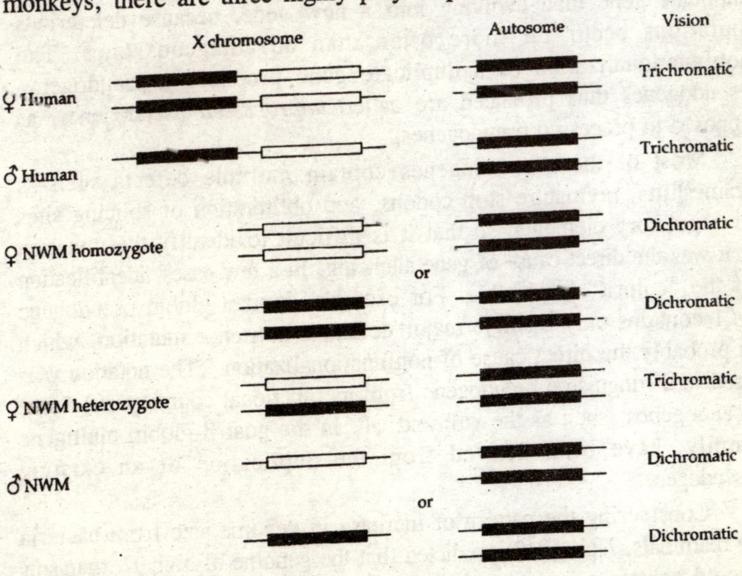

Fig. 10.7. Molecular basis of trichromatic vision in humans and New World monkey (NWM).

locus. Two of these alleles have spectral sensitivity peaks similar to those of human red and green colour pigments, respectively, while the third one has an intermediate spectral peak. For this reason, a female monkey heterozygous for two of the three alleles is trichromatic, like a normal human, though male monkeys, who carry only one X chromosome, are *dichromatic*, like a colour-blind human, i.e., they can only distinguish between blue and green or blue and red, but not between green and red. Thus, in the case of humans and Old World monkeys, trichromatic vision is achieved by a mechanism akin to isozymes (i.e., two distinct proteins encoded by different loci), while in heterozygous female squirrel monkeys, the same end is achieved through the use of two "allozymes" (i.e., two distinct allelic forms at a single locus). If trichromatic vision confers a selective advantage on its carriers, as commonly thought, then the long-term maintenance of the three colour-sensitive alleles at a locus in New World monkeys is achieved through overdominant selection.

Nonfunctionalization of Redundant Duplicates

A duplicate gene may accumulate deleterious mutations and become nonfunctional, as long as the other duplicate gene is functioning normally. Indeed, this is a much more likely fate for a redundant duplicate gene than evolving into a new gene, because deleterious mutations occur far more often than advantageous ones. The nonfunctionalization of a duplicate gene produces a pseudogene. Pseudogenes thus produced are called *unprocessed* pseudogenes, as opposed to processed pseudogenes.

Most of these pseudogenes contain multiple defects such as frameshifts, premature stop codons, and obliteration of splicing sites or regulatory elements, so that it is difficult to identify the mutation that was the direct cause of gene silencing. In a few cases, identification of the "culprit" is possible. For example, human globin pseudogene $\psi\zeta 1$ contains only a single major defect, a nonsense mutation, which is probably the direct cause of nonfunctionalization. (The notation ψ is used to distinguish a pseudogene from its functional counterpart.) Some pseudogenes, such as the $\psi\beta^x$ and $\psi\beta^z$ in the goat β-globin multigene family, have been derived from the duplication of an existing pseudogene.

Considering the pattern of increase in genome size from bacteria to mammals, Nei (1969) predicted that the genome of higher organisms would contain a large number of nonfunctional genes. There is now indeed ample evidence that loss of duplicate genes has occurred repeatedly in the long history of the evolution of life. For example, electrophoretic studies suggest that salmonid and catostomid fishes, which are of tetraploid origin, have lost duplicate gene expression in about 50% of the cases. Furthermore, pseudogenes have been found in almost every gene family that has been examined in detail e.g., the 5S rRNA genes, the α- and β-globin gene clusters, and the immunoglobulin genes.

The rate of loss of a duplicate gene from a population or the rate of fixation of a pseudogene in a population depends on population size as well as on the rate of null mutation. This is because the process occurs mainly by random genetic drift. In an extremely large population, the effect of genetic drift is negligible, so that defective duplicate genes may never be fixed in the population. In finite populations, fixation of pseudogenes can occur and many authors have studied the rate of fixation, assuming that there are two duplicate loci and that mutation occurs irreversibly from the normal to the nonfunctional or null state. In the case of no linkage, Watterson (1983) showed that the mean time until fixation of the null allele at one of the two loci is roughly given by

$$T = N \log N - N\psi(2N\upsilon) + 2.53N \qquad \ldots(1)$$

where $\psi(.)$ denotes the digamma function, υ is the rate of mutation from the normal to the null state per gene per generation, and N is the effective population size. It is assumed that the population is initially free of null mutants and that double-null homozygotes are lethal but all other genotypes are normal.

There are several interesting properties. First, tight linkage ($r=0.001$) has only a minor effect on T, as long as the population size is not extremely large. Second, T is small in a small population but increases with population size. That is because as population size increases the effect of random drift becomes weaker and selection becomes effective. When N is 10^6, it takes approximately 10 million generations for a pseudogene to be fixed in the population. Third, in a small population T is largely determined by υ, the null mutation rate. Indeed, when N is of the order of 100, T is roughly given by $1/(2\upsilon)$. For example, when $\upsilon=10^{-5}$, $N=100$, and $r=0.5$, T is approximately 50,900 generations. In a large population, however, an increase in υ has only a minor effect on T. For instance, if $N=10^5$ and $r=0.5$, T decreases from 1.16×10^6 to 0.97×10^6 when υ increases from 10^{-5} to 10^{-4}. Fourth, the standard deviation is as large as the mean. Thus, the fixation time is subject to large random errors.

The preceding computation shows that the rate of fixation of pseudogenes is highly dependent on population size and mutation rate. At the present time, however, there is still no data to test the details of these relationships. In practice, the rate of fixation would depend on several other factors such as the effect of null alleles on the fitness of heterozygotes, attainment of disomic segregation at meiosis in the case of tetraploidization, and differentiation of the regulatory systems of the duplicate genes. All these factors would complicate the relationships among T, N, and υ. In general, however, these factors have an effect of prolonging the fixation time.

In a multigene family, the gene number may fluctuate from time to time as a result of unequal crossing-over. If the number becomes larger than the optimal number, some of the genes will be free to become pseudogenes. In this case, the mean time for the first pseudogene to become fixed in the population is expected to be considerably shorter than that for the case of two duplicate genes, because any of the multiple genes can become nonfunctional. This seems to be in agreement with actual data, because most of the multigene families include some pseudogenes at present.

Dating Gene Duplications

Two genes are said to be *paralogous* if they are derived from a duplication event, but *orthologous* if they are derived from a speciation event. For example, genes α and β were derived from duplication of an ancestral gene and are therefore paralogous, while gene α from species 1 and gene α from species 2 are orthologous, and so are gene

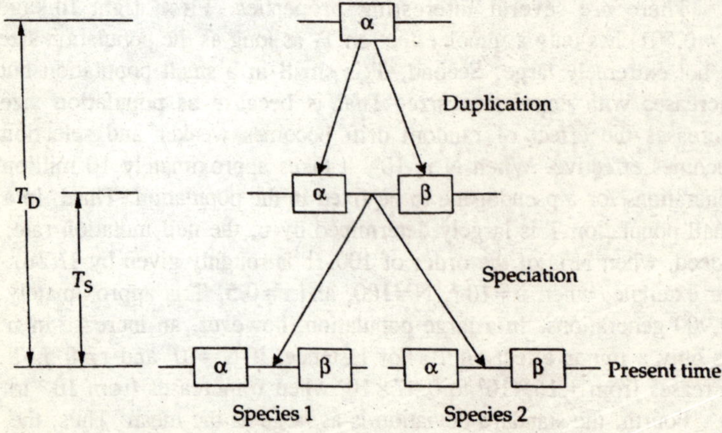

Fig. 10.8. Model for estimating the time of a gene duplication event (T_D).

β from species 1 are gene β from species 2. We can estimate the date of duplication, T_D, from sequence data, if the rate of substitution in genes α and β is known. The rate of substitution can be estimated from the number of substitutions between the orthologous genes in conjunction with knowledge of the time of divergence, T_S, between species 1 and 2. An estimate of T_D can be obtained as follows.

For gene α, let K_α be the number of substitutions per site between the two species. Then, the rate of substitution is gene α is estimated by

$$r_\alpha = K_\alpha / (2T_S) \quad \text{...(2)}$$

The rate of substitution in gene β, r_β, can be obtained in a similar manner. The average substitution rate for the two genes is given by

$$r = (r_\alpha + r_\beta) / 2 \quad \text{...(3)}$$

To estimate T_D, we need to know the number of substitutions per site between genes α and β ($K_{\alpha\beta}$). This number can be estimated from four pairwise comparisons: (1) gene α from species 1 and gene β from species 2, (2) gene α from species 2 and gene β from species 1, (3) both genes from species 1, and (4) both genes from species 2. From these four estimates we can compute the average value ($\overline{K}_{\alpha\beta}$) for $K_{\alpha\beta}$, from which we can estimate T_D as

$$T_D = \overline{K}_{\alpha\beta} / (2r) \quad \text{...(4)}$$

Note that in the case of protein-coding genes, by using the numbers of synonymous and nonsynonymous substitutions separately, we can obtain two independent estimates of T_D. The average of these two estimates may be used as the final estimate of T_D. However, if the number of substitutions per synonymous site between genes α and β is large, say larger than 1, then the number of synonymous substitutions cannot be estimated accurately, and so synonymous substitutions may not provide a reliable estimate of T_D. In such cases, only the number of nonsynonymous substitutions should be used.

In the above, we have assumed rate constancy. This assumption can be tested by the four pairwise comparisons mentioned above. The assumption fails if an approximate equality does not hold among the four comparisons. As will be discussed in the next chapter, problems due to concerted evolutionary events may also arise and complicate the estimation of T_D.

Another method for dating gene-duplication events is to consider the phylogenetic distribution of genes in conjunction with paleontological data pertinent to the divergence date of the species in question. For example, all vertebrates with the exception of jawless fish (Agnatha) encode α- and β-globin chains. There are two possible explanations for this observation. One is that the duplication event producing the α and β globins occurred in the common ancestor of the Agnatha and the other vertebrates, but all the Agnatha species have lost one of the two duplicates. This is possible but not likely, because such a scenario would require that the losses occurred independently in many evolutionary lineages. The other explanation is that the duplication event occurred after the divergence of jawless fish from the ancestor of all other vertebrates but before the radiation of the other vertebrates from each other (450-500 million years ago). This latter explanation is thought to be more plausible, and the duplication date is commonly taken to be 450-500 million years ago.

Obviously, the above methods can only provide us with rough estimates of duplication dates, and all estimates should be taken with caution. Note that in comparison with estimation of species divergence dates, estimation of duplication dates is more difficult because in the former case many genes can be used whereas in the latter only the sequence data of the duplicate genes can be used, so that the estimate is subject to larger standard errors.

Globin Superfamily of Genes

In this section the globin superfamily and in the next section the homeobox genes will be used as two examples to illustrate how divergent duplicate genes may contribute to the physiological or structural complexity of higher organisms.

The globin superfamily has experienced all the possible evolutionary pathways that can occur in families of duplicated genes: (1) retention of original function, (2) acquisition of new function, and (3) loss of function in some duplicates. In humans, the globin superfamily consists of three families: the myoglobin family, whose single member is located

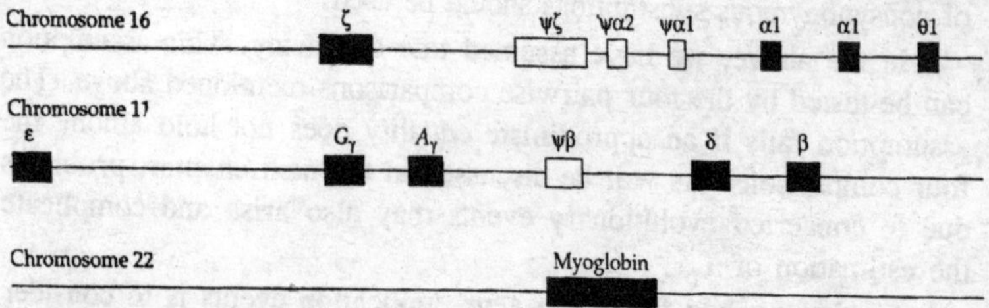

Fig. 10.9. The chromosomal arrangement of the three gene families belonging to the globin superfamily of genes in humans: the α-globin family on chromosome 16; β-globin family on chromosome 11; and myoglobin on chromosome 22.

on chromosome 22, the α-globin family on chromosome 16, and the β-globin family on chromosome 11.

The three families of globin genes produce two types of functional proteins: myoglobin and hemoglobin. The two proteins diverged about 600-800 million years ago and have become specialized in several respects. In terms of tissue specificity, myoglobin became the oxygen-storage protein in muscles, whereas hemoglobin become the oxygen carrier in blood. In terms of quarternary structure, myoglobin retains a monomeric structure, while hemoglobin has become a tetramer composed of two α and two β chains. In terms of function, myoglobin evolved a higher affinity for oxygen than did hemoglobin, while the function of hemoglobin has become more refined and diversified. Mammalian hemoglobin, for instance, has the capabilities of (1) transporting H^+ and CO_2 in addition to O_2, (2) binding four oxygen molecules cooperatively, (3) responding to tissue acidity (the Bohr effect), and (4) regulating its oxygen affinity through the level of organic phosphate in blood. Apparently, the heteromeric structure has facilitated the refinement of the function of hemoglobin. In fact, isolated α chains are monomers and do not have properties (2)-(4) and although β chains can form tetramers, they do not show cooperativity and thus, like the α chain and myoglobin, would not be a suitable oxygen carrier in blood.

The hemoglobin in humans and the vast majority of vertebrates is made up of two types of chains, one encoded by an α family member, the other by a member of the β family. As discussed above, the α and α families diverged about 450-500 million years ago. Since jawless fish contain only one type of monomeric hemoglobin, polymerization of hemoglobin in vertebrates probably occurred close to the time of the α-β divergence.

In humans, the α family consists of four functional genes: ζ, α_1, α_2 and θ1. It also contains three pseudogenes: ψζ, ψα1, and ψα2. The β family consists of five functional genes: ε, $^G\gamma$, $^A\gamma$, β, and δ, and one pseudogene, ψβ. The two families have diverged in both physiological properties and ontological regulation. In fact, distinct globins appear at different developmental stages: $\zeta_2\varepsilon_2$ and $\alpha_2\varepsilon_2$ in the embryo, $\alpha_2\gamma_2$ in the fetus, and $\alpha_2\beta_2$ and $\alpha_2\delta_2$ in adults; the time at which θ1 is expressed is not known. Furthermore, differences in oxygen-binding affinity have evolved among these globins. For example, the fetal hemoglobin $\alpha_2\gamma_2$ has a higher oxygen affinity than either adult hemoglobin, enabling the fetus to receive enough oxygen from the maternal blood (the high affinity occurs because the γ chain has serine

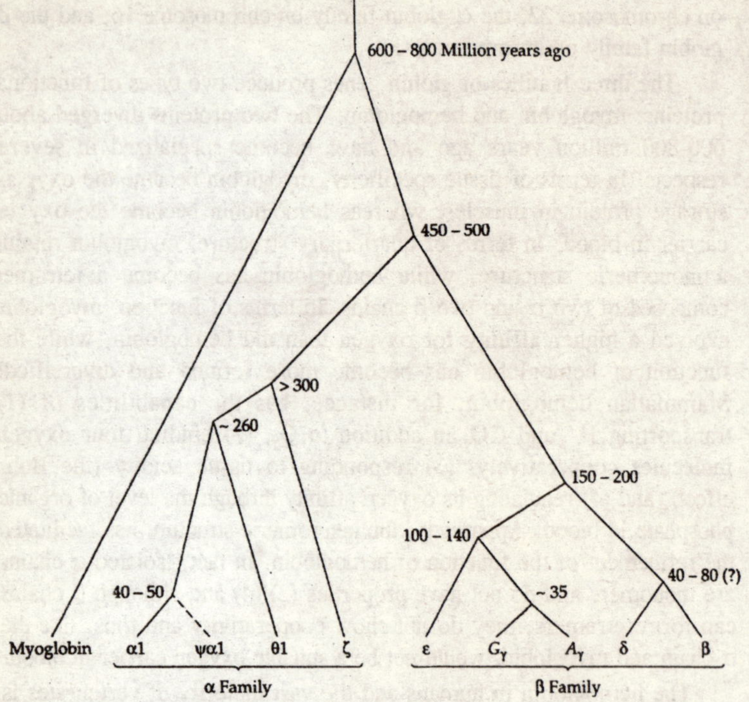

Fig. 10.10. Evolutionary history of human globin genes. The broken line denotes a pseudogene lineages.

at position 143 instead of histine as in the β chain). This phenomenon exemplifies again the fact that gene duplication can result in refinements of a physiological system.

Among the α family members the embryonic type ζ is the most divergent, having branched off more than 300 million years ago. The θ1 globin branched off about 260 million years ago. Because the divergence time between the two α genes is uncertain, only the α_1 gene is shown in the figure. The α_1 and α_2 genes have almost identical DNA sequences and produce an identical polypeptide. This similarity would seem to indicate a very recent divergence time. However, the similarity could also be the result of concerted evolution. The two genes are present in humans and all the apes and so could have diverged from each other more than 20 million years ago.

Among the β family members, the adult types (β and δ) and the nonadult types (γ and ε) diverged about 155-200 million years ago. The ancestor of the two γ genes diverged from the ε gene about 100-

140 million years ago. The duplication that created $^{G}\gamma$ and $^{A}\gamma$ occurred after the separation of the human and Old World monkey lineage from the New World monkey lineage, about 35 million years ago. The divergence between the δ and β genes was previously estimated to be 40 million years ago, but more recent DNA sequence data suggest that it may have occurred even before the eutherian radiation, about 80 million years ago. We note from the above discussion that, in both families, there is a good correlation between the time of divergence and the degree of functional or regulatory divergence between genes.

HOMEOBOX GENES

Homeotic (*Hox*) genes are genes in which mutations can transform part of a body segment, or an entire body segment, into the corresponding part of another segment. For example, in the *Antennapedia* mutant of *Drosophila* the antennae on the head of the fly are transformed into an additional pair of the second legs. The first homeotic mutant, bithorax, was discovered by Bridges and Morgan (1923) and later shown by E.B. Lewis to be part of a gene cluster, the bithorax complex. The first molecular cloning of bithorax genes was done by Bender et al. (1983). Later, McGinnis et al. (1984) and Scott and Weiner (1984) simultaneously discovered that homeotic genes share a common sequence element of 180 bp. This element has been known as the *homeobox*, and all genes containing this box, including nonhomeotic genes, are called homeobox genes. The discovery of the homeobox has generated much excitement, for it not only facilitates the study of the function of homeotic genes, but also serves as a probe for fishing out and identifying other homeobox genes. As homeotic genes are master control genes that specify the body plan and regulate development of higher organisms, understanding their regulation, structure, and function is key to the study of genetic control of development in higher organisms. These genes are also essential for understanding the evolution of body, plans, which has been a major issue in evolution.

Today, more than 350 homeobox genes and their relatives have been identified and sequenced in fungi, plants, and animals, and they can be grouped into at least 30 different classes in the past many gene names were very confusing but a new nomenclature has been adopted. The evolution of homeobox genes has been studied by many authors. Here we shall focus on the Antennapedia (Antp)-class homeobox genes, which are the best studied and are of special importance, because they specify the body segments along the anterior-posterior axis of the animal embryo and their developmental patterning.

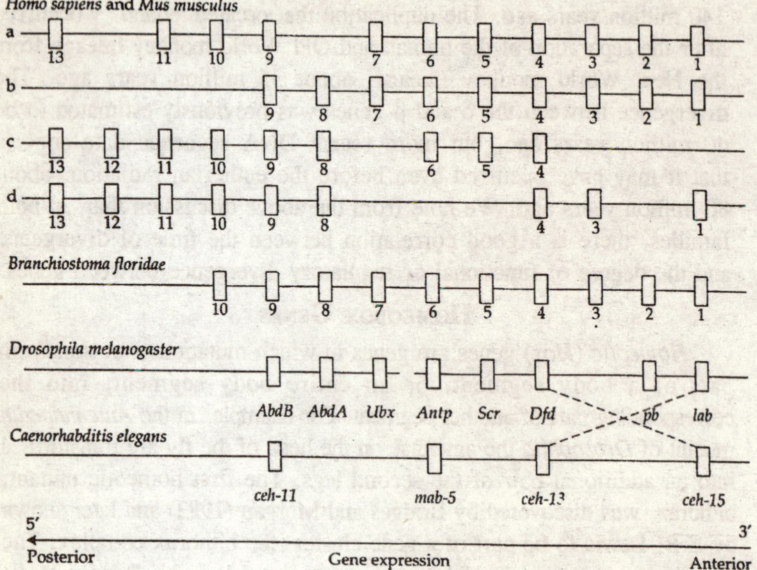

Fig. 10.11. Genomic organizations of Antennapedia-class homeobox genes in the human (Homo sapiens), mouse (Mus musculus), amphioxus (Branchiostoma floridae), Drosophila (D. melanogaster), and nematode (Caenorhabditis elegans).

In vertebrates (with the possible exceptions of primitive vertebrates) there are four clusters of Antp-class genes located on four different chromosomes. The genes within and between the clusters are related and can be classified into 13 cognate gene groups; however, none of the clusters has all 13 cognate genes. The amphioxus (*Branchiostoma floridae*), which is thought to be a sister group of vertebrates, has one cluster of at least 10 cognate genes. There are eight Antp-class homeobox genes in *Drosophila melanogaster*, which are located on the same chromosome but in two separate clusters, the Antennapedia complex and the bithorax complex. The nematode (*Caenorhabditis elegans*) has one cluster of four Antp-class homeobox genes. Interestingly, the temporal order of expression of the gene in each cluster is from 3' to 5', i.e., along the 3'-to-5' direction, each successive gene expresses later in the developmental process and more posterior along the anterior-posterior axis of the animal embryo.

There are studies of Antp-class homeobox genes in many other animals such as hydras, flatworms, annelids, crustaceans, acorn worms, and others. For simplicity, these genes are not included in the following discussion.

The following discussion is based mainly on Zhang and Nei (1996). It should be noted that the relationship inferred in any of these studies are not very certain because the homeobox sequence, which has been the basis of the influence, is only 180 bp or 60 codons long. Therefore, the inferences discussed below should be taken with caution, especially for those groupings with a low bootstrap value.

The gene in the tree are clustered into two major groups, I and II. Group I consists of subgroups A, B, and C; this subgrouping is tentative because the bootstrap proportions (BP), except that for subgroup B, are not high. Subgroup A includes cognate genes 1 and 2 of the mouse, human, and amphioxus; pb (proboscipedia) and lab (labial) of *Drosophila*; and ceh-13 and of the nematode. Inclusion of ceh-13 in this subgroup is reasonable, because ceh-13 and ceh-15 of the nematode have apparently been inverted and ceh-13 seems to be orthologous to cognate gene 1. Subgroup B includes cognate gene 3 of the mouse, human, and amphioxus. Subgroup C consists of cognate genes 4-8 of the mouse, human, and amphioxus; Dfd (Deformed), Scr (Sex combs reduced), Antp, AbdA (Abdominal-A), and Ubx (Ultrabithorax) of *Drosophila*; and mab-5 and ceh-15 of the nematode. ceh-15 and mab-5 C are supposed to be orthologous with cognate gene groups 4 and 6, respectively, but they form a cluster separate from other genes in subgroup C. This cluster should be taken with caution because it has a low BP value. In figure, cognate groups 5-8 do not form monophyletic groups, but this result could be due to sampling errors.

Group-II genes are divided into subgroups D and E, though each grouping is supported by only a low BP value. Subgroup D includes cognate genes 9 and 10 of the mouse, human, and amphioxus. The Hox-9 gene of the amphioxus clusters with cognate group 10 genes, but this may be due to sampling errors, especially in view of the low BP value for this clustering. Also, whether cognate group 10 should be in subgroup D or E is uncertain, because when only human and mouse genes were used in the analysis, the group was included in subgroup E, not in subgroup D. With this caveat in mind, figure suggests that subgroup E consists of cognate genes 11-13 of the mouse and human, and AbdB (Abdominal-B) of *Drosophila*. The *Drosophila* AbdB gene is supposed to belong to cognate gene group 9, and in view of the low BP value for the E subgroup, its inclusion in subgroup E seems to be due to sampling errors. The nematode gene ceh-11, which was previously aligned with cognate group 9 genes, seems to be an outgroup to all other Antp-class genes; however, this is uncertain because it is supported by a low BP value.

The divergence of nematode ceh-13, *Drosophila-lab*, and chordates cognate group 1 genes postdated the divergence of cognate gene groups 1 and 2 and that the last common ancestor of the nematode, *Drosophila*, and chordates already had cognate gene groups 1 and 2. Similarly suggests that the last common ancestor of he nematode, *Drosophila*, and chordates also had cognate gene 3, the ancestral gene of subgroup C, and the ancestral gene group II. In other words, there were at least five cognate genes in the last common ancestor of Pseudocoelomates and Coelomates (insects and chordates). In the nematode, however, cognate genes 2 and 3 seem to have been lost later, and the ancestral genes of cognate genes 4-8 apparently gave rise to mab-5 and ceh-15. The positions of ceh-15 and ceh-13 are inverted according to the sequence comparison and functional analysis of Burglin and Ruvkun (1993), probably due to a chromosome inversion in the nematode.

The divergence of subgroups D and E occurred earlier than that of the amphioxus and vertebrate cognate group 10 genes. Therefore, if the tree is correct on this point, the amphioxus must have had cognate group 11-13 genes previously. These genes have important functions in the limb formation in the mouse. It is possible that they have been lost, consistent with the fact that the fins of amphioxus, which may be homologous to the limbs of tetrapods, are poorly developed. Or it is possible that these genes have not yet been found in the amphioxus.

Fig. 10.12 shows a scenario proposed by Zhang and Nei (1996) for the evolution of the Antp-class genes through gene duplications. In the beginning, duplication of a single ancestral gene produced the ancestral genes of groups I and II, and the mutational change of these genes initiated the step for the anterior and posterior parts of the animal embryo to be controlled by separate genes. Then, a series of duplications of the ancestral group I gene generated subgroups A, B, and C. Subgroups A and B are involved in the segmentation of the anterior parts, whereas subgroup C genes control the intermediate part along the anterior-posterior axis of the animal embryo. In the case of group II genes, the first duplication seems to have produced the ancestral genes of cognate groups 9-10 and the others, and the subsequent duplication generated the remaining cognate genes. An ancestral cluster of at least 10 cognate genes was probably formed before the divergence between the amphioxus and vertebrate lineages. Gene clusters a, b, c, and d in vertebrates were probably generated by two events of genome duplication in the early stage of vertebrate evolution, and later a few genes in each cluster were lost.

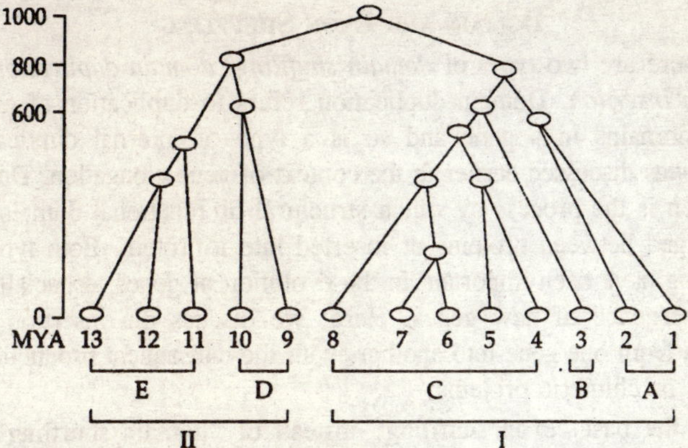

Fig. 10.12. Evolutionary scenario of the 13 cognate groups of the Antp-class homeobox gene family. Branch lengths are not proportional to the evolutionary time. I and II are two major groups and A, B, C, D and E are five subgroups.

On the basis of amino acid divergence and under the assumption of a molecular clock, the time of divergence between groups I and II has been estimated to be about 100 million years ago, suggesting that the Antp-class homeobox gene family evolved in the very early stage of metazoan evolution. Similarly, subgroups D and E and subgroups B and C were estimated to have diverged about 800 and 600 million years ago, respectively. These estimates suggest that the five subgroups were formed by 600 million years ago, predating the divergence of Pseudocoelomates and Coelomates.

Of course, many of the above inferences are uncertain and much further research is needed to clarify the evolutionary history of these genes. Also, the above discussion is concerned mainly with the evolutionary history of the Antp-class homeobox genes. For an excellent review of the relationships between the evolution of homeotic genes and the evolution of body plans and body parts, Carroll (1995), who suggested tat the generation of arthropod and tetrapod diversity has largely involved regulatory changes in the expression of homeotic genes and the evolution of interactions between homeotic genes and the genes they regulate, though changes in the number and function of homeotic genes have also been important. Presently, the Antp-class homeobox genes are the best studied, yet our understanding of their evolution is limited. In view of the existence of a large number of homeobox genes. Their evolution is a subject that will be require many years of effort by both molecular biologists and evolutionists.

Domain and Exon Shuffling

There are two types of *domain shuffling*: *domain duplication* and *domain insertion*. Domain duplication refers to duplication of one or more domains in a gene and so is a type of internal duplication, which was discussed earlier in the context of gene elongation. Domain insertion is the process by which structural or functional domains are exchanged between proteins or inserted into a protein. Both types of shuffling have been important in the evolution of genes, especially for the emergence of new genes. Here, we discuss the insertion of a domain from one gene into another, with the consequent production of mosaic or chimeric proteins.

In the past "exon shuffling" instead of "domain shuffling" has been commonly used but the latter seems to be a better or, at least, broader term. Shuffling of a domain is more likely to succeed than shuffling of an exon, because a domain is a functional or structural unit whereas an exon is often not. Of course, an exon may be successfully shuffled even if it does not encode a domain, but such a shuffling may have little functional or evolutionary significance. Note further that in the case where a multiexonic domain is shuffled, domain rather than exon is the unit of shuffling. Moreover, domain shuffling can occur without the involvement of any exon. In fact, no intron has yet been found in prokaryotic protein coding genes, but many cases of domain shuffling have already been discovered in bacteria. Nevertheless, the existence of introns has greatly promoted domain shuffling, especially in vertebrate genes. For this and for the historic reasons, we shall also discuss exon shuffling.

There has been a controversy regarding the following questions: How many kinds of domain exist and what was their origin? One view is that there were only a few small polypeptides in early stages of life and that most contemporary proteins are descended from them via duplication and modification. The opposing view is that a large number of polypeptides existed early in the history of living systems and that all proteins, past and present, are the result of shuffling these primordial structural units. This view is based on the "intron early" notion which assumes that primitive proteins were encoded by "minigenes" that were spliced together, with coding regions (exons) being separated by introns. The intron early hypothesis is a subject of intense debate.

Mosaic Proteins

One of the first mosaic proteins discovered was tissue plasminogen activator (TPA). Plasminogen is converted by TPA into its active form, plasmin, which dissolves fibrin, a soluble fibrous protein in blood clots. The conversion is greatly accelerated by the presence of fibrin itself, the substrate of plasmin. Fibrin binds both plasminogen and TPA, thus aligning them for catalysis. This mode of molecular alignment allows plasmin production only in the proximity of fibrin, thus conferring fibrin specificity to plasmin. By contrast, urokinase (UK), a urinary plasminogen activator, lacks fibrin specificity. A comparison of the amino acid sequences of TPA and UK showed that TPA contains a 43-residue sequence at its amino-terminal end that has no counterpart in UK. This segment can form a finger-like structure and is homologous to the finger domains responsible for the fibrin affinity of another protein, fibronectin (FN), which is a large glycoprotein present in plasma and cell surfaces that promotes cellular adhesion. Deletion of this segment leads to a loss of the fibrin affinity of TPA. The homology

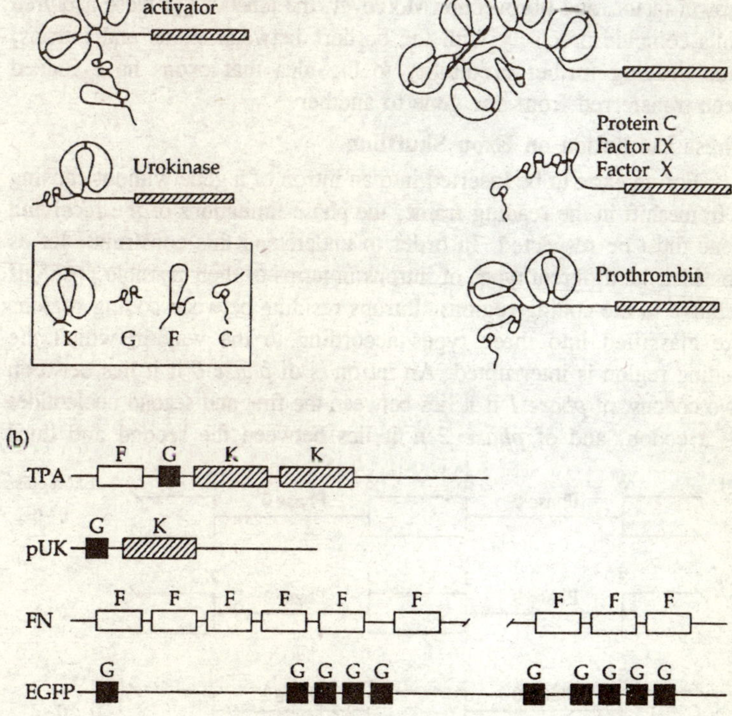

Fig. 10.13. (a) Various structural modules observed in tissue plasminogen activator (TPA) and some other proteins. (b) The origin of the modules acquired through exon insertion in the tissue plasminogen activator (TPA) protein.

of TPA with FN is restricted to this finger domain. Thus, domain (exon) shuffling was probably responsible for the acquisition of this domain by TPA from either FN or a similar protein.

TPA also contains a segment homologous to epidermal, growth factor (EGF) and the growth-factor-like regions of other proteins, such as Factor IX and Factor X, which are blood-clotting enzymes in the blood-coagulation pathway. In addition, the carboxy-terminal regions of TPA are homologous to the proteinase parts of trypsin and other trypsin-like serine proteinases, which are enzymes that hydrolyze proteins into peptide fragments. Finally, the nonproteinase parts of TPA contain two structures similar to the kringles of plasminogen. (A kringle is a cysteine-rich sequence that contains three internal disulfide bridges and forms a pretzel-like structure resembling the Danish cake bearing this name.) Thus, during its evolution, TPA captured at least four DNA segments from at least three other genes: plasminogen, epidermal growth factor, and fibronectin. Moreover, the junctions of these acquired units coincide precisely with the borders between exons and introns, thus, leading further credibility to the idea that exons have indeed been transferred from one gene to another.

Phase Limitation on Exon Shuffling

For an exon to be inserted into an intron of a gene without causing a frameshift in the reading frame, the phase limitations of the receiving gene must be respected. In order to understand this constraint, let us consider the different types of introns in terms of their possible positions relative to the coding regions. Introns residing between coding regions are classified into three types according to the way in which the coding region is interrupted. An intron is of *phase 0* if it lies between two codons, of *phase 1* if it lies between the first and second nucleotides of a codon, and of *phase 2* if it lies between the second and third

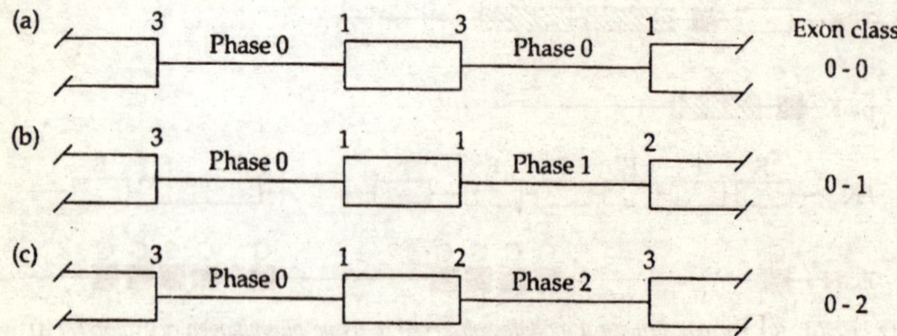

Fig. 10.14. Phases of introns and classes of exons.

nucleotides of a codon. Exons are grouped into classes according to the phases of their flanking introns. For example, the middle exon is flanked by a phase-0 intron at its 5' end and by a phase 1 intron at its 3' end and is said to be class 0-1. An exon that is flanked by introns of the same phase at both ends is called a *symmetrical exon*, otherwise it is *asymmetrical*. For example, the middle exon is symmetrical. Of the nine possible classes of exons, three are symmetrical (0-0, 1-1, and 2-2), and six are asymmetrical.

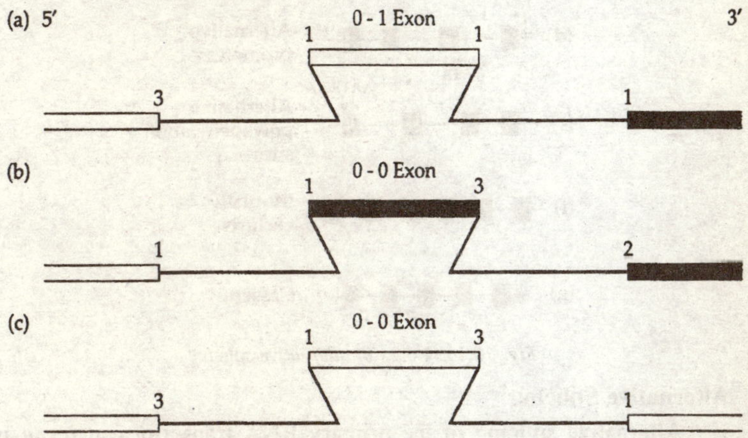

Fig. 10.15. Consequences of exon insertion into introns. Shaded boxes indicate frameshifts. (a) Insertion of a 0-1 asymmetrical exon into a phase 0 intron; (b) insertion of a 0-0 symmetrical exon into a phase-1 intron; (c) insertion of a 0-0 symmetrical exon into a phase-0 intron.

Only symmetrical exons can be inserted into introns. For example, the insertion of a 0-1 exon into a phase-0 intron causes a frameshift in all the subsequent exons. Moreover, insertion of symmetrical exons is also restricted. A 0-0 exon can be inserted only into introns of phase 0, and 1-1 exon can be inserted only into introns of phase 1, and a 2-2 exon can be inserted only into introns of phase 2. For example, the insertion of a 0-0 exon into an intron of phase 1 causes a frameshift in the inserted exon and in all exons on the 3' side, while the insertion of a 0-0 exon into an intron of phase 0 causes no frameshift. Interestingly, the vast majority of known movable exons are of the 1-1.

Alternative Pathways for Producing New Functions

In addition to gene duplication and domain shuffling, there are other mechanisms for producing new genes or polypeptides. For such mechanisms are discussed in the next four sections.

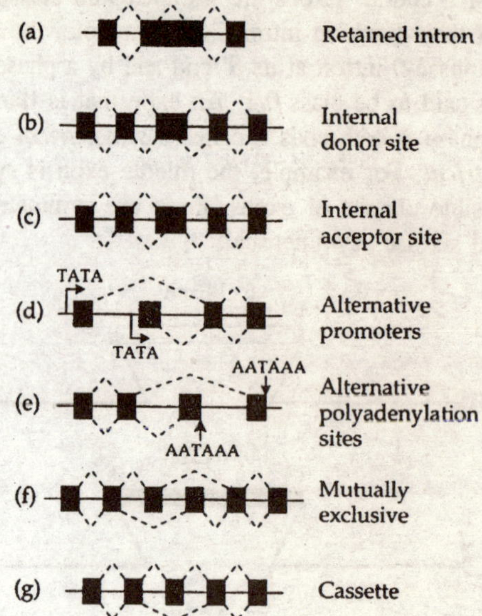

Fig. 10.16. Modes of alternative splicing.

Alternative Splicing

Alternative splicing of the primary RNA transcript can result in the production of different polypeptides from the same DNA segment. This phenomenon has been found in numerous eukaryotic genes and in a number of eukaryotic transposons and animal viruses.

The classification of the modes of alternative splicing summarized by Smith et al. (1989):

Retained introns

Perhaps the simplest form involves introns that can either be spliced or retained in the mature RNA. Failure to splice the intron can result in the insertion of a peptide segment if the open reading frame is maintained. Alternatively, premature termination of the reading frame or frameshift may result in functionally different products or may effectively shut off gene expression if the unspliced RNA produces no functional product. Examples of such retained introns are found in the genes for fibronectin and *Drosophila* P element transposase.

Alternative 5' donor sites and 3' acceptor sites

The use of alternative donor or acceptor sites for a given exon results in the excision of introns of different lengths with complementary variations in exon size. Again, this can give rise to insertion of small alternative peptide segments, frameshifting, or premature termination of reading frames. Such use of competing splicing sites is common in viral transcription units such as the adenovirus Ela and SV40 T/t units, and is also used in eukaryotic genes such as the *Drosophila* ultrabithorax and transformer genes.

Alternative promoters and cleavage/polyadenylation sites

In some instances, alternative 5' and 3' end sequences arise through the differential utilization of promoters or cleavage/poly(A) sites, each associated with their own exons. This can lead to variability in the amino acid sequence of the N or C terminal of protein, respectively, if the variable segments extend into the protein-coding region. The variations in 5' and 3' untranslated sequences may lead to differential regulation of gene expression via effects on RNA stability, transport, or translational efficiency. Alternative promoter selection dictates the splicing pattern of the myosin light chain 1/3 (MLC1/3) gene, while examples of alternative 3' end exons with associated poly(A) addition sites are found in the α and β tropomyosins, the immunoglobulin μ heavy chain, and the calcitonin/CGRP genes.

Internal mutually exclusive exons

Internal exons are sometimes used mutually exclusively. One member of the pair is always spliced into the mRNA, but never both. The two exons are thus never spliced together, nor are both skipped. These exons code for alternative peptide segments within an otherwise constant flanking amino acid sequence. Such exons are found in the genes for the contractile proteins α and β tropomyosin, troponin-T, and myosin light chain 1/3, and in the pyruvate kinase gene.

Cassette exons

Some cassette exons can be included or excluded independently of other exons, and usually the same reading frame is maintained whether the exon is spliced in or out, though frameshift or premature termination can result from inclusion. When several cassette exons are present in a gene a very high degree of diversity can be generated, as in the NCAM and troponin-T genes. The observed hypervariability within the N-terminal region of troponin-T region is based on the presence of five cassette exons at the 5' end of the gene, these in conjunction with a pair of mutually exclusive exons in theory allow the generation of up to 64 different isoforms.

Alternative splicing has often been used as a means of developmental regulation. A very intriguing situation is seen in several genes involved in the process of sex determination in *Drosophila*

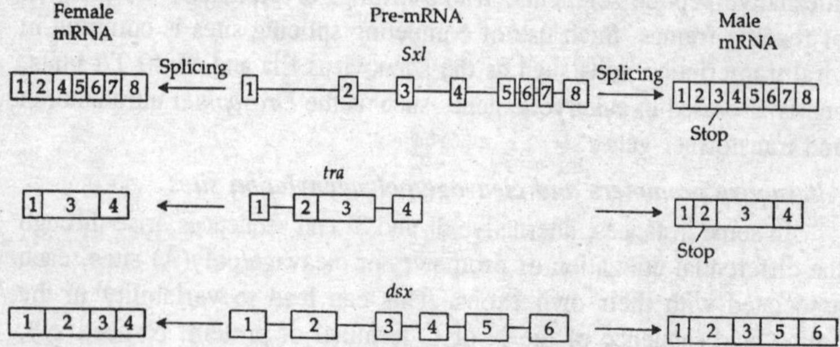

Fig. 10.17. The patterns of splicing of the sexlethal (Sxl), transformer (tra), and doublesex (dsx) genes in Drosophila melanogaster females (left) and males (right).

melanogaster. At least three genes, *Sexlethal* (Sxl), *transformer* (tra), and *doublesex* (dsx), are spliced differently in males and females. In the case of *dsx*, the gene has six exons, and exons 1, 2, 3, and 4 are used in the female, whereas exons 1, 2, 3, 5 and 6 are used in the male. In the cases of *Sxl* and *tra*, the products of the alternative splicing in males contain premature termination codons and are therefore nonfunctional. For example, exon 3 is *Sxl* contains an in-frame stop codon, but the mRNA in females does not contain this exon.

A special instance of alternative splicing is illustrated by the case of intron-encoded proteins. In such cases, the intron contains an open reading frame that encodes a protein or part of a protein that is completely different in function from the protein encoded by the flanking exons. In some cases, the open reading frame is an extension of the upstream exon, e.g., intron a14α in the yeast mitochondrial gene *cox I*. In other cases the intron includes not only a free-standing protein-coding gene, but also the necessary signals for transcription initiation and termination. Intron a14α is intriguing, since it encodes a maturase also functions as an endonuclease in DNA recombination.

The evolution of alternative splicing requires that an alternative splice junction site be created *de novo*. Since splicing signals are

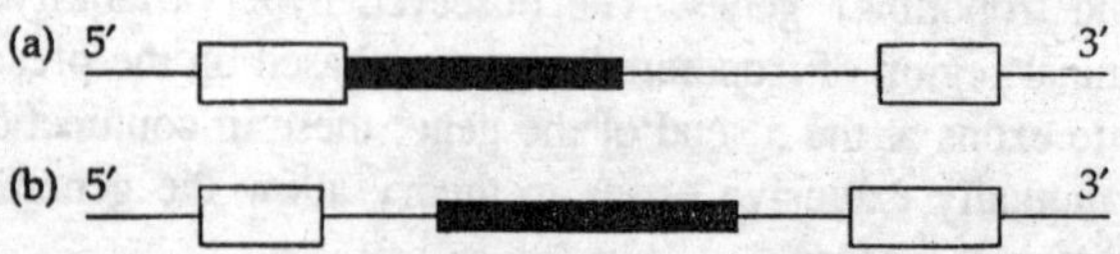

Fig. 10.18. Examples of intron-encoded proteins: (a) an open reading frame (shaded) that is an extension of the upstream exon; (b) a free-standing open reading frame, the transcription initiation and termination signals of which reside within the intron.

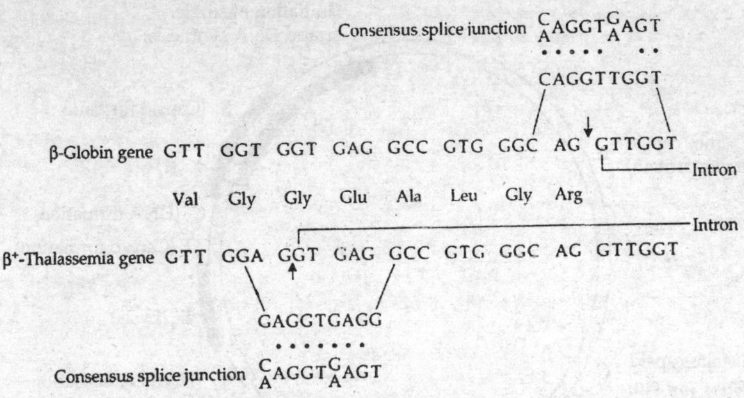

Fig. 10.19. The nucleotide sequences at the border between exon 1 and intron 1 in the β-globin gene from a normal individual and a patient with β⁺-thalassemia.

usually 5-10 nucleotides long, it is possible that such sites are created with an appreciable frequency by mutation. Indeed, many such examples are known in the literature. For example, a case in which a synonymous substitution in a glycine codon turned a coding region into a splice junction. In cases of pathological manifestations such as the β⁺-thalassemia, the new splice site is used more frequently than the old splice site (i.e., most of the mRNA synthesized after such a mutation occurs is of the altered type). Such a mutation will obviously have deleterious effects and is not expected ever to become fixed in the population. However, if he newly created splice site is much weaker, then most mRNA will be of the original type and only small quantities of the new mRNA will be made. Such a change will not obliterate the old function, and yet will create an opportunity to produce a new protein.

Overlapping Genes

It has been found that a DNA segment can code for more than one gene by using different reading frames. This phenomenon is widespread in viruses, organelles, and bacteria. Fig. 10.20 shows the genetic map of φX174, which is a single stranded DNA bacteriophage. Several overlapping genes are observed. For example, gene B is completely contained inside gene A, while gene K overlaps gene A on the 5' end and overlaps gene C on the 3' end.

Overlapping genes can also arise by the use of the complementary strands of a DNA sequence. For example, the gene specifying $tRNA^{Ile}$ and $tRNA^{Gln}$ in the human mitochondrial genome are located on different strands, and there is a three nucleotide overlap between them that

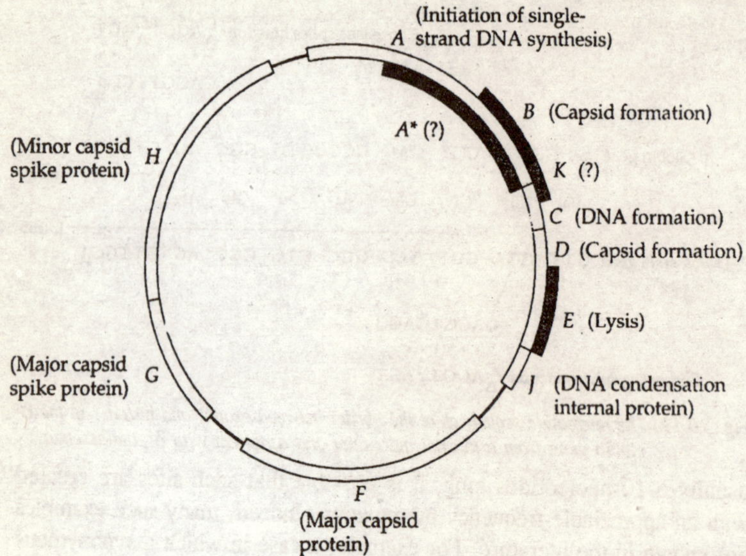

Fig. 10.20. Genetic map of the circular genome of the single-stranded DNA bacteriophage ϕX174.

reads 5'–CTA–3' in the former and 5'–TAG–3' in the latter. The question arises as to how overlapping genes may come into existence during evolution. To answer this question, we note that open reading frames abound throughout the genome. Therefore, it is possible that potential coding regions of considerable length exist in either a different reading frame of an existing gene or on the complementary strand. Because only three of 64 possible codons are termination codons, even a random DNA sequence might contain open reading frames hundreds of nucleotides long. If by chance such a reading frame contains an initiation codon and a transcription initiation site, or if such sites are created by mutation, an additional mRNA will be transcribed and subsequently translated into a new protein. Whether the new product has a beneficial function or not is another matter, but if it does, the trait may become fixed in the population.

The rate of evolution is expected to be slower in stretches of DNA encoding overlapping genes than in similar DNA sequences using only one reading frame. The reason is that the proportion of nondegenerate sites is higher in overlapping genes than in nonoverlapping genes, thus reducing the proportion of synonymous mutations out of the total number of mutations.

Gene Sharing

An extremely intriguing situation of creating a new function is the recruitment of a gene product to serve an additional function without any changes in its amino acid sequence. This phenomenon has been termed *gene sharing*. Gene sharing means that a gene acquires and maintains a second function without duplication and without loss of the primary function. Gene sharing may, however, require a change in the regulation system of tissue specificity or developmental timing.

Gene sharing was first discovered in crystallins, which are used in the eye lens to maintain transparency and proper light diffraction. The first finding was that the ε crystallin from birds and crocodiles is identical in its amino acid sequence with lactate dehydrogenase B and possesses identical LDH activity. Subsequent work showed that the two "proteins" are encoded by the same gene. A second crystallin, δ, which exists in all birds and reptiles, has also been shown to be identical in sequence with another enzyme, argininosuccinate lyase, which catalyzes the conversion of argininosuccinate into the amino acid arginine, and these proteins are likewise encoded by the same gene. Similarly, τ crystallin in lampreys, bony fishes, reptiles, and birds has been shown to be identical to and encoded by the same gene as α-enolase, a glycolytic enzyme converting 2- phosphoglycerate into phosphoenolpyruvate. Thus, δ, ε, and τ crystallins illustrate instances of gene sharing, whereby a gene acquires additional roles without being duplicated. The α, β, and γ crystallins, on the other hand, are classical examples of proteins that evolved by means of gene duplication and subsequent sequence divergence from ancestral genes specifying different proteins (e.g., heat-stock genes, which encode proteins expressed following exposure to excessive heat).

The finding that the same polypeptide can serve both as an enzyme and as a structural protein blurs the traditional distinction between enzymes and nonenzymatic proteins. Note also that once an enzyme has been recruited to serve as a structural protein in lens, it is subject to two sets of evolutionary pressure. This constraint may not allow sequence changes that enhance its function as a crystallin but that are not of benefit to the enzymatic role. A solution to this adaptive conflict may result if a gene duplication occurs. Such an example is provided by the argininosuccinate lyase (ASL)/δ crystallin family. Of the two tandemly arranged chicken ASL/δ crystallin genes, δ1 is specialized for lens expression and produces >95% of the lens δ crystallin mRNA; by contrast, the δ2 gene, which appears to encode the enzymatically active ASL, produces most of the ASL/δ crystallin mRNA in nonlens tissues, although it is still much more abundant in the lens than in other tissues.

RNA Editing

RNA editing is defined as the transcriptional or posttranscriptional modification of an RNA molecule that changes its coding specificity by a process other than splicing. The first example of RNA editing is the insertion and deletion of uridine (U) residues within the coding regions of many mitochondrial transcripts of kinetoplastid trypanosomes. Many types of RNA editing are now known.

One of the common types of RNA editing is C-to-U conversion in the RNA transcript. This conversion may occur partially or completely in some tissues but not in the others, leading to different variants of a protein in different tissues or multiple variants in the same tissue. Occasionally, it can produce a new protein with a quite different function, such as in the case of the synthesis of apolipoprotein B-48 mRNA. Apolipoprotein (apo) B is the largest protein species of apolipoproteins, which are the major carriers of various lipids in human blood. ApoB is heterogeneous but exists primarily in two forms: apoB-100 and apob-48. In humans, apoB-100 is synthesized primarily by the liver and is the major protein constituent of very low density lipoproteins (VLDL) and low-density lipoproteins (LDL), whereas apoB-48 is synthesized by the intestine and is found in chylomicrons and chylomicron remnants. ApoB-48, unlike apoB-100, does not bind to the LDL receptor and has a molecular weight about 48% of that of apoB-100, which is a huge monomer protein, having 4536 amino acid residues. Traditionally, one would have guessed that apoB-48 is either produced by posttranslational cleavage of apoB-100 or encoded by a different gene. Unexpectedly, it was fond that apoB-48 is translated from an intestinal mRNA with an in-frame UAA stop codon resulting from a C-to-U change in the codon CAA encoding Gln^{2153} in apoB-100 mRNA. Thus, although there is only one apoB gene in humans and rodents, two quite different proteins can be produced from the same gene.

INDEX

A
Agnatha, 117
Allozymes, 297
Amphioxus, 144
Analogous, 111
Ancient conserved regions, 120, 121, 123
Antennapedia, 307
Aotus azarae, 194
Apoptosis, 2
Aquifex aeolicus, 34
Arabidopsis, 21, 22
Arabidopsis thaliana, 21
Archain, 119
Asymmetrical, 315
Atonal, 119

B
B. bufo, 51
Biochemical functions, 31
Birth-and-death, 126
Block, 2
Bone fide, 210
Borrelia burgdorferi, 34
Bos indicus, 30
Bos taurus, 30
Bradyrhizobium japonicum, 36
Branchiostoma floridae, 308
Bufo bufo, 51

C
C. elegans, 19, 20, 21, 22, 120
Caenorhabditis, 119, 120
Caenorhabditis elegans, 18, 19, 120, 213, 308
Callithrix jacchus, 153
Cebus apella, 194
Coding DNA, 50
Concerted evolution, 45, 125
Conservative, 60
Contig, 9
Contig map, 6
Contiguous clone map, 9
Convergent evolution, 178, 193
Copia, 62
Crystallin genes, 80
Cul-de-sacs, 64
Culex, 295

D
D. melanogaster, 20, 21, 22
De novo, 103, 196, 318
Dichromatic, 300
Dictyostelium, 87, 119, 121

Differential phylogenetic footprinting, 186
Dishevelled, 119
Divergent evolution, 126, 178
Dodo, 119
Domain, 111, 286
Domain duplication, 312
Domain insertion, 312
Domain shuffling, 312
Dorsal, 119
Dose repetitions, 295
Dot-matrix, 257
Doublesex, 318
Drosophila, 21, 41, 62, 89, 94, 104, 119, 120, 121, 132, 165, 170, 213, 216, 286, 307, 309, 310, 316, 317
Drosophila melanogaster, 18, 20, 308, 317
Drosophila-lab, 310
Durum, 17

E

E. coli, 9, 11, 12, 13, 14, 35, 36, 122, 123, 187
Ectopic, 97
Einkorn, 17
Emmer, 17
Enhancer of split, 119
Escherichia coli, 165, 296
Exon, 2
Exon shuffling, 285
Exon theory of genes, 99
Expressed sequence tag, 10

F

Farro, 17
Folds, 111
Fourfold degenerate, 251
Fugu, 88, 89, 91
Fugu rubripes, 88, 93, 94
Functional, 179
Functional domain, 286

G

Gain, 27
Galago, 104
Galago crassicaudatus, 162
Gap penalty, 259
Gaps, 256
Gene clusters, 40
Gene conversion, 48
Gene elongation, 287
Gene family, 40, 295
Gene recruitment, 132
Gene sharing, 132, 321
Genes, 31
Genetic fingerprinting, 60

H

H. influenzae, 33, 34
Haemophilus influenzae, 32
Hedgehog, 216
Highly conserved regions, 197
Homeobox, 307
Homeotic, 307
Homologous, 111
Hylobates lar, 176

I

Ido, 179
Illegitimate, 97
in vitro, 80, 85, 185, 188, 200, 216
In vivo, 199, 216, 235
Indels, 256
Intron, 2
Intron sliding, 100, 220
Introns early, 99
Introns late, 99
Isoaccepting groups, 156
Isozymes, 297

K

Kruppel, 174

L
LINE element-mediated recombination, 105
Lymnea stagnalis, 120
M
M. genitalium, 15, 16, 33, 34
M. jannaschii, 14, 15
M. pahari, 195, 198
M. tuberculosis, 36
Mariner, 89
Mature, 3
Mechanistic, 179
Methanococcus, 36
Methanococcus jannaschii, 14, 32
Minibrain, 119
Minimal organism, 15, 33
Module, 111, 286
Motifs, 111
mRNA surveillance, 97
Multigene family, 295
Multiple hits, 274
Mus domesticus, 195, 198
Mus pahari, 198
Mycoplasma genitalium, 15, 33
Myxococcus, 131
N
Non-coding, 50
Non-synonymous, 31
Nondegenerate, 251
O
Oncorhynchus mykiss, 219
One-parameter model, 264
Open reading frames, 10
Operons, 2, 11
ORFs, 10
Orthologous, 17, 302
Overdominant selection, 152
P
Paralinear, 246
Parallel substitution, 274
Paralogous, 17, 302
Peptide-binding region, 149
Per nucleotide site, 243
Per se, 201
Phylogenetic footprinting, 185
Physarum, 131
Plasmodium, 87
Pointed, 216
Positional cloning, 6
Post-translational modifications, 3
Promoter shuffling, 184
Proteins, 31
Proteome project, 3
R
Radiation hybrids, 8
Relative-rate test, 222
Replicative, 60
Restriction fragment length polymorphisms, 9
Restriction map, 6
Reverse genetics, 6
Rolling circle replication, 56
S
S. cerevisiae, 121
S. griseus, 35
Saccharomyces cerevisiae, 32, 120, 297
Salmonella, 35
Scattered repeats, 51
Secale cereale, 17
Selfish DNA, 54
Semi-processed, 75
Sequence convergence, 180
Sequence tagged site, 10
Sexlethal, 318
Short tandem repeat polymorphisms, 9
Single copy pseudogenes, 81
Slippage, 55
Slowpoke, 119

Son of sevenless, 119
Sonic, 216
Spelt, 17
Structural, 179
Structural domain, 286
Sulculus, 179
Sulculus diversicolor, 89, 179
Superfamily, 295
Symmetrical, 106
Symmetrical exon, 315
Synonymous, 31

T

T. aestivum, 17
T. dicoccum, 17
T. spelta, 17
T. turgidum, 17
Tandem, 51
Tandem repeats, 41
Taxon-specific, 132
Tetrahymena, 297
Total number, 244
Trans-species polymorphism, 152
Transformer, 318
Transposition, 60
Triticale, 17
Triticum monococcum, 17
Triticum turgidim, 17
Twofold degenerate, 251

U

Unequal crossing over, 56
Unequal crossing over, 42
Unprocessed, 300

V

Variable number tandem repeats, 8
Vice versa, 27, 119, 190, 216

W

Wingless, 214

X

X. borealis, 46
X. laevis, 44, 46
Xenopus, 44, 46, 90, 93, 194
Xenopus laevis, 41, 296
Xenopus tropicalis, 44